Plant Gene Research

Basic Knowledge and Application

Edited by

E. S. Dennis, Canberra

B. Hohn, Basel

Th. Hohn, Basel (Managing Editor)

P. J. King, Basel

J. Schell, Köln

D. P. S. Verma, Montreal

Springer-Verlag Wien New York

Plant DNA Infectious Agents

Edited by Th. Hohn and J. Schell

Springer-Verlag Wien New York

Dr. Thomas Hohn
Friedrich Miescher-Institut, Basel

Dr. Joszef Schell
Max-Planch-Institut für Züchtungsforschung, Köln

With 76 Figures

Softcover reprint of the hardcover 1st edition 1987

Library of Congress Cataloging-in-Publication Data

Plant DNA infectious agents.

(Plant gene research)
Bibliography: p.
Includes index.
1. Plant genetic engineering. 2. Plant molecular genetics. 3. Insertion elements, DNA.
4. Plant viruses.
I. Hohn, Thomas, 1938– . II. Schell, Joszef S.
III. Series.
QK981.5.P56 1987 581.87'3282 87-23554

ISBN-13: 978-3-7091-7458-6

ISSN 0175-2073
ISBN-13: 978-3-7091-7458-6 e-ISBN-13: 978-3-7091-6977-3
DOI: 10.1007/978-3-7091-6977-3

Preface

There has been recent rapid progress in the transformation of plants with foreign DNA, making use either of the natural routes of genetic invasion that viruses and bacteria have developed, or of chemical, mechanical and electrical tricks to make plant protoplast membranes permeable to nucleic acids. Genes integrated into plant virus genomes can be carried systemically from the initial site of infection into the rest of the plant. Genes placed between the borders of *Agrobacterium tumefaciens* T-DNA can be transferred into single cells or plant tissue, which then divides to produce wound calli, or as in the case of an *Agrobacterium rhizogenes* infection, grow out into new roots. Calli and roots can be grown into whole plants. If virus genomes are placed between the T-DNA borders, a very effective infectious route, termed „agroinfection“, is established. Once inside a protoplast, DNA finds its way into the nucleus where it can finally integrate into the resident chromosome and be expressed. Whether it can also find its way into chloroplasts is not yet clear, but at least translation products can be targeted into this organelle.

Regeneration of whole organisms from single cells is a special feature of plants and offers a unique tool to study genes in a multicellular organism. In addition, as in animal cells, transcription and translation of transforming genes can be studied in plant cells during „transient expression“. Some chapters in this book deal with these practical aspects, and others describe the strategies of natural nucleic acid invasion into plants, by itself a fascinating topic. The next volume of this series will deal with attempts to exploit the technologies described to unravel plant strategies for developmental and environmental gene control.

Basel and Köln, 1987 — Thomas Hohn and Joszef Schell

Contents

Chapter 1

The Molecular Biology of Cauliflower Mosaic Virus and Its Application as Plant Gene Vector

Bruno Gronenborn

Max-Planck-Institut für Züchtungsforschung
Abt. Genetische Grundlagen der Pflanzenzüchtung
D-5000 Köln 30
Federal Republic of Germany

With 3 Figures

Contents

I. Introduction

The artificial introduction of new traits into living cells by means of molecular gene transfer techniques has been developed into a routine procedure, applicable to both prokaryotic and various eukaryotic organisms. For higher plants, nature itself has evolved an elegant and sophisticated gene transfer system in the Ti-plasmid of *Agrobacterium tumefaciens*. It ist now widely used to introduce and stably anchor new genes in the genome of susceptible host plants. (See Chapter 7 by S. Rogers, this volume.)

As an alternative to a gene vector which always leads to the integration of the newly introduced genes into the genome of the respective host plant, as is the case for T-DNA mediated gene transfer, an autonomously replicating multicopy vector analogous to bacterial or yeast plasmids would be of interest.

Viruses fulfill these criteria and consequently a wide variety of bacterial and animal viruses were developed into efficient expression and transduction vectors. The vast majority of plant viruses, however, contain RNA as the genetic material (Matthews, 1981). Since RNA is not as readily accessible to genetic engineering techniques as is DNA, plant viruses containing a DNA genome have attracted far more interest with respect to their potential use as gene vectors. The only two groups of plant viruses to have a DNA genome are the caulimoviruses and the geminiviruses.

The biology of the geminiviruses has been reviewed by Goodman (1981); Harrison (1985) and Stanley (1985) and is well covered by Davies (Chapter 2; this volume).

Caulimoviruses have been reviewed extensively by Shepherd (1976; 1979) and Shepherd and Lawson (1981). The molecular biology of caulimoviruses has been covered by Howell (1985); Covey (1985) and Maule (1985b). The potential of DNA plant viruses, in particular cauliflower mosaic virus, as vectors has been addressed by Hohn *et al.*, 1982); Hohn and Hohn (1982); Howell (1982); Hull and Davies (1983) and Gronenborn (1984).

The intention of this article is to briefly summarize some details of the life cycle of cauliflower mosaic virus (CaMV) and to highlight the progress which has been made as well as the difficulties that were encountered in trying to develop cauliflower mosaic virus as a gene vector for higher plants.

II. The Biology of Cauliflower Mosaic Virus

Cauliflower mosaic virus (CaMV) is the type member of the caulimovirus group, small isometric plant viruses with a double-stranded DNA genome (Shepherd *et al.*,1968; Matthews, 1981). Members of the same group are Carnation etched ring virus (CERV), Dahlia mosaic virus (DaMV), Figwort mosaic virus (FMV), Mirabilis mosaic virus (MMV), Strawberry vein banding virus (SVBV), Thistle mottle virus (ThMV) and Soybean chlorotic mottle virus (SoyCMV) (Maule, 1985b).

All members of the caulimovirus group have a fairly restricted host range; CaMV for instance is naturally restricted to the *cruciferae* but may experimentally be transmitted to *Datura stramonium* or *Nicotianana cleve-*

landii (Lung and Pirone, 1972; Hills and Campbell, 1968). Recently an aphid transmissible strain of CaMV has been characterized which is able to systematically infect *Datura stramonium* and a variety of *Nicotiana* species (Schoelz *et al.*, 1986 a).

The standard experimental host plant for CaMV, however, is *Brassica campestris, var. rapa,* a turnip. Following infection, this host plant exhibits symptoms of disease ranging from severe stunting and necrosis of the leaves to almost an invisibly mild vein clearing without any gross growth interference. The extent of these symptoms varies between the different isolates of CaMV and may be influenced by environmental conditions such as the growth temperature of the host plants (Yamaoka *et al.*, 1982).

In nature CaMV is spread by aphids *(Myzus persicae)* in a semi-persistent manner. In addition, the virus is easily transmitted mechanically by rubbing it onto the leaves of susceptible plants. The successful transmission by an aphid vector is dependent on a virus gene encoding the so called aphid acquisition factor protein (Lung and Pirone, 1973; Lung and Pirone, 1974). Two variants of CaMV have been isolated from natural virus populations which are no longer transmissible by aphids. They may, however, be transmitted in mixed populations together with wildtype virus (Lung and Pirone, 1973). One of these isolates (CM 4-184) is a deletion mutant of 421 base pairs in the aphid acquisition factor gene (Hull, 1980; Howarth *et al.*, 1981).

A. The Virus Particle

The virion of CaMV is an icosahedron with a diameter of about 50 nm, and a molecular weight of 22.8 million daltons (Hull *et al.*, 1976 a). In preparations stained with uranylacetate the virus particles often appear empty. Based on neutron diffraction data Chauvin *et al.*, 1979) proposed that the viral DNA is tightly associated with the inner surface of the capsid. In recent years it has gradually become apparent that this seemingly "empty" particle does in fact harbour a variety of proteins.

The coat protein(s) of CaMV were found to have a molecular weight of 58 K and 44 K or 42 K (Hahn and Shepherd, 1980; Hahn and Shepherd, 1982; Al Ani *et al.*, 1979). Both species are phosphorylated, the 58 K species being the precursor of the 44 K virion protein. Whether the conversion of the 58 K protein into the 44 K species is a natural step in the assembly of the virus particle is not known.

Recently Menissier-De Murcia *et al.* (1986) and Martinez-Izquierdo and Hohn (1987) reported the association of a protein kinase activity with purified CaMV particles, the biological role of this protein kinase, however, remains unknown.

In addition to the phosphorylation, a glycosylation of the capsid protein(s) has been reported (Du Plessis and Smith, 1981). Furthermore, the association of a DNA-polymerase activity and a small-sized DNA-binding protein with the CaMV particle have been reported (Menissier *et al.*, 1984; Giband *et al.*, 1986). (See part II. C. 1).

B. The DNA of CaMV

The genome of CaMV is a double-stranded DNA molecule of about 8 kb in size. DNA isolated from virus particles appears as a variety of entangled and twisted circular molecules which are not covalently closed (Shepherd *et al.*, 1970; Shepherd and Wakeman, 1971; Hull and Shepherd, 1977; Civerolo and Lawson, 1978). Rather, they contain discontinuities which are sensitive to S1-nuclease in both strands of the DNA. These discontinuities, originally called "nicks" or "gaps", have been mapped on the CaMV DNA by Volovitch *et al.* (1978) and Hull and Howell (1978). Later, the first DNA sequence of a CaMV strain (isolate Cabb-S) by Franck *et al.* (1980) and the further analysis by Richards *et al.* (1981) revealed, that the DNA forms short triple-stranded overlaps at the positions of the "gaps". Complete copies of the CaMV genome, cloned in plasmid vectors and propagated in *E. coli*, do not contain these peculiar structures anymore since they are eliminated during replication in the bacteria. Nevertheless, cloned CaMV DNA, covalently closed at the former dicontinuities, is fully infectious (Howell *et al.*, 1980; Lebeurier *et al.*, 1980), yet following passage through plants the virus progeny has acquired the discontinuities at their fixed positions again.

These remarkable structures in the DNA of CaMV are a common feature of all caulimoviruses (Hull and Donson, 1982; Richins and Shepherd, 1983; Donson and Hull, 1983). How these overlap structures are introduced and what their biological function might be remained obscure for some time, but their origin and role during the life cycle of the virus became immediately evident when the involvement of a reverse transcription step in the replication of the virus was proposed. (See section II, D.)

C. The Genetic Organisation of CaMV

The DNA sequence of three different isolates of CaMV has been determined: Cabb-S (8024 bp) by Franck *et al.* (1980); CM 1841, a non aphid-transmissible strain (8031 bp) by Gardner *et al.* (1981); and the strain D/H (8016 bp), a high virulence strain by Balasz *et al.* (1982). The three isolates show only 5 % sequence divergence which is almost exclusively restricted to the coding regions. Most of the base changes occur at the third positions of codons, so they are neutral for the corresponding proteins. A comparison shows that the Cabb-S and CM 1841 strains are more related to each other than to the D/H strain.

From the DNA sequence six open reading frames (ORFs) have been deduced (Franck *et al.*, 1980), (Fig. 1). Two additional small ORFs were pointed out by Hohn *et al.* (1982). These two reading frames are the least conserved between the three isolates.

All reading frames are exclusively located on the plus-strand of the viral DNA. The first five are closely linked: only one base separates ORF I from II and II from III. The ORFs III and IV overlap for a short

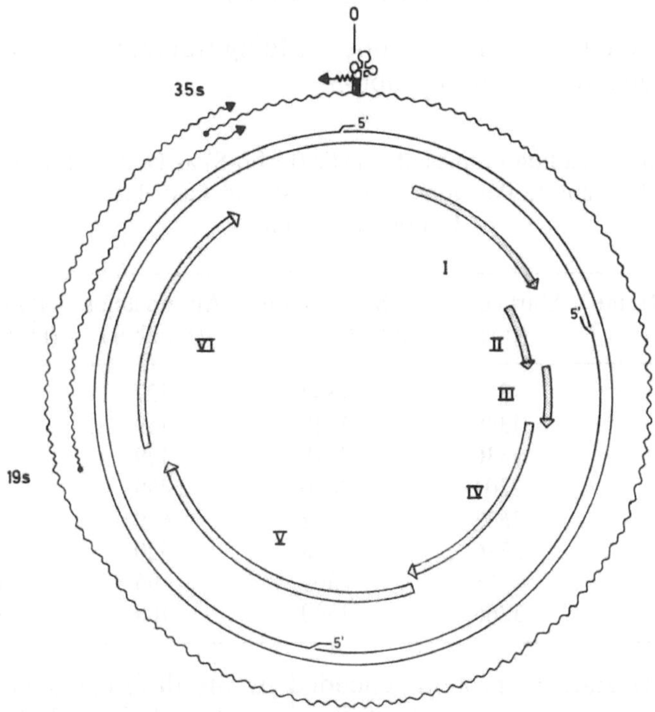

Fig. 1. The genome organisation of cauliflower mosaic virus. The CaMV chromosome is represented by the *circular double line.* Three short overlap structures (discontinuities or "gaps") are indicated. The single discontinuity of the minus DNA strand is defined as the zero point of the map; the base pair counting is clockwise. The two discontinuities of the plus DNA strand are located in ORF II and ORF V, only the discontinuity in ORF V is essential for replication. The minus strand primer binding site spans nucleotide position 1—14, the Met initiator tRNA primer is symbolized by its *cloverleaf* structure. The two major transcripts are shown as wavy lines, their start points are indicated by 19 S and 35 S respectively. The synthesis of the minus DNA strand by reverse transcriptase is primed by the tRNA hybridizing to the 35 S transcript and proceeds counterclockwise as indicated by the *zigzag arrow.* The genes (open reading frames) are shown as *dotted arrows* and numbered *I* through *VI.* The two promoters are located in the small intergenic region between ORF V and ORF VI (19 S promoter) and just after ORF VI in the large intergenic region (35 S promoter) as indicated

distance, as do ORFs IV and V (assuming that the first possible AUG codon deduced from the DNA sequence corresponds to the start codon of the respective gene). A small intergenic region of about 100 base pairs separates ORFs V and VI, a large intergenic region of about 1 kb is located after ORF VI.

A strikingly similar genetic organisation has been recently reported for the chromosome of carnation etched ring virus (CERV) (Hull *et al.,* 1986); possible implications of this similarity will be discussed later.

1. The Genes of CaMV

Table 1 shows the locations of the open reading frames and the size of the potential proteins for the strain Cabb-S.

Table 1. The Open Reading Frames of CaMV (Cabb-S) as Deduced from the DNA Sequence. The Nucleotide Positions of the First and Last Base of the Respective Codon are Given.

Open reading frame	Start codon (position on Cabb-S)	Stop codon (position on Cabb-S)	Amino acids (number)	Polypeptide (mol. Weight)
I	364	1347	327	37 K
II	1349	1828	159	18 K
III	1830	2219	129	15 K
IV	2201	3670	489	57 K
V	3633	5672	680	79 K
VI	5776	7338	520	61 K
VII	13	302	96	11 K
VIII	3266	3590	108	12 K

The identification of proteins encoded by the different open reading frames is almost complete. In the following, ORF I through VI will be called "genes" since the corresponding proteins have been identified.

The protein encoded by gene I very recently has been found as being part of the inclusion body in infected plants (Harker *et al.*, 1987; Martinez-Izquierdo *et al.*, 1987). Using antibodies directed against the ORF I product synthesized in *E. coli*, Harker *et al.* (1987) were able to identify polypeptides of 46 K, 42 K and 38 K molecular weights respectively. The apparently higher molecular weight as compared to the theoretical coding capacity of ORF I (36.9 K) might be due to post-translational modifications, e. g. glycosylation. Employing similar techniques Martinez-Izquierdo *et al.* (1987) identified a 41 K polypeptide encoded by gene I. The same polypeptide was synthesized when SP 6 derived mRNA comprising ORF I was translated *in vitro* in a rabbit reticulocyte lysate.

However, the function of the gene I product has not yet been clarified. Amino acid sequence comparisons yielded some limited homology with the 30 K protein of tobacco mosaic virus (TMV) (Hull *et al.*, 1986) and with the ATP-binding domains of some protein kinases (Martinez-Izquierdo *et al.*, 1987). Whether the gene I protein is involved in cell to cell movement, by analogy to TMV, or whether it plays a role in the process of virus replication remains a matter of speculation.

Gene II encodes the aphid acquisition factor as was suggested by Howarth *et al.* (1981). By deletion analysis and the construction of hybrid CaMV genomes between aphid transmissible and non-aphid transmissible strains Woolston *et al.* (1983) and Armour *et al.* (1983) have shown that the ability of the virus to be transmitted by aphids is specified by the 18 K polypeptide encoded in ORF II. This protein is known to affect the structure of the inclusion bodies or viroplasm (Givord *et al.,* 1984), but how this relates to aphid transmission ist still an open question.

Gene III encodes a 15K polypeptide found in nuclear fractions of CaMV infected turnip leaves as well as in inclusion body preparations (Xiong *et al.,* 1984). Recently Giband *et al.* (1986) were able to show that the gene III product (p15) is a DNA binding protein with a preference for double stranded DNA. The expression in *E. coli* of p15 in the form of a fusion protein with the first ten amino acids of β-galactosidase (Mesnard *et al.,* 1984) yields also a double-stranded DNA binding protein. *In planta,* this polypeptide is additionally processed to a 11K form which becomes associated with the virus particle and no longer binds DNA. Since it still reacts with an antiserum against a synthetic aminoterminus of p15, the DNA binding potential resides in the carboxy-terminal part of p15.

The coat protein of the virus has been tentatively assigned to gene IV on the basis of its amino acid composition in comparison with the amino acid sequence deduced from the first DNA sequence (Franck *et al.,* 1980). Later, Daubert *et al.* (1982) were able to prove this assignment by the finding that a protein expressed from ORF IV in *E. coli* reacted with antisera raised against purified capsid protein.

The identification of gene V as the virus polymerase gene soon followed the proposition of a replication model for CaMV involving a reverse transcription step (Pfeiffer and Hohn, 1983; Hull and Covey, 1983b). By comparison of the amino acid sequences of the reverse transcriptases of Rous sarcoma virus (RSV), Moloney murine leukemia virus (Mo-MuLV) and the predicted polymerase sequence of hepatitis B virus with the deduced amino acid sequence of CaMV ORF V Toh *et al.* (1983) and Volovitch *et al.* (1984) detected striking homologies between these proteins. Only ORF V or CaMV has the potential to code for a protein similar in size to retoviral reverse transcriptases.

Several groups have partially purified the CaMV gene V protein from infected cells or from engineered bacteria or yeast cells (Volovitch *et al.,* 1984; Pfeiffer *et al.,* 1984; Ziegler *et al.,* 1985; Takatsuji *et al.,* 1986). Taken together, the results provide strong evidence that the gene V of CaMV encodes a viral polymerase with reverse transcriptase activity.

Gene VI was the first gene of CaMV to have its product assigned. Odell and Howell (1980) have shown that a virus specific polyadenylated m-RNA (19S RNA) directs the synthesis of 66K protein in an *in vitro* translation system. By hybrid arrested translation experiments they and Al Ani *et al.* (1980) mapped the RNA to the region of ORF VI on the CaMV genome. Hybrid release translation experiments by Covey and Hull (1981) and aminoterminal sequencing of the *in vitro* translation product, combined with immunological reaction with an antiserum against the inclusion body protein carried out by Xiong *et al.* (1982), provided final proof that gene VI encodes the inclusion body or viroplasm protein.

The viroplasm or inclusion body has been found to be the intracellular site of virus replication (see II. D.). In addition to that, Schoelz *et al.*

(1986b) have identified the aminoterminal part of the inclusion body protein as being responsible for the systemic spread of the virus in various host plants. They exchanged a part of gene VI from the strains Cabb-B and CM 1841 which are restricted to necrotic local lesion formation in *Datura stramonium* (hypersensitive response) for the homologous region of the systemic strain D4. The development of either a local lesion or a systemic response on *Datura stramonium* and *Nicotiana bigelovii* correlated with the presence of the 5′ half of the inclusion body gene (VI) from the restricted or systemic strains respectively.

No polypeptides or functions have been assigned to the open reading frames VII and VIII. Although frameshift and deletion mutations in ORF VII did not interfere with virus multiplication (Howell *et al.*, 1981; Daubert *et al.*, 1983; Dixon *et al.*, 1983), ORF VII probably is translated into a protein, the function of which is still a matter of speculation (Dixon *et al.*, 1986b). Gene VII might serve to regulate the initiation of translation of a polycistronic messenger RNA comprising the following genes I, II and III which it precedes (see Chapter II. B.).

2. Transcripts of CaMV

Only one strand of the CaMV chromosome, the *alpha* or (−) strand, is transcribed into RNA (Howell and Hull, 1978; Hull *et al.*, 1979). Two major polyadenylated transcripts have been isolated from infected leaf tissue. A small (19S) transcript covers ORF VI (Odell *et al.*, 1981) and is the message of p66, the matrix protein of the viroplasm (Odell *et al.*, 1980; Al Ani *et al.*, 1980; Covey and Hull, 1981). This RNA can be translated *in vitro* and is the only subgenomic RNA which has been unequivocally shown to be a messenger of a single individual gene of CaMV.

A large (35S) transcript covers the entire genome and is terminally redundant by 180 bases (Covey *et al.*, 1981; Guilley *et al.*, 1982). Detailed mapping by S1 nuclease and primer extension confined the start point of the 19S transcript 11 (Covey *et al.*, 1981) or 12 bases (Guilley *et al.*, 1982) in front of the first ATG codon of ORF VI. Its 3′ end was mapped to position 7615, the same position as the 3′ end of the major 35S transcript (Guilley *et al.*, 1982; Covey *et al.*, 1981). This implies that the 19S and the 35S transcripts coterminate. Al polyadenylation signal (AATAAA) is located 18 bases upstream of the common 3′ end of the two transcripts.

The 5′ end of the major 35S transcript has been mapped to position 7435 on the CaMV genome (Guilley *et al.*, 1982). Both, the start points of the 19S and the 35S transcript are preceded by typical TATAAA and CAAT boxes characteristic for eukaryotic RNA polymerase II initiation sites (Breathnach and Chambon, 1981).

Thus, a moderately strong promoter is located in the small intergenic region preceding ORF VI (the 19S promoter) and a very strong one at the 5′ extremity of the large intergenic region (the 35S promoter) (Odell *et al.*, 1985). See Fig. 1.

This is in agreement with the results of Guilfoyle (1980), who has shown that in isolated nuclei from CaMV infected leaf cells RNA polymerase II synthesizes virus specific RNA. In addition, Olszewski *et al.* (1982) were able to isolate the DNA of CaMV as transcriptionally active covalently closed minichromosomes from the nuclei of infected turnip leaves.

The discovery of the covalently closed form of CaMV DNA resolved the puzzle of how RNA polymerase II was able to transcribe past the discontinuity in the minus-strand of the CaMV DNA. Non-encapsidated covalently closed molecules of CaMV DNA were also found by Menissier *et al.* (1982) and Hull and Covey (1983 a).

Other transcripts than the 19 S and the 35 S RNAs have been described by Condit *et al.* (1983 a; 1983 b), but their biological significance remains unclear.

However, Plant *et al.* (1985) were able to hybrid select a subgenomic RNA species that roughly covers ORF V of CaMV and can be translated *in vitro* into a polypeptide of 75 K, a size similar to that of the proposed reverse transcriptase. So far, this remains the only report on a subgenomic messenger RNA of CaMV apart from the 19 S RNA.

D. *The Replication of CaMV*

Kamei *et al.* (1969) and later Favali *et al.* (1973) presented evidence that the cytoplasmic inclusion bodies are the major site of CaMV replication. These results were disregarded for a while following the discovery of supercoiled forms of DNA and the minichromosome of CaMV in the nuclei of infected cells. However, the proposal by Summers and Mason (1982) of a reverse transcription step in hepatitis B virus replication inspired a rethinking of the replication mechanism of CaMV.

Piecing together the peculiar overlap structures of the CaMV chromosome and the terminally redundant genomic transcript (35 S RNA), several authors proposed a model for the replication of cauliflower mosaic virus (Pfeiffer and Hohn, 1983; Guilley *et al.*, 1983; Hull and Covey, 1983 b; Marco and Howell, 1984). The model is based on the involvement of reverse transcription converting the genomic 35 S RNA into a DNA copy. It elegantly explains the formation of the formerly quite enigmatic gap structures in the virion DNA of CaMV.

The 35 S transcript is the key replicative intermediate in the life cycle of CaMV, and a methionine initiator-tRNA is suggested to serve as a primer for the synthesis of the first (minus-DNA) strand by the reverse transcriptase (Pfeiffer & Hohn, 1983; Guilley *et al.*, 1983). In the region where the 35 S RNA is terminally redundant, the reverse transcriptase template switch its from the 5′ end of the 35 S RNA to the 3′ end of the 35 S RNA. Following this step, the synthesis of the (−) strand DNA may procede until a full length cDNA copy of the 35 S RNA has been completed.

An intermediate, indicative of this priming mechanism, consisting of a small DNA molecule of about 625 bases covalently attached to about 100 bases of RNA at its 5′ end (sa-DNA) has been found by Covey *et al.* (1983) and Turner and Covey (1984). This molecule is equivalent to the "strong-stop DNA" of retroviruses (Varmus, 1982). In addition to its occurence as free DNA in the cell, it is found tightly associated (encapsidated?) with mature virus particles (Turner and Covey, 1984; Maule, 1985a; Pietrzak and Hohn, 1985).

Initiation of the (+) strand DNA synthesis occurs at a site almost 180 degrees from the tRNA primer binding site on the circular genome of CaMV. How initiation at this site occurs has not been resolved. The initiation of the (+) strand synthesis at the position of the third discontinuity has been studied by Pietrzak and Hohn (1985). They found that the orientation of the respective signal sequence determines whether a discontinuity is introduced or not. The actual sequence which determines the initiation of (+) strand synthesis and thereby leads to the formation of a discontinuity is not precisely known. There is, however, some homology to the consensus sequence for the initiation of retroviral (+) strand synthesis (Pfeiffer and Hohn, 1985; Varmus, 1982; Covey, 1986).

This asymmetric mode of replication explains the formation of the overlap structures in either strand of the viral DNA as being the result of strand displacement at the termination sites of either DNA strand synthesis. The gapped molecules formed by some hepatitis viruses are analogous to the "overlapping" DNA molecules of CaMV (Summers and Mason, 1982; Marsh *et al.,* 1985; Mazzolini *et al.,* 1985).

Taken together, all these results reveal a striking similarity between caulimoviruses, hepatitis B viruses, and retroviruses (Pfeiffer and Hohn 1983; Varmus 1983). Furthermore, Toh *et al.* (1983) and Volovitch *et al.* (1984) found stretches of strong homology at the protein level between retroviral reverse transcriptases, hepatitis B virus polymerases, and the potential protein of CaMV ORF V. Subsequently, Volovitch *et al.* (1984); Pfeiffer *et al.* (1984); Ziegler *et al.* (1984); Mazzolini *et al.* (1985); Thomas *et al.* (1985) and Takatsuji *et al.* (1986) detected and partially purified an RNA-dependent DNA polymerase activity from either infected *Brassica* leaves or from engineered bacterial or yeast strains expressing the ORF V protein of CaMV.

Additional evidence for the involvement of a reverse transcriptase was obtained by Hirochika *et al.* (1985) who analysed and sequenced a deletion mutant of 856bp in genes I and II. This intrinsically non viable mutant occurs rather frequently in virion DNA preparations of strain Cabb-S (Japan) but not in CM 1841. The deletion end points match the consensus sequence of splice donor and acceptor sites. *In vitro* generated point mutations in the donor site led to the occurrence of altered deletions upon passage through the plants. They resulted from the activation of cryptic splice donor sites near the mutated original one. The occurence of site specific deletions of this kind in the DNA of CaMV can only be explained by the reverse transcription of spliced CaMV RNA. Whether spliced RNAs, however, play any role in the expression of the CaMV genome remains an open question.

Based on the „retroid virus" model for the life cycle of CaMV B. Hohn *et al.* (1986) designed an elegant experiment to indirectly prove the involvement of the 35S RNA as a replicative intermediate. They con-

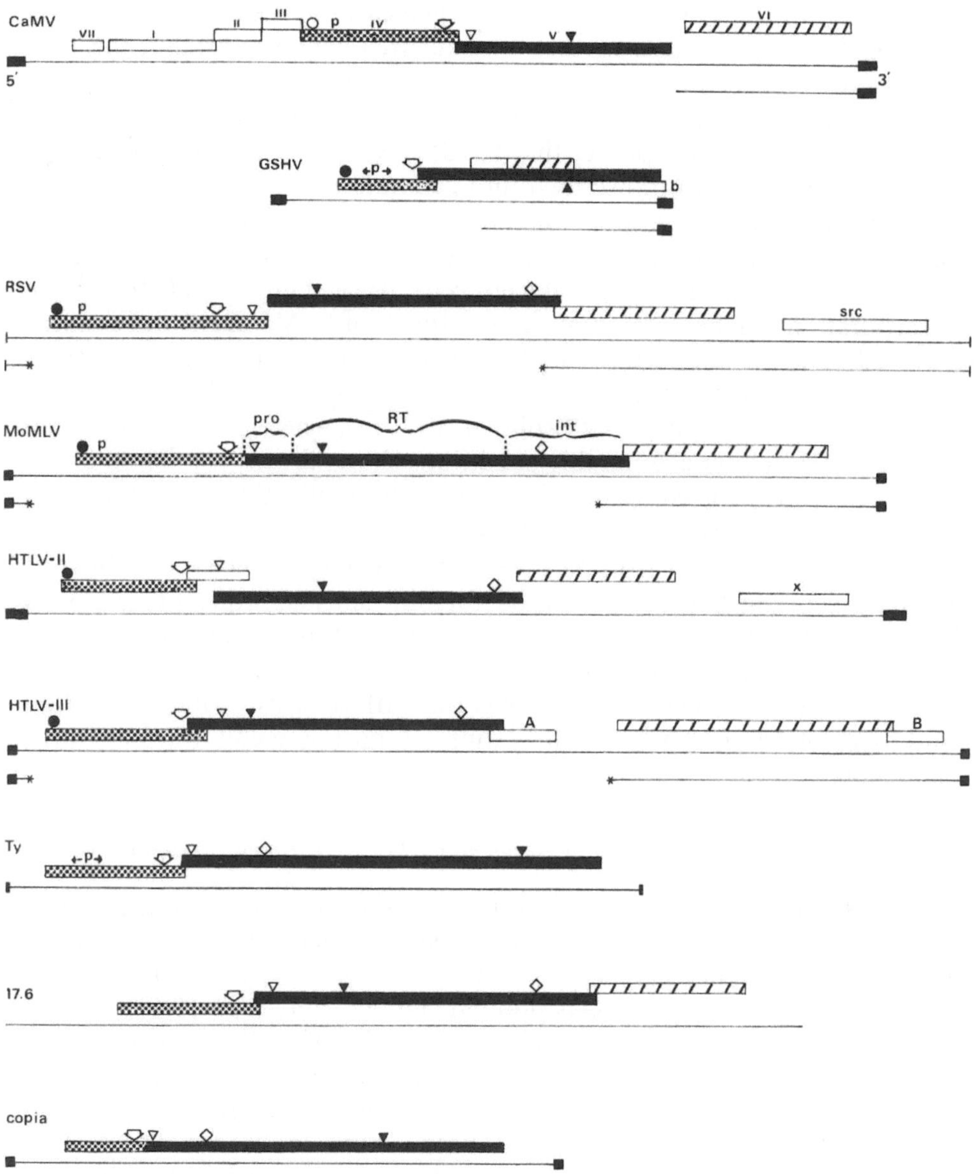

Fig. 2. Genome organization of cauliflower mosaic virus (CaMV), ground squirrel hepatitis virus (GSHV), Rous sarcoma virus (RSV), Moloney murine leukaemia virus (MoMLV), human T cell lymphotropic viruses II and III (HTLV-II and HTLV-III), yeast Ty element, *Drosophila copia*-like element 17.6 and *Drosophila copia* element. Each is shown on the genome-length RNA (—), the terminal repeats being indicated by (■ – ■). the subgenomic mRNA for the *gag* gene or its equivalent is shown below the genomic RNA; * indicates a splice. The *gag* (▒), *pol* (▬) and *env* (▨) genes are highlighted. Genes I to VII of CaMV, the b gene of GSHV, the *src* gene of RSV, the X gene of HTLV-II and the A and B genes of HTLV-III are noted. The extents of the protease (pro), the reverse transcription (RT) and the integrase (int) domains of MoMLV and various features and regions common to all or most genomes are shown: ○, Hydrophilic, ●, hydrophobic; p, phosphoprotein; ⬠, nucleic acid binding domain; ▽, protease; ▼ reverse transcriptase; ◇ integrase. (Taken from Hull and Covey (1986); with permission)

structed *in vitro* a derivative of CaMV, which carries an intron in the region of gene II. After passage of this mutant virus genome through the host plants they indeed found, that the intron sequence was removed correctly at the respective splice donor and acceptor sites. The intronless CaMV genome thus recovered, proves that the 35 S RNA is processed like a messenger RNA and retro-transcribed into genomic DNA.

It is still an open question whether, in addition to the retroid mode of replication, CaMV might be able to replicate via its supercoiled minichromosome in the nucleus of infected cells. Ansa *et al.* (1982) reported the incorporation of label into open circular forms of CaMV DNA in *bona fide* purified nuclei of infected tissue. Under the rather artificial conditions of tissue culture, the persistance of cauliflower mosaic virus DNA as supercoiled form has been observed (Rollo *et al.*, 1985), whereas figworth mosaic virus (FMDV) under comparable conditions shows the accumulation of the same replicative intermediates found in cells replicating CaMV via the reverse transcription pathway (Rollo *et al.*, 1986).

An accumulation of virus particles in the nuclei of infected cells instead of the cytoplasm has been reported for one unusual strain of CaMV in *Nicotiana clevelandii* (Gracia *et al.*, 1985) as well as for carnation etched ring virus (CERV) in *Saponaria vaccaria* (Hearon and Lawson, 1981). It is however not clear, whether this behaviour reflects a true viral replication and assembly step, or whether it is simply a post-assembly transport phenomenon.

E. Structural Homologies Between CaMV and Retroid Elements

An intriguing similarity in structural organization between cauliflower mosaic virus, ground squirrel hepatitis virus, some retroviruses and the transposing elements Ty from yeast and the *copia* and *copia*-like element 17.6 from *Drosophila* has been pointed out by Toh *et al.* (1985), Covey (1986) and Hull and Covey (1986). According to the alignment shown in Fig. 2, the coat protein of CaMV would be equivalent to the *gag* gene of the retroviruses. In the carboxy-terminal part of these proteins the amino acid motif . .-C-X_2-C-X_4-H-X_4-C-. . has been found to be conserved, at a position within a region that encodes a highly basic nucleic acid binding protein (NBP). The same sequence is found to be encoded at the end of the coat protein gene of CaMV.

The similarities between the CaMV gene V, the hepatitis virus *pol* gene and the retroviral reverse transcriptases have been mentioned in the previous chapter. The positions of the protease domain and the reverse transcriptase domain within ORF V of CaMV and the respective *pol* regions of different retroviruses and reverse transcribing elements are indicated in Fig. 2.

A more detailed and comprehensive illustration of the close relationship between CaMV and other reverse transcribing elements is given by Hull and Covey (1986), Fuetterer and Hohn (1987) and Mason *et al.* (1987), to whom the reader is kindly referred.

III. The Development of CaMV into a Plant Gene Vector

A. *Mutants of CaMV*

1. *Mutations in ORF II*

In order to use cauliflower mosaic virus as a vector for the introduction and expression of foreign genes in plants, it was necessary to identify sites or regions on the CaMV genome, into which additional genes might be inserted or which may be replaced by the genes of interest. The only region that seemed to be dispensable for the replication of the virus was the area of the genome later defined as ORF II, the aphid acquisition factor gene. A deletion in the genome of the natural mutant strain CM 4-184 was mapped in this region of the CaMV chromosome (Hull, 1980).

We reasoned that those parts of the genome which can tolerate a deletion, and hence do not code for any function vital to viral propagation, might also be replaced by some additional DNA. Therefore we started to modify a unique restriction site (*Xho* I) in strain CM 1841 which is located in the part of the genome, deleted in strain CM 4-184.

Virion DNA of CM 1841 was cut with *Xho* I, filled *in vitro* by polymerase I, ligated and used to infect turnip plants. Indeed, *Xho* I resistant mutants could be recovered from the virus progeny. The DNA sequence of these mutants revealed, however, that the mutations were not the anticipated frame-shift mutations, but rather were in-frame additions of only three instead of four bases at the *Xho* I site (Gronenborn *et al.,* 1980). Later, the determination of the first DNA sequence of a CaMV strain by Franck *et al.* (1980) identified an open reading frame (ORF II) covering this site of manipulation. (See part III. B. for further explanation of these results.)

These initial experiments were carried out using uncloned virion DNA of cauliflower mosaic virus. A prerequisite, however, for an efficient *in vitro* manipulation and mutagenesis of the CaMV genome are infectious molecules of the viral genome derived from DNA cloned in *E. coli.* Howell *et al.* (1980) were the first to demonstrate that cloned copies of CaMV DNA retain their infectivity after they are released from the bacterial vector and used to infect turnip plants. Thereby the earlier report of Szeto *et al.* (1977), who stated that the gap structures might be necessary for infectivity of the CaMV DNA, was disproven. Working with another isolate of CaMV (Cabb-S) Lebeurier *et al.* (1980) were also able to show infectivity of the cloned CaMV DNA.

Using a cloned copy of the CaMV strain CM 1841 (Gardner *et al.,* 1981) we found that small sized insertions (65 bp or 256 bp) were tolerated by CaMV when inserted into ORF II (Gronenborn *et al.,* 1981). There is, however, a limitation to the size of the additional DNA which can be stably maintained and propagated within the viral genome during its systemic spread throughout a plant. An additional 531 bp fragment could not be recovered from encapsidated viral DNA. Similar results have been obtained by Daubert *et al.* (1983) and Dixon *et al.* (1983). The largest additional piece of DNA which has been found to be encapsidated has a length of 351 base pairs (Gronenborn, unpublished).

2. *Mutations in Other Regions of the CaMV Genome*

The effects of small sized (10 to 30 bp) insertions at various sites in the genome of CaMV were studied by Howell *et al.* (1981); Daubert *et al.* (1983); Dixon *et al.* (1983); Choe *et al.* (1985) and Melcher *et al.* (1986a). Only ORF II and the large intergenic region were found to tolerate both small sized and larger (up to 250 bp) insertions.

Two other open reading frames maintained small insertions, but only at particular positions: 12 bp in frame insertions at two different sites in ORF VI leading to the addition of four amino acids were found to be infectious (Daubert *et al.,* 1983). This part of the ORF VI, the gene coding for the inclusion body protein, is the most variable between the different CaMV isolates sequenced (Balazs *et al.,* 1982).

A peculiar mutation in gene III has been obtained by Melcher *et al.* (1986a). They changed the sequence at position 2041 (Cabb-S) by deleting four nucleotides and inserting additional 16 base pairs, resulting in a change of the amino acid sequence from .-T-. to .-P-G-F-R-A-. . at position 72 of the gene III protein. This mutant produces normal symptoms of infection but is unable to coexist with the wildtype Cabb-S strain. The reason for this behaviour is unknown.

A 10bp frameshift linker inserted at the extreme end of ORF IV alters the carboxyterminal sequence of the coat protein and also shortens it by three amino acids. The mutation turned out to be viable but leads to a delay in the appearance of disease symptoms (Dixon *et al.,* 1983). It is interesting to note that this particular insertion mutation causes a frameshift in ORF V between the first and second ATG codon deduced from the DNA sequence. Therefore the first ATG codon predicted by analysis of the DNA sequence might not be the actual initiation codon for protein synthesis of ORF V. The intercistronic distance between the termination codon of ORF IV and the second ATG codon of ORF V is only seven base pairs.

B. *Translational Polarity*

In order to define the maximal size of additional DNA that can be propagated by CaMV and still lead to a systemic spread of the hybrid virus, we carried out a more extended study of the effect of "medium-sized" (351bp) insertions into ORF II of strain CM 1841 and into the truncated short ORF II of strain CM 4-184 (Howarth *et al.,* 1981). Based upon the sequence analysis of a variety of insertion induced deletions, which had arisen during passage of the hybrid viral DNA through the plants, we found an influence of nonsense mutations in ORF II on the stability of the mutant CaMV genome (Sieg and Gronenborn, 1982). The same was true for single frameshift mutations in ORF II.

This behaviour strongly resembles the polar effects of nonsense and frameshift mutations observed in polycistronic bacterial operons.

These results led us to hypothesize that the genes I, II and III of CaMV are translated from a polycistronic mRNA. The first three reading frames are closely linked with an intercistronic space of only one base. In CaMV strain D/H the stop codon of ORF I and the start codon of ORF II overlap in the sequence motif "ATGA" (Balazs *et al.*, 1982), in which the first three bases are the start codon of ORF II, the last three form the stop codon of ORF I.

A mechanism by which eukaryotic ribosomes are able to translate a polycistronic messenger bearing closely linked reading frames is described by the "relay-race" model of translation (Sieg and Gronenborn, 1982; Sieg and Gronenborn, in preparation). After finishing protein synthesis at the stop condon of one open reading frame, the ribosomal subunits do not immediately dissociate from the messenger RNA but continue the process of scanning for a certain distance. If an AUG codon occurs within about the next 50 nucleotides, translation is resumed again. Open reading frames interlinked by only small intercistronic distances as for instance the ORFs I, II and III of CaMV, will be passed on to the same ribosome like the bar in a relay-race.

In this context, it is interesting to note that all open reading frames on the genome of carnation etched ring virus (CERV), another caulimovirus whose DNA sequence has been determined recently, are linked by the sequence motiv "ATGA" (Hull *et al.*, 1986). A similar arrangement of reading frames is found also in bacteriophage *lambda,* where 13 open reading frames are linked in this way (Kröger and Hobom, 1982).

Furthermore, the DNA sequence analysis of virus progeny derived from infections with the CaMV deletion mutant SA 103 of the aphid acquisition factor gene (ORF II, Armour *et al.*, 1983) revealed that an additional base had become deleted by selection *in planta,* and hence a contiguous truncated open reading frame II is restored (Melcher *et al.*, 1986a).

A different model for the expression mechanism of a potential 35 S- or 35 S-derived polycistronic messenger RNA of CaMV has been proposed by Hull (1984). He found stretches of complementarity between the 18 S ribosomal RNA and a leader sequence located between base position 270 and 280 on the CaMV genome. This sequence is followed by a "selector sequence" which in turn is complementary to sequences preceeding the individual open reading frames. It is suggested that a polycistronic messenger RNA of CaMV is folded in such a way as to bring the initiation codons of the open reading frames into close proximity with a ribosome by hybridizing with the selector sequence.

However, the existence of fully viable deletion mutants in which both the complementary sequence to the ribosomal 18 S RNA and the selector sequence are missing (Dixon and Hohn, 1984; Gronenborn, unpublished) casts doubt on the involvement of the proposed mechanism in translation initiation.

It is an intrinsic feature of the relay-race model described previously, that any open reading frame replacing the original ORF II of CaMV has to be translated into protein, since translation of a reading frame preceeding

ORF III is necessary to ensure the efficient continuation of translation of ORF III. Gene III encodes a protein indispensable for the multiplication of the virus.

In conclusion, the relay-race model of translation makes a clear-cut prediction on how foreign genes are to be inserted into the CaMV chromosome, in order not to interfere with the specific translation mechanism of the virus, and to become expressed and remain stably incorporated within the virus genome:

1. The insertion of genes replacing ORF II of CaMV should not cause any polar effects on the expression of the genes downstream of ORF II.
2. The intercistronic distance between ORF I, any additional new gene and ORF III should be as small as possible; preferably a few bases only.

C. Transducing Cauliflower Mosaic Virus Variants

The exact application of the predictions made by the relay-race model of translation on the design of transducing CaMV variants led to the first viable and stable transducing cauliflower mosaic virus expressing a foreign gene in plants (Brisson *et al.*, 1984). The gene coding for a methotrexate insensitive dihydrofolate reductase (DHFR), derived from the R67 R-factor of *E. coli* (O'Hare *et al.*, 1981), was tailored in such a way as to precisely fit into a deletion of the entire ORF II. The DHFR gene is now the new ORF II; it has a length of 240 bp. The 5′ intercistronic distance between ORF I ist 9 base pairs and the 3′ intercistronic distance to ORF III is one base pair as in the wildtype virus. In a second construction which is still suboptimal with regard to the relay-race model, and which has a longer intercistronic distance between ORF I and the start codon of the DHFR gene (31 bp plus a small part of reading frame II), the genome of the hybrid virus is less stable and the insert is gradually lost during repeated passages of infection. In the tight construction described, first the insert is stably maintained even after repeated passages through plants. The DHFR protein was readily detected in western blots; its activity was assayed by measuring the methotrexate insensitive in corporation of ^{32}P-labeled phosphate into total plant DNA (Brisson *et al.*, 1984). Both transducing cauliflower mosaic viruses render infected plants partially resistant to methotrexate.

Besides the DHFR gene, the coding sequence for a human interferon (IFN-*alpha*-D) has been inserted into the CaMV genome replacing ORF II according to the rules of the "relay-race". It is maintained fairly stably and leads to the expression of active interferon (Penswick and Hohn, *pers. comm.)*.

D. Defective Complementing Mutants of CaMV

Although it has been shown that CaMV may be used to transfer and express new genes in plants, the size limitations imposed on the additional DNA that is encapsidated into virus particles is still unsatisfactory. Furthermore, the exact tailoring of foreign genes replacing ORF II, which is necessary to avoid interference with the relay-race expression of the virus genome, makes CaMV difficult to use.

Therefore the development of a complementation system between a helper virus and a defective transducing virus or between two defective virus mutants is most desirable.

In the past, numerous attempts to establish complementation between defective mutants of CaMV were started, but have so far always failed. In all cases, only wild-type recombinants were recovered (Howell *et al.*, 1981; Lebeurier *et al.*, 1982; Walden and Howell, 1982; Walden and Howell, 1983; Choe *et al.*, 1985 and Melcher *et al.*, 1986b). This apparent lack of complementation has been attributed to either efficient homologous recombination or to the continuous template switching activity of the reverse transcriptase during replication of the viral genome. The fact that template switching of the reverse transcriptase is a frequent event in the life of CaMV, and that it even might contribute to the evolution of CaMV by generating diversity between different strains has been shown by Dixon *et al.* (1986a) and Grimsley *et al.* (1986b).

In order to avoid the deleterious influence of both recombination and template switching on the coexistence of two defective mutant CaMV genomes in the same plant, we constructed pairs of overlapping deletions which lack sequence homology (Fig. 3).

These mutants spanning ORF I plus II and ORF II and III failed to cause symptoms of infection on turnip plants when inoculated separately. However, when they were used to co-infect plants, the symptoms of disease appeared with almost normal speed and severity. Analysis of the viral progeny showed no recombinant virus that was infectious on its own.

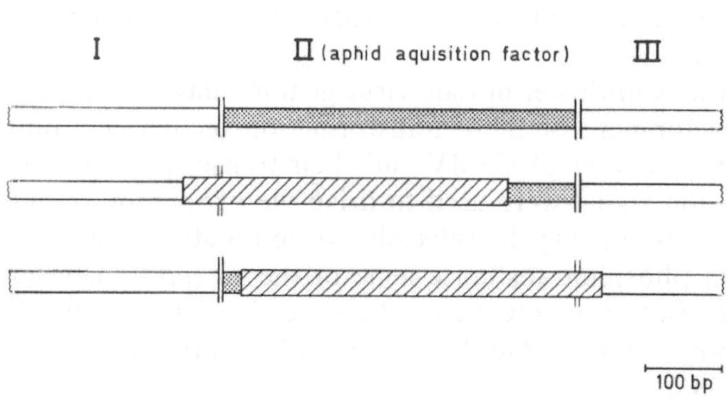

Fig. 3. Defective complementing deletions of CaMV.
Fig. 3 shows a pair of deletion mutations which are able to complement each other without recombining to nondefective virus progeny. I, II, III, indicate the genes I, II, III of CaMV. the non-essential gene II (the aphid acquisition factor gene) is highlighted as dotted area. The extent of the respective deletions is given by the hatched bar. WT = wildtype; I^-, II^-, III^+ and I^+, II^-, III^- denote the genotype of the mutants

Using this approach we have created CaMV variants with artificial bipartite genomes that are able to fully complement one another and spread systemically throughout the host plant as defective viruses (Gronenborn *et al.*, to be published).

Taking advantage of overlapping deletion mutants, is seems feasible to accommodate larger genes into the virus genome and to achieve the systemic spread of defective transducing particles.

IV. Elements of CaMV as Tools in Plant Genetic Engineering

In recent years considerable progress has been made in the manipulation of plant cells at the level of protoplasts. Transformation of plant protoplasts without the help of *Agrobacterium* mediated gene transfer has become a routine procedure for several plant species.

Two different techniques for introducing DNA into plant cells devoid of their cell wall have been developed: A treatment of the protoplasts with chemicals that neutralize the negative charge of the protoplast membrane, e. g. polyethylene glycol in combination with divalent cations (Krens *et al.*, 1982; Hain *et al.*, 1985; Meyer *et al.*, 1985) or an electrical pulse treatment to discharge the membrane for a short period of time (Zimmermann and Vienken, 1982; Fromm *et al.*, 1985).

Both methods, which had been previously used for the transformation of animal cells (Graham and van der Eb, 1973; Neumann *et al.*, 1982) and the fusion of plant protoplasts (Zimmermann and Scheurich, 1981) can be applied as well to force unprotected DNA out of solution through the membrane of a protoplast into the cytoplasm of the cell. In an as yet not understood process this DNA finds its way into the nucleus and may become expressed either in a transient fashion (Fromm *et al.*, 1985) or become a stably incorporated new constituent of the cell genome (Paszkowski *et al.*, 1984).

Elements of the cauliflower mosaic virus genome have been invaluable tools for the development of these transformation techniques. Both, the 19S and the 35S promoter of CaMV and their transcription termination and polyadenylation signal were used to drive the expression of chimeric gene constructions employing the selectable drug resistance gene (APH 3′ II), the neomycin phosphotransferase gene of the *E. coli* transposon *Tn5* (Beck *et al.*, 1982). Both promoters are active in cells of monocotyledonous and dicotyledonous plants (Potrykus *et al.*, 1985; Fromm *et al.*, 1986; Pietrzak *et al.*, 1986).

During the establishment of the "naked DNA" transformation technique, the CaMV promoters played the same important role as did the nopaline synthase promoter of *Agrobacterium tumefaciens* (Depicker *et al.*, 1982) for the development of the first chimeric genes for plant transformation (Herrera-Estrella *et al.*, 1983).

Structural and functional analyses of the CaMV 35S promoter, and the rbcS gene E9 promoter of pea, defined the sequences necessary for promoter function and revealed the very high relative strength of the constitutive 35 S promoter in comparison to a light regulated rbcS promoter of pea and the nopaline synthase promoter of *Agrobacterium tumefaciens* (Odell *et al.*, 1985; Morelli *et al.*, 1985).

This very high level of constitutive transcription found in a wide variety of monocot and dicot plants led to the 35 S promoter of cauliflower mosaic virus becoming one of the most commonly used promoters for the non-regulated high expression of new traits in chimeric gene constructions (Bevan *et al.*, 1985; Powell Abel *et al.*, 1986; Ecker *et al.*, 1986).

An interesting further application of cauliflower mosaic virus as a whole is its use in studying the process of transfer of the T-DNA from *Agrobacterium tumefaciens* into the genome of different acceptor plants, as described recently by Grimsley *et al.* (1986 a) and Hille *et al.* (1986). The term "agroinfection" was coined for this way of surmounting the natural host range barriers of viruses and *Agrobacterium tumefaciens*. The same technique had been used before by Shewmaker *et al.* (1985) to transfer CaMV into tobacco and by Gardner and Knauf (1986) for the transfer of viroid cDNA sequences into tomato plants.

A more detailed description of the "agroinfection" and its applications is given by Grimsley and Bisaro in Chapter 4 of this volume.

V. Vectors Based on Other Plant Viruses

The development of plant viruses other than CaMV into autonomously replicating vectors is still in its infancy. Out of the large group of RNA viruses, only brome mosaic virus (BMV) has been successfully engineered to replicate and express a foreign gene (French *et al.*, 1986).

The genome of brome mosaic virus consists of three different RNAs of messenger polarity. RNA 3 is dicistronic and its second cistron encodes the coat protein of BMV. In infected cells a subgenomic mRNA species (RNA 4) comprising only the coat protein cistron accumulates to high amounts (Ahlquist *et al.*, 1981).

For the *in vitro* transcription of viral cDNAs by *E. coli* RNA polymerase Ahlquist and Janda (1984) designed a special vector which they have used to produce large amounts of the brome mosaic virus RNAs. In subsequent infection experiments with barley protoplasts and plants all three BMV RNAs synthesized *in vitro* were shown to replicate (Ahlquist *et al.*, 1984).

Using RNA of clones in which they had replaced the coat protein cistron of RNA 3 by the bacterial chloramphenicol acetyltransferase gene (CAT), French *et al.* (1986) were able to detect replication and expression of the new gene in the barley protoplasts. It has not been reported whether or not this hybrid RNA becomes encapsidated when it is used together with wildtype RNA 3 in mixed infections on whole plants.

Recently, the second group of DNA plant viruses, the geminiviruses, have been attracting considerable interest as potential candidates for the development of independently replicating viral vectors. Unlike the caulimoviruses, geminiviruses infect a broad range of host species both within the mono- and dicotyledonous plants.

We have cloned the DNA of wheat dwarf virus (WDV). WDV is a geminivirus infecting grasses, which has a monopartide genome and is transmitted by leafhoppers (Lindsten *et al.*, 1970; Lindsten *et al.*, 1980). We have cloned and studied its replication and expression in *Triticum monococcum* and *Zea mays* suspension culture cells. Protoplasts of the respective cells were transfected by double-stranded WDV DNA either as a linear molecule of total genome size or in the form of a dimeric WDV unit in a bacterial plasmid. Both forms of cloned WDV DNA led to the formation in protoplast derived cells of the same replicative intermediates, and viral DNAs as are found in infected plants. In addition to the replication of the genome of wildtype WDV, we were able to obtain replication and expression of different foreign genes replacing the coding region of the viral capsid protein (Matzeit *et al.*, in preparation). The genes that have been successfully replicated and expressed in this way by the engineered wheat dwarf virus are the neomycin phosphotransferase gene, the chloramphenicol acetyltransferase gene and the β-galactosidase gene of *E. coli.*

These results make us feel rather optimistic about the further development of these viruses into useful autonomously replicating viral vectors.

VI. Conclusion and Outlook

Efforts at development of cauliflower mosaic virus as a viral vector for plant genetic engineering were accompanied by the constant emergence of new and unanticipated insights in the actual complexity of CaMV biology. The asymmetric replication of the CaMV genome via reverse transcription, and its peculiar mode of gene expression involving an as yet unidentified, but most probably polycistronic messenger RNA, are obstacles to a simple and straightforward use of CaMV as a vector. Nevertheless, with our current knowledge of the life cycle of the virus, it should be possible to increase the list of non-defective transducing CaMV variants and to establish a successful complementation between defective transducing CaMV variants.

Whether engineered RNA viruses will lead to applicable vector systems has been the subject of a recent and controversial discussion (van Vloten-Doting *et al.*, 1985; Siegel, 1985). The development of the first successful transducing brome mosaic virus (French *et al.*, 1986), however, definitely supports the more optimistic assessment of RNA virus vectors as it was expressed by Siegel (1985).

The very recent progress in our understanding of the biology of the second group of DNA plant viruses, the geminiviruses, will undoubtedly result in a plethora of new viral vectors for plant genetic engineering.

Acknowledgements

I thank B. Baker, P. Czernilofsky, I. Moore and S. Sieg for discussion and critically reading the manuscript.

Support came from the Bundesminister für Forschung und Technologie (Zentrum für Molekulare Gen- und Zelltechnologie: BCT 0365 2-1).

VII. References

Ahlquist, P., Janda, M., 1984: cDNA cloning and *in vitro* transcription of the complete brome mosaic virus genome. Mol. Cell. Biol. **4,** 2876—2882.

Ahlquist, P., French, R., Janda, M., Loesch-Fries, S., 1984: Multicomponent RNA plant virus infection derived from cloned viral cDNA. Proc. Natl. Acad. Sci. U.S.A. **81,** 7066—7070.

Ahlquist, P., Lucknow, V., Kaesberg, P., 1981: Complete nucleotide sequence of brome mosaic virus RNA 3. J. Mol. Biol. **153,** 23—38.

Al Ani, R., Pfeiffer, P., Lebeurier, G., 1979: The structure of cauliflower mosaic virus. Part 2: Identity and location of the viral polypeptides. Virology **93,** 188—197.

Al Ani, R., Pfeiffer, P., Whitechurch, O., Lesot, A., Lebeurier, G., Hirth, L., 1980: A virus specified protein produced upon infection by cauliflower mosaic virus. Ann. Virol. (Inst. Pasteur) **131 E** 33—53.

Ansa, O. A., Bowyer, J. W., Shepherd, R. J., 1982: Evidence for replication of cauliflower mosaic virus DNA in plant nuclei. Virology **121,** 147—156.

Armour, S. L., Melcher, U., Pirone, T. P., Lyttle, D. J., Essenberg, R. C., 1983: Helper component for aphid transmission encoded by region II of cauliflower mosaic virus DNA. Virology **129,** 25—30.

Balazs, E., Guilley, H., Jonard, G., Richards, K., 1982: Nucleotide sequence of DNA from an altered-virulence isolate D/H of the cauliflower mosaic virus. Gene **19,** 239—249.

Beck, E., Ludwig, G., Auerswald, E. A., Reiss, B., Schaller, H., 1982: Nucleotide sequence and exact localization of the neomycin phosphotransferase gene from transposon Tn 5. Gene **19,** 327—336.

Bevan, M. W., Mason, S. E., Goelet, P., 1985: Expression of tobacco mosaic virus coat protein by a cauliflower mosaic virus promoter in plants transformed by *Agrobacterium.* EMBO J. **4,** 1921—1926.

Breathnach, R., Chambon, P. 1981: Organisation and expression of eukaryotic split genes. Ann. Rev. Biochem. **50,** 349—383.

Brisson, N., Paszkowski, J., Penswick, J. R., Gronenborn, B., Potrykus, I., Hohn, T., 1984: Expression of a bacterial gene in plants by using a viral vector. Nature **310,** 511—514.

Chauvin, C., Jacrot, B., Lebeurier, G., Hirth, L., 1979: The structure of cauliflower mosaic virus. A neutron diffraction study. Virology **96,** 640—641.

Choe, I. S., Melcher, U., Richards, K., Lebeurier, G., Essenberg, R. C., 1985: Recombination between mutant cauliflower mosaic virus DNAs. Plant Mol. Biol. **5,** 281—289.

Civerolo, E. L., Lawson, R. H., 1978: Topological forms of cauliflower mosaic virus nucleic acid. Phytopathology **68,** 101—109.

Condit, C., Hagen, T. J., McKnight, T. M., Meagher, R. B., 1983 a: Characterization and preliminary mapping of cauliflower mosaic virus transcripts. Gene **25,** 101—108.

Condit, C., Meagher, R. B., 1983b: Multiple, discrete 35S transcripts of cauliflower mosaic virus. J. Mol. Appl. Genet. **2,** 301—314.

Covey, S. N., 1985: Organization and expression of cauliflower mosaic virus. In: Davies, J. W. (ed.), Molecular Plant Virology (Vol. 2), pp. 121—159. Boca Raton: CRC Press.

Covey, S. N., 1986: Amino acid sequence homology in the *gag* region of reverse transcribing elements and the coat protein gene of cauliflower mosaic virus. Nucl. Acids Res. **14,** 623—633.

Covey, S. N., Hull, R., 1981: Transcription of cauliflower mosaic virus DNA. Detection of transcripts, properties, and location of the gene encoding the virus inclusion body protein. Virology **111,** 463—474.

Covey, S. N., Lomonossoff, G. P., Hull, R., 1981: Characterisation of cauliflower mosaic virus DNA sequences which encode major polyadenylated transcripts. Nucl. Acids Res. **9,** 6735—6747.

Covey, S. N., Turner, D., Mulder, G., 1983: A small DNA molecule containing covalently-linked ribonucleotides originates from the large intergenic region of the cauliflower mosaic virus genome. Nucl. Acids Res. **11,** 251—264.

Daubert, S., Richins, R., Shepherd, R. J., Gardner, R. C., 1982: Mapping of the coat protein gene of cauliflower mosaic virus by its expression in a prokaryotic system. Virology **122,** 444—449.

Daubert, S., Shepherd, R., Gardner, R., 1983: Insertional mutagenesis of the cauliflower mosaic virus genome. Gene **25,** 201—208.

Depicker, A., Stachel, S., Dhaese, P., Zambryski, P., Goodman, H., 1982: Nopaline synthase: transcript mapping and DNA sequence. J. Mol. Appl. Genet. **1,** 561—574.

Dixon, L. K., Koenig, I., Hohn, T., 1983: Mutagenesis of cauliflower mosaic virus. Gene **25,** 189—199.

Dixon, L. K., Hohn, T., 1984: Initiation of translation of the cauliflower mosaic virus genome from a polycistronic mRNA: evidence from deletion mutagenesis. EMBO J. **3.** 2731—2736.

Dixon, L., Nyffenegger, T., Delley, G., Martinez-Izquierdo, J., Hohn, T., 1986a: Evidence for replicative recombination in cauliflower mosaic virus. Virology **150,** 463—468.

Dixon, L., Jiricny, J., Hohn, T., 1986b: Oligonucleotide directed mutagenesis of cauliflower mosaic virus DNA using a repair-resistant nucleoside analogue: identification of an anogene initiation codon. Gene **41,** 225—231.

Donson, J., Hull, R., 1983: Physical mapping and molecular cloning of caulimovirus DNA. J. Gen. Virol. **64,** 2281—2288.

Du Plessis, D. H., Smith, P., 1981: Glycosylation of the cauliflower mosaic virus capsid polypeptide. Virology **109,** 403—408.

Ecker, J. R., Davis, R. W., 1986: Inhibition of gene expression in plant cells by expression of antisense RNA. Proc. Natl. Acad. Sci. U.S.A. **83,** 5372—5376.

Favali, M. A., Bassi, M., Conti, G. G., 1973: A quantitative autoradiographic study of intracellular sites of replication of cauliflower mosaic virus. Virology **53,** 115—119.

Franck, A., Guilley, H., Jonard, G., Richards, K., Hirth, L., 1980: Nucleotide sequence of cauliflower mosaic virus DNA. Cell **21,** 285—294.

French, R., Janda, M., Ahlquist, P., 1986: Bacterial gene inserted in an engineered RNA virus: Efficient expression in monocotyledonous plant cells. Science **231,** 1294—1297.

Fromm, M., Taylor, L. P., Walbot, V., 1985: Expression of genes transferred into monocot and dicot plant cells by electroporation. Proc. Natl. Acad. Sci. U.S.A., **82,** 5824—5828.

Fromm, M. E., Taylor, L. P., Walbot, V., 1986: Stable transformation of maize after gene transfer by electroporation. Nature **319,** 791—793.

Fuetterer, J. and Hohn, T., 1987: Involvement of nucleocapsids in reverse transcription: a general phenomenon? Trends Biochem. Sci. **12,** 92—95.

Gardner, R. C., Howarth, A. J., Hahn, P., Brown-Luedi, M., Shepherd, R. J., Messing, J., 1981: The complete nucleotide sequence of an infectious clone of cauliflower mosaic virus by M13 mp7 shotgun sequencing. Nucl. Acids Res. **9,** 2871—2888.

Gardner, R. C., Knauf, V. C., 1986: Transfer of *Agrobacterium* DNA to plants requires a T-DNA border but not the *vir*E locus. Science **231,** 725—727.

Giband, M., Mesnard, J. M., Lebeurier, G., 1986: The gene III product (P15) of cauliflower mosaic virus is a DNA-binding protein while an immunologically related P11 polypeptide is associated with virions. EMBO J. **5,** 2433—2438.

Givord, L., Xiong, C., Giband, M., Koenig, I., Hohn, T., Lebeurier, G., Hirth, L., 1984: A second cauliflower mosaic virus gene product influences the structure of the viral inclusion body. EMBO J. **3,** 1423—1427.

Goodmann, R. M., 1981: Geminiviruses. In: Kurstak, E. (ed.), Handbook of Plant Virus Infections and Comparative Diagnosis, pp. 883—910. Amsterdam: Elsevier North-Holland Biomedical Press.

Gracia, O., Shepherd, R. J., 1985: Cauliflower mosaic virus in the nucleus of *Nicotiana.* Virology **146,** 141—145.

Graham, F. L., van der Eb, A. J., 1973: A new technique for the assay of human adenovirus 5 DNA. Virology **52,** 456—467.

Grimsley, N., Hohn, B., Hohn, T., Walden, R., 1986a: „Agroinfection", an alternative route for viral infection of plants by using the Ti-plasmid. Proc. Natl. Acad. Sci. U.S.A. **83,** 3282—3286.

Grimsley, N., Hohn, T., Hohn, B., 1986b: Recombination in a plant virus: template-switching in cauliflower mosaic virus. EMBO J. **5,** 641—646.

Gronenborn, B., Gardner, R., Shepherd, R., 1980: Introduction of mutations into the genome of cauliflower mosaic virus *in vitro.* Abstracts of the 6th EMBO Ann. Symp. Heidelberg.

Gronenborn, B., Gardner, R. C., Schaefer, S., Shepherd, R. J., 1981: Propagation of foreign DNA in plants using cauliflower mosaic virus as vector. Nature **294,** 773—776.

Gronenborn, B., 1984: Cauliflower mosaic virus: A plant gene vector. In: Schell, J. S., Starlinger, P. (eds.), The impact of gene transfer techniques in eukaryotic cell biology, pp. 108—118. Berlin - Heidelberg - New York: Springer.

Guilfoyle, T. J., 1980: Transcription of the cauliflower mosaic virus genome in isolated nuclei from turnip leaves. Virology **107,** 71—80.

Guilley, H., Dudley, R. K., Jonard, G., Balazs, E., Richards, K. E., 1982: Transcription of cauliflower mosaic virus DNA: detection of promoter sequences and characterization of transcripts. Cell **30,** 763—773.

Guilley, H., Richards, K. E., Jonard, G., 1983: Observations concerning the discontinuous DNAs of cauliflower mosaic virus. EMBO J. **2,** 277—282.

Hahn P., Shepherd, R. J., 1980: Phosphorylated proteins in cauliflower mosaic virus. Virology **107,** 295—297.

Hahn, P., Shepherd, R. J., 1982: Evidence for a 58-kilodalton polypeptide as precursor of the coat protein of cauliflower mosaic virus. Virology **116,** 480—488.

Hain, R., Stabel, P., Czernilofsky, A. P., Steinbiß, H. H., Herrera-Estrella, L.,

Schell, J., 1985: Uptake, integration, expression and genetic transmission of a selectable chimaeric gene by plant protoplasts. Mol. Gen. Genet. **199,** 161—168.

Harker, C. L., Mullineaux, P. M., Bryant, J. A., Maule, A. J., 1987: Detection of CaMV gene I and gene VI protein products *in vivo* using antisera raised to COOH-terminal β-galactosidase fusion proteins. Plant Mol. Biol. **8,** 275—287.

Harrison, B. D., 1985: Advances in geminivirus research. Ann. Rev. Phytopathol. **23,** 55—82.

Hearon, S. S., Lawson, R. H., 1981: Effects of light intensity, photoperiod, and temperature on symptom expression and host and virus ultrastructure in *Saponaria vaccaria* infected with carnation etched ring virus. Phytopathology **71,** 645—652.

Herrera-Estrella, L., Depicker, A., Van Montagu, M., Schell, J., 1983: Expression of chimaeric genes transferred into plant cells using a Ti-plasmid-derived vector. Nature **303,** 209—213.

Hille, J., Dekker, M., Luttighuis, H. O., van Kammen, A., Zabel, P., 1986: Detection of T-DNA transfer to plant cells by *A. tumefaciens* virulence mutants using agroinfection. Mol. Gen. Genet. **205,** 411—416.

Hills, G. J., Campbell, R. N., 1968: Morphology of broccoli necrotic yellows virus. J. Ultrastruct. Res. **24,** 134—144.

Hirochika, H., Takatsuji, H., Ubasawa, A., Ikeda, J.-E., 1985: Site-specific deletion in cauliflower mosaic virus DNA: possible involvement of RNA spicing and reverse transcription. EMBO J. **4,** 1673—1680.

Hohn, T., Richards, K., Lebeurier, G., 1982: Cauliflower mosaic virus on its way to become a plant vector. Curr. Top. Microbiol. Immunol. **96,** 193—236.

Hohn, B., Hohn, T., 1982: Cauliflower mosaic virus: a potential vector for plant genetic engineering. In: Kahl, G., Schell, J. (eds.), Molecular Biology of Plant Tumors, pp. 549—560. New York: Academic Press.

Hohn, T., Hohn, B., Pfeiffer, R., 1985: Reverse transcription in CaMV. Trends in Biochem. Sci. **10,** 205—209.

Hohn, B., Balazs, E., Rüegg, D., Hohn, T., 1986: Splicing of an intervening sequence from hybrid cauliflower mosaic viral RNA. EMBO J. **5,** 2759—2762.

Howarth, A. J., Gardner, R. C., Messing, J., Shepherd, R. J., 1981: Nucleotide sequence of naturally occurring deletion mutants of cauliflower mosaic virus. Virology **112,** 678—685.

Howell, S. H., 1982: Plant Molecular Vehicles: Potential vectors for introducing foreign DNA into plants. Ann. Rev. Plant Physiol. **33,** 609—650.

Howell, S. H., 1985: The molecular biology of plant DNA viruses. CRC Critical Reviews in Plant Sciences **2,** 287—316.

Howell, S. H., Hull, R., 1978: Replication of cauliflower mosaic virus and transcription of its genome in turnip leaf protoplasts. Virology **86,** 468—481.

Howell, S. H., Walker, L. L., Dudley, R. K., 1980: Cloned cauliflower mosaic virus DNA infects turnips *(Brassica rapa).* Science **208,** 1265—1267.

Howell, S. H., Walker, L. L., Walden, R. M., 1981: Rescue of *in vitro* generated mutants of the cloned cauliflower mosaic virus genome in infected plants. Nature **293,** 483—486.

Hull, R., Shepherd, R. J., Harvey, J. D., 1976: Cauliflower mosaic virus: An improved purification procedure and some properties of the virus particles. J. Gen. Virol. **31,** 93—100.

Hull, R., Shepherd, R. J., 1977: The structure of cauliflower mosaic virus genome. Virology **79,** 216—230.

Hull, R., Howell, S. H., 1978: Structure of the cauliflower mosaic virus genome. II.

Variation in DNA structure and sequence between isolates. Virology **86,** 482—493.

Hull, R., Covey, S. N., Stanley, J., Davies, J. W., 1979: The polarity of the cauliflower mosaic virus genome. Nucl. Acids Res. **7,** 669—677.

Hull, R., 1980: Structure of the cauliflower mosaic virus genome. III. Restriction endonuclease mapping of thirty-three isolates. Virology **100,** 76—90.

Hull, R., Donson, J., 1982: Physical mapping of the DNAs of carnation etched ring and figwort mosaic viruses. J. Gen. Virol. **60,** 125—134.

Hull, R., Covey, S. N., 1983a: Characterization of cauliflower mosaic virus DNA forms isolated from infected turnip leaves. Nucl. Acids Res. **11,** 1881—1895.

Hull, R., Covey, S. N., 1983b: Does cauliflower mosaic virus replicate by reverse transcription? Trends Biochem. Sci. **8,** 119—121.

Hull, R., Davies, J. W., 1983: Genetic engineering with plant viruses, and their potential as vectors. Adv. Virus Res. **28,** 1—33.

Hull, R., 1984: A model for the expression of CaMV nucleic acid. Plant Mol. Biol. **3,** 121—125.

Hull, R., Covey, S. N., 1986: Genome organization and expression of reverse transcribing elements: variations and a theme. J. Gen. Virol. **67,** 1751—1758.

Hull, R., Sadler, J., Longstaff, M., 1986: The sequence of carnation etched ring virus DNA: comparison with cauliflower mosaic virus and retroviruses. EMBO J. **5,** 3083—3090.

Kamei, T., Rubio-Huertos, M., Matsui, C., 1969: Thymidine-^{3}H uptake by X-bodies associated with cauliflower mosaic virus infection. Virology **37,** 506—508.

Krens, F. A., Molendijk, L., Wullems, G. J., Schilperoort, R. A., 1982: *In vitro* transformation of plant protoplasts with Ti-plasmid DNA. Nature **296,** 72—74.

Kröger, M., Hobom, G., 1982: A chain of interlinked genes in the *nin*R region of bacteriophage lambda. Gene **20,** 25—38.

Lebeurier, G., Hirth, L., Hohn, T., Hohn, B., 1980: Infectivities of native and cloned DNA of cauliflower mosaic virus. Gene **12,** 139—146.

Lebeurier, G., Hirth, L., Hohn, B., Hohn, T., 1982: *In vivo* recombination of cauliflower mosaic virus DNA. Proc. Natl. Acad. Sci. U.S.A. **79,** 2932—2936.

Lindsten, K., Vacke, J., Gerhardson, B., 1970: A preliminary report on three cereal virus diseases new to Sweden spread by *Macroteles* and *Psammotettix* leafhoppers. Natl. Swed. Inst. Plant Prot. Contrib. **14,** 281—297.

Lindsten, K., Lindsten, B., Abdelmoeti, M., Juntti, N., 1980: Purification and some properties of wheat dwarf virus. Proceedings of the 3rd Conference on Virus Diseases of *Gramineae* in Europe, pp. 27—31. Harpenden: Rothamsted Experimental Station.

Lung, M. C. Y., Pirone, T. P., 1972: *Datura stramonium,* a local lesion host for certain isolates of cauliflower mosaic virus. Phytopathology **62,** 1473—1474.

Lung, M. C. Y., Pirone, T. P., 1973: Studies on the reason for differential transmissibility of cauliflower virus isolates by aphids. Phytopathology **63,** 910—914.

Lung, M. C. Y., Pirone, T. P., 1974: Acquisition factor required for aphid transmission of purified cauliflower mosaic virus. Virology **60,** 260—264.

Marco, Y., Howell, S. H., 1984: Intracellular forms of viral DNA consistent with a model of reverse transcriptional replication of the cauliflower mosaic virus genome. Nucl. Acids Res. **12,** 1517—1528.

Martinez-Izquierdo, J., Hohn, T., 1987: Cauliflower mosaic virus coat protein is phosphorylated *in vitro* by a virion-associated protein kinase. Proc. Natl. Acad. Sci. U.S.A. **84,** 1824—1828.

Martinez-Izquierdo, J. A., Fütterer, J., Hohn, T., 1987: Protein encoded by ORF I of cauliflower mosaic virus is part of the viral inclusion body. Virology, in press.

Marsh, L., Kuzj, A., Guilfoyle, T., 1985: Identification and characterization of cauliflower mosaic virus replication complexes — analogy to hepatitis B viruses. Virology **143,** 212—223.

Mason, W. S., Taylor, J. M., Hull, R., 1987: Retroid virus genome replication. Adv. Virus Res. **32,** 35—96.

Matthews, R. E. F., 1981: Plant Virology. 2nd ed. New York: Academic Press.

Maule, A. J., 1985a: Partial characterization of different classes of viral DNA and kinetics of DNA synthesis in turnip protoplasts infected with cauliflower mosaic virus. Plant Mol. Biol. **5,** 25—34.

Maule, A. J., 1985b: Replication of caulimoviruses in plants and protoplasts. In: Davies, J. W. (ed.), Molecular Plant Virology (Vol. 2), pp. 161—190. Boca Raton: CRC Press.

Mazzolini, M., Bonneville, M., Volovitch, M., Magazin, M., Yot, P., 1985: Strand-specific viral DNA synthesis in purified viroplasms isolated from turnip leaves infected with cauliflower mosaic virus. Virology **145,** 293—303.

Melcher, U., Steffens, D. L., Lyttle, D J., Lebeurier, G., Lin, H., Choe, I. S., Essenberg, R. C., 1986a: Infectious and non-infectious mutants of cauliflower mosaic virus DNA. J. Gen. Virol. **67,** 1491—1498.

Melcher, U., Choe, I. S., Lebeurier, G., Richards, K., Essenberg, R. C., 1986b: Selective allele loss and interference beween cauliflower mosaic virus DNAs. Mol. Gen. Genet. **203,** 230—236.

Menissier, U., Lebeurier, G., Hirth, L. 1982: Free cauliflower mosaic virus supercoiled DNA in infected plants. Virology **117,** 322—328.

Menissier, J., de Murcia, G., Lebeurier, G., Hirth, L., 1983: Electron microscopic studies of the different topological forms of the cauliflower mosaic virus DNA: knotted encapsidated DNA and nuclear minichromosome. EMBO J. **2,** 1067—1071.

Menissier, J., Laquel, P., Lebeurier, G., Hirth, L., 1984: A DNA polymerase activity is associated with cauliflower mosaic virus. Nucl. Acids Res. **12,** 8769—8778.

Menissier-De Murcia, J., Geldreich, A., Lebeurier, G., 1986: Evidence for a protein kinase activity associated with purified particles of cauliflower mosaic virus. J. Gen. Virol. **67,** 1885—1891.

Mesnard, J.-M., Geldreich, A., Xiong, C., Lebeurier, G., Hirth, L., 1984: Expression of a putative plant viral gene in *Escherichia coli.* Gene **31,** 39—47.

Meyer, P., Walgenbach, E., Bussmann, K., Hombrecher, G., Saedler, H., 1985: Synchronized tobacco protoplasts are efficiently transformed by DNA. Mol. Gen. Genet. **201,** 513—518.

Modjtahedi, N., Volovitch, M., Sossountzov, L., Habricot, Y., Bonneville, J. M., Yot, P., 1984: Cauliflower mosaic virus induced viroplasms support viral DNA synthesis in a cell-free system. Virology **133,** 289—300.

Morelli, G., Nagy, F., Fraley, R. T., Rogers, S. G., Chua, N.-H., 1985: A short conserved sequence is involved in the light inducability of a gene encoding ribulose 1.5-bisphosphate carboxylase small subunit of pea. Nature **315,** 200—204.

Neumann, E., Schaefer-Ridder, M., Wang, Y., Hofschneider, P. H., 1982: Gene transfer into mouse lyoma cells by electroporation in high electric fields. EMBO J. **1,** 841—845.

Odell, J. T., Howell, S. H., 1980: The identification, mapping and characterization of mRNA for p66, a cauliflower mosaic virus-encoded protein. Virology **102,** 349—359.

Odell, J. T., Dudley, R. K., Howell, S. H., 1981: Structure of the 19 S RNA transcript encoded by the cauliflower mosaic virus genome. Virology **111**, 377—385.

Odell, J. T., Nagy, F., Chua, N.-H., 1985: Identification of DNA sequences required for activity of the cauliflower 35 S promoter. Nature **313**, 810—812.

O'Hare, K., Benoist, C., Breathnach, R., 1981: Transformation of mouse fibroblasts to methotrexate resistance by a recombinant plasmid expressing a prokaryotic dihydrofolate reductase. Proc. Natl. Acad. Sci. U.S.A. **78**, 1527—1531.

Olszewski, N. E., Hagen, G., Guilfoyle, T. J., 1982: A transcriptionally active, covalently closed minichromosome of cauliflower mosaic virus DNA isolated from turnip leaves. Cell **29**, 395—402.

Paszkowski, J., Shillito, R. D., Saul, M., Mandak, V., Hohn, T., Hohn, B., Potrykus, I., 1984: Direct gene transfer to plants. EMBO J. **3**, 2717—2722.

Pfeiffer, P., Hohn, T., 1983: Involvement of reverse transcription in the replication of the plant virus CaMV: a detailed model and test of some aspects. Cell **33**, 781—784.

Pfeiffer, P., Laquel, P., Hohn, T., 1984: Cauliflower mosaic virus replication complexes: characterization of the associated enzymes and of the polarity of the DNA snythesized *in vitro*. Plant Mol. Biol. **3**, 261—270.

Pietrzak, M., Hohn, T., 1985: Replication of the cauliflower mosaic virus: role and stability of the cloned $\Delta 3$ discontinuity sequence. Gene **33**, 169—179.

Pietrzak, M., Shillito, R. D., Hohn, T., Potrykus, I., 1986: Expression in plants of two bacterial antibiotic resistance genes after protoplast transformation with a new plant expression vector. Nucl. Acids Res. **14**, 5857—5868.

Plant, A. L., Covey, S. N., Grierson, D., 1985: Detection of a subgenomic mRNA for gene V, the putative reverse transciptase gene of cauliflower mosaic virus. Nucl. Acids Res. **13**, 8305—8321.

Potrykus, I., Saul, M., Petruska, J., Paszkowski, J., Shillito, R. D., 1985: Direct gene transfer into protoplasts of a graminaceous monocot. Mol. Gen. Genet. **199**, 183—188.

Powell Abel, P., Nelson, R. S., De, B., Hoffmann, N., Rogers, S. G., Fraley, R. T., Beachy, R. N., 1986: Delay of disease development in transgenic plants that express the tobacco mosaic virus coat protein gene. Science **232**, 738—743.

Richards, K. E., Guilley, H., Jonard, G., 1981: Further characterization of the discontinuities of cauliflower mosaic virus DNA. FEBS Lett. **134**, 67—70.

Richins, R. D., Shepherd, R. J., 1983: Physical maps of the genomes of Dahlia mosaic virus and mirabilis mosaic virus — two members of the caulimovirus group. Virology **124**, 208—214.

Rollo, F., Covey, S. N., 1985: Cauliflower mosaic virus DNA persists as supercoiled forms in cultured turnip cells. J. Gen. Virol. **66**, 603—608.

Rollo, F., Covey, S. N., Amici, A., 1986: Figwort mosaic virus DNA replicates in cultured *Datura stamonium* cells. J. Gen. Virol. **67**, 2227—2232,

Schoelz, J., Shepherd, R. J., Daubert, S., 1986a: Region VI of cauliflower mosaic virus encodes a host range determinant. Mol. Cell. Biol. **6**, 2632—2637.

Schoelz, J. E., Shepherd, R. J., Richins, R. D., 1986b: Properties of an unusual strain of cauliflower mosaic virus. Phytopathology **76**, 451—454.

Shepherd, R. J., Wakeman, R. J., Romanko, R. R., 1968: DNA in cauliflower mosaic virus. Virology **36**, 150—152.

Shepherd, R. J., Bruening, G. E., Wakeman, R. J., 1970: Double-stranded DNA from cauliflower mosaic virus. Virology **41**, 339—347.

Shepherd, R. J., Wakeman, R. J., 1971: Observation on the size and morphology of cauliflower mosaic virus deoxyribonucleic acid. Phytopathology **61**, 188—193.

Shepherd, R. J., 1976: DNA viruses of higher plants. In: Lauffer, M., Bang, F. B., Maramorosch, K., Smith, K. M. (eds.), Advances in Virus Research (Vol. 20), pp. 305—339. New York: Academic Press.

Shepherd, R. J., 1979: DNA plant viruses. Ann. Rev. Plant Physiol. **30,** 405—423.

Shepherd, R. J., Lawson, R. H., 1981: Caulimoviruses. In: Kurstak, E. (ed.), Handbook of Plant Virus Infections and Comparative Diagnosis. pp. 848—878. Amsterdam: Elsevier/North-Holland Biomedical Press.

Shewmaker, C. K., Caton, J. R., Houck, C. M., Gardner, R. C., 1985: Transcription of cauliflower mosaic virus integrated into plant genomes. Virology **140,** 281—288.

Sieg, K. Gronenborn, B., 1982: Introduction and propagation of foreign DNA in plants using cauliflower mosaic virus as vector. Abstract. NATO/FEBS Advanced course: Structure and function of plant genomes. p. 154.

Siegel, A., 1985: Plant-virus-based vectors for gene transfer may be of considerable use despite a presumed high error frequency during RNA synthesis. Plant Mol. Biol. **4,** 327—329.

Stanley, J., 1985: The molecular biology of geminiviruses. In: Maramorosch, K., Murphy, F. A., Shatkin, A. J. (eds.), Advances in Virus Research (Vol. 30), pp. 139—177. New York - London: Academic Press.

Stanley, J., Davies, J. W., 1985: Structure and function of the DNA genome of geminiviruses. In: Davies, J. W. (ed.) Molecular Plant Virology (Vol. 2), pp. 191—218. Boca Raton: CRC Press.

Summers, J., Mason, W., 1982: Replication of the genome of a hepatitis B-like virus by reverse transcription of an RNA intermediate. Cell **29,** 403—415.

Szeto, W. W., Hamer, D. H., Carlson, P. S., Thomas, C. A., 1977: Cloning of cauliflower mosaic virus (CaMV) DNA in *Escherichia coli.* Science **196,** 210—212.

Takatsuji, H., Hirochika, H., Fukushi, T., Ikeda, J.-E., 1986: Expression of cauliflower mosaic virus reverse transcriptase in yeast. Nature **319,** 240—243.

Thomas, C. M., Hull, R., Bryant, J. A., Maule, A. J., 1985: Isolation of a fraction from cauliflower mosaic virus-infected protoplasts which is active in the synthesis of (+) and (−) strand viral DNA and reverse transcription of primed RNA templates. Nucl. Acids Res. **13,** 4557—4576.

Toh, H., Hayashida, H., Miyata, T., 1983: Sequence homology between retroviral reverse transcriptase and putative polymerases of hepatitis B virus and cauliflower mosaic virus. Nature **305,** 827—829.

Toh, H., Kikuno, R., Hayashida, H., Miyata, T., Kugimiya, W., Inouye, S., Yuki, S., Saigo, K., 1985: Close structural resemblance between putative polymerase of a *Drosophila* transposable genetic element 17.6 and *pol* gene product of Moloney murine leukemia virus. EMBO J. **4,** 1267—1272.

Turner, D. S., Covey, S. N., 1984: A putative primer for the replication of cauliflower mosaic virus by reverse transcription is virion associated. FEBS Letters **165,** 285—289.

Van Vloten-Doting, L., Bol, J.-F., Cornelissen, B., 1985: Plantvirus-based vectors for gene transfer will be of limited use because of the high error frequency during viral RNA synthesis. Plant. Mol. Biol. **4,** 323—326.

Varmus, H. E., 1982: Form and function of retroviral proviruses. Science **216,** 812—820.

Varmus, H. E., 1983: Reverse transcription in plants? Nature **304,** 116—117.

Volovitch, M., Drugeon, G., Yot, P., 1978: Studies on the single-stranded discontinuities of the cauliflower mosaic virus genome. Nucl. Acids Res. **5,** 2913—2925.

Volovitch, M., Modjtahedi, N., Yot, P., Brun, G., 1984: RNA-dependent DNA polymerase activity in cauliflower mosaic virus-infected plant leaves. EMBO J. **3,** 309—314.

Walden, R. M., Howell, S. H., 1982: Intergenomic recombination events among pairs of defective cauliflower mosaic virus genomes in plants. Mol. Appl. Genet. **1,** 447—456.

Walden, R. M., Howell, S. H., 1983: Uncut recombinant plasmids bearing nested cauliflower mosaic virus genomes infect plants by intragenomic recombination. Plant Mol. Biol. **2,** 27—31.

Woolston, C. J., Covey, S. N., Penswick, J. R., Davis, J. W., 1983: Aphid transmission and a polypeptide are specified by a defined region of the cauliflower mosaic virus genome. Gene **23,** 15—23.

Xiong, C., Muller, S., Lebeurier, G., Hirth, L., 1982: Identification by immunoprecipitation of cauliflower mosaic virus *in vitro* major translation product with a specific serum aganist viroplasm protein. EMBO J. **1,** 971—976.

Xiong, C., Lebeurier, G., Hirth, L. 1984: Detection *in vivo* of a new gene product (gene *III*) of cauliflower mosaic virus. Proc. Natl. Acad. Sci. U.S.A. **81,** 6608-6612.

Yamaoka, N., Morita, T., Furusawa, I., Yamamoto, M., 1982: Effect of temperature on the multiplication of cauliflower mosaic virus. J. Gen. Virol. **61,** 283—287.

Ziegler, V., Laquel, P., Guilley, H., Richards, K., Jonard, G., 1985: Immunological detection of cauliflower mosaic virus gene V protein produced in engineered bacteria or infected plants. Gene **36,** 271—279.

Zimmermann, U., Scheurich, P., 1981: High frequency fusion of plant protoplasts by electric fields. Planta **151,** 26—32.

Zimmermann, U., Vienken, J., 1982: Electric field-induced cell-to-cell fusion. J. Membr. Biol. **67,** 165—182.

Note added in proof:

After this chapter had been completed the potential use of caulimoviruses as gene vectors was also reviewed by R. J. Shepherd (1986): Caulimoviruses as potential gene vectors for higher plants. In: Setlow, J. K., Hollaender, A., (eds.), Genetic Engineering: Principles and Methods. pp. 241—276. New York - London: Plenum Press.

Chapter 2

The Structure, Expression, Functions and Possible Exploitation of Geminivirus Genomes

J. W. Davies, R. Townsend and J. Stanley

John Innes Institute, Colney Lane, Norwich NR4 7UH, U. K.

With 4 Figures

Contents

I. Introduction

The geminiviruses are single-stranded (ss) DNA viruses with genomes comprising one or two circles of 2.5—3.0 Kb. The biological and molecular properties of some members of the group have been reviewed recently (Harrison, 1985; Stanley and Davies, 1985; Stanley, 1985). The number of known viruses is increasing as a consequence of intensifying interest in the group. Three main areas of research offer particularly exciting possibilities.

1) Geminiviruses as model plant genes. They provide simple genomes which are relatively easy to prepare and have evolved to be adapted to the host cells, utilizing much of the plant's gene expression machinery. Unlike the double-stranded (ds) DNA caulimoviruses, some geminiviruses infect monocotyledonous plants and are therefore of particular interest in studies of the regulation mechanisms of gene expression in cereals.

2) Molecular pathology. Geminiviruses collectively infect a wide range of plant species, although some individual members have rather limited ranges. Some isolates of the same virus show different symptom or transmission properties. Some geminiviruses are transmitted by leafhoppers, others by whiteflies; most show a degree of specificity for the type of cell

they infect. The availability of fundamental molecular biological data, including several complete nucleic acid sequences of geminivirus genomes, now makes the group an attractive one for the investigation of functions such as symptom expression, cell-to-cell spread, host range and transmission at the molecular level.

3) Plant gene vectors. A number of important properties make the group of special interest in the construction of gene vectors. Geminiviruses replicate in the nucleus of infected cells via dsDNA replicative forms (RFs) which are apparently produced in very high copy numbers. A feature of the replication of some geminiviruses is the presence of defective DNA circles which are able to replicate if complemented by a normal genome but lack about half the molecule. Furthermore, the monocotyledonous hosts include several which are of economic importance, including rice, wheat, maize, sorghum, millet and sugarcane.

In order to exploit the potential of the geminiviruses it is essential to gain an understanding of the organisation and expression of their genomes and assign functions to the various genes. Much of this information is now becoming available and forms the basis of this review. We have chosen to concentrate on three geminiviruses with which we are currently working; cassava latent virus (CLV) (sometimes known as African cassava mosaic virus or ACMV), maize streak virus (MSV) and beet curly top virus (BCTV), which are representative of the group as a whole. We have also taken this opportunity to speculate about some of the developments in plant gene vectors, based on geminiviruses, which could be achieved in the near future.

II. Genome Organisation

A. Coding Regions

The first geminivirus to be sequenced was CLV (Stanley and Gay, 1983). The nucleotide sequence revealed a bipartite genome comprising two ssDNA circles, component 1 (2779 nucleotides) and component 2 (2724 nucleotides). In the field CLV is transmitted by whiteflies but cloned DNA proved infectious when mechanically inoculated to plants. However, both components of the genome were required to produce a systemic infection (Stanley, 1983). the two DNAs had a total of twelve open reading frames (ORFs) with a potential to code for proteins of up to ~ 40,000 Da in either the virion (+) sense sequence or its complement (–). Subsequently, two other whitefly-transmitted geminiviruses were sequenced. Both tomato golden mosaic virus (TGMV) (Hamilton *et al.,* 1984) and bean golden mosaic virus (BGMV) (Howarth *et al.,* 1985) are serologically related to CLV and both proved to have similar genome organisation to CLV but only six ORFs were conserved in terms of approximate position, size (Fig. 1) and homology (Table 1) between the three viruses. The major degree of conservation is found between the coat proteins which range in size between 27,000 and 30,000 Da and are encoded in the virion (+) sense of

Table 1. Comparison of Potential Gene Products of Some Geminiviruses of Dicotyledonous Hosts

DNA circle	Gene product size (kDA)				Direct amino acid homology (%)		
	CLV	TGMV	BGMV	BCTV	CLV-TGMV	CLV-BGMV	CLV-BCTV
1+	30.2	28.6	27.7	29.6	71	68	15
1−	40.3	40.3	40.2	40.9	63	64	61
	15.1	14.9	19.6	20.1	51	49	30
	15.8	15.7	15.6	16.1	55	47	38
2+	29.3	29.3	29.7	—	34	29	—
2−	33.7	21.1	33.1	—	51	46	—

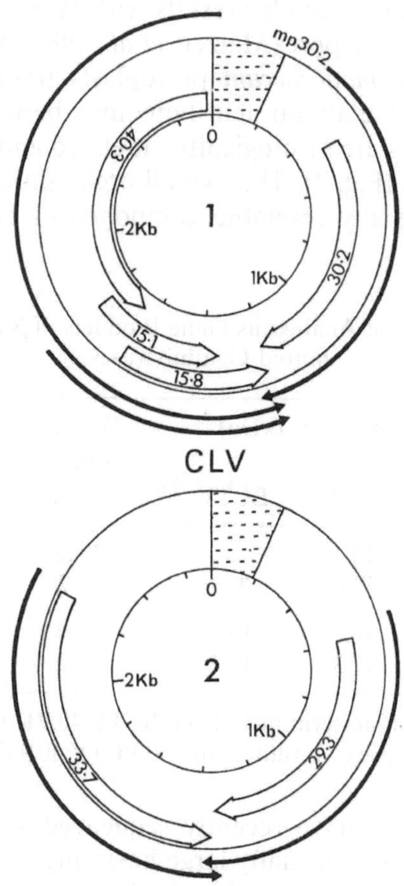

Fig. 1. Organisation of the genome of CLV (Kenyan). The open reading frames ORFs) indicated (⇨) are those which are conserved in the other whitefly-transmitted geminiviruses BGMV and TGMV (see Table 1). These all correspond to regions to which RNA transcripts have been mapped (➡). Transcript mp 30.2 is the mRNA for coat protein. The intergenic "common region" is represented by the stippled section

DNA 1. Among the other putative products, that of ~ 40,000 Da, encoded on the complementary strand of DNA 1, shows more than 60 % direct amino acid homology with the corresponding product of CLV. Presumably those ORFs which are not common to all three viruses, particularlay those which overlap conserved ORFs in the opposite orientation (e. g. the 10,600, 21,000 and 13,500 Da ORFs in CLV DNA 1^- which overlap the 32,000 Da coat protein gene in DNA 1^+ are fortuitous and not expressed (Stanley and Davies, 1985).

In contrast to the whitefly-transmitted geminiviruses which infect dicotyledonous hosts, exhaustive sequencing of virion DNAs of the leafhopper-transmitted geminiviruses MSV (Mullineaux *et al.*, 1985; Howell, 1984; 1985) and wheat dwarf virus (WDV) (McDowell *et al.*, 1985) has revealed only one circle of ssDNA. It is probable that the leafhopper-transmitted geminiviruses which infect monocotyledonous plants are monopartite; convincing evidence of the infectivity of cloned copies of the MSV DNA has now been presented (Grimsley, Hohn, Davies and Hohn, 1987; see also Chapter 4 of this volume). Similar results with WDV may be forthcoming in view of a preliminary report (Matzeit *et al.*, 1985) of the synthesis of ds viral RFs in *Triticum monococcum* protoplasts transfected with cloned WDV. There is extensive amino acid homology between MSV and WDV (Table 2) although they are serologically unrelated and they share a similar arrangement of ORFs (Fig. 2). The overall organisation of the genomes of MSV and WDV generally resemble component 1 of the whitefly-transmitted geminiviruses.

Table 2. Comparison of Analogous Gene Products of Some Leafhopper-Transmitted Geminiviruses

DNA polarity	Gene product size (kDa)			Direct amino acid homology (%)		
	MSV	WDV	BCTV	MSV-WDV	MSV-BCTV	WDV-BCTV
+	10.9	10.1	—	32 .6	—	—
	27.0	29.4	29.6	35.2	25.0	22.3
−	31.4	30.2	40.9	41.0	32.8[a]	33.8[a]
	17.8	17.3	40.9	52.0	25.0[b]	34.5[b]

[a] homologies with the amino-terminus of the BCTV 40.9 K ORF
[b] homologies with the carboxy-terminus of the BCTV 40.9 K ORF

Stanley *et al.* (1986) have recently sequenced an infectious clone of BCTV. This virus has an unusually large host range among different dicotyledonous genera but is transmitted by leafhoppers and has a single DNA. The arrangement of ORFs in BCTV is very similar to CLV DNA 1 (Fig. 3) and there is extensive homology between the putative products of the complementary (−) strand ORFs of the two viruses. In contrast, the products of the virion strand ORFs bear little resemblance to those of CLV

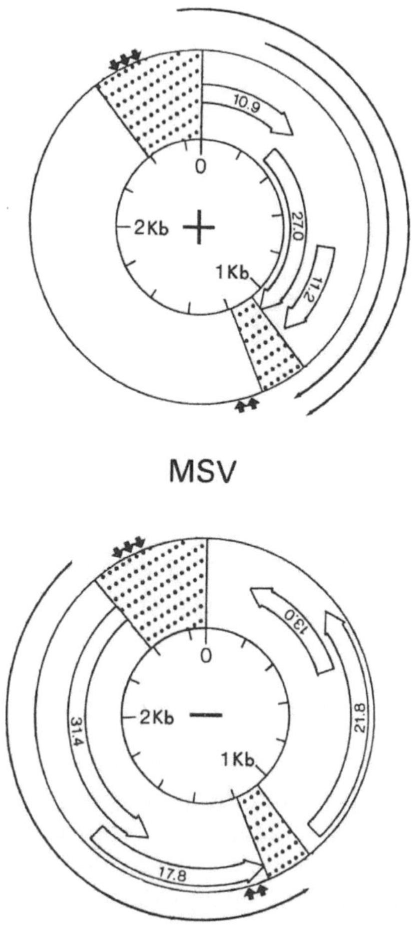

Fig. 2. Organisation of the genome of MSV (Nigerian) showing the coding and potential coding regions of the virion sense (+) DNA and its complement (−). The open reading frames (⇨) assume that the protein synthesis intiates at the first ATG and that potential products are those of at least 10 K molecular weight. The narrow headed outer arrows (→) represent the approximate mapping positions of the RNA transcripts. Those on the (+) DNA are the 1.05 and 0.9 Kb RNAs, overlapping at their 5′ termini (nucleotides 163 and 2682) and co-terminating at nucleotide 1114. The thick arrows (⬇) indicate the positions of five of the nine possible stem-loops, and the stippled regions represent the large intergenic region (L) and small intergenic region (S)

except at the carboxy-terminus of the putative coat protein. Comparison with MSV and WDV suggests that the BCTV coat protein more closely resembles that of leafhopper-transmitted geminiviruses (Tables 1 and 2) apparently reflecting a strong association between the virus particle and its insect vector (see section on gene function).

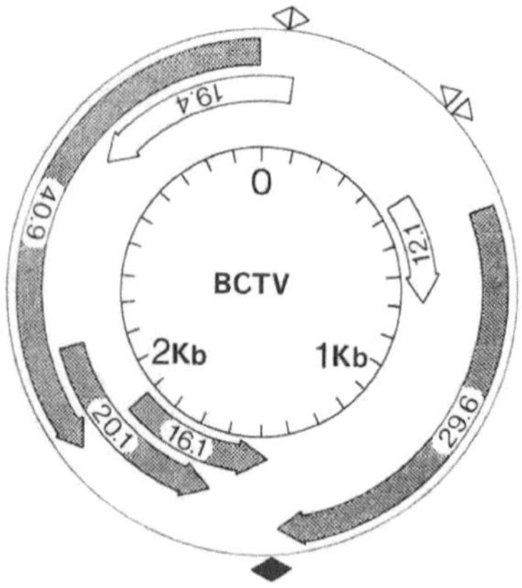

Fig. 3. Organisation of the genome of BCTV, showing the potential coding regions. The shaded ORFs represent those which have equivalent homologous ORFs in other geminiviruses (see Tables 1 and 2). On the basis of comparison with other geminiviruses the 29.6 K ORF is the coat protein gene. Consensus TATA sequences are indicated (△) and also potential polyadenylation signals (▲)

B. Non-coding Regions

Non-coding or intergenic regions account for ~ 16% of CLV DNA 1 and ~ 39% of DNA 2 and within these regions is a stretch of conserved sequence ~ 200 nucleotides long. The presence of the same sequence in both DNAs strongly suggests that it plays an important role in functions such as genome replication or encapsidation. "Common regions" are a feature of all other bipartite geminiviruses which have been examined and all have the capacity to form a stable loop structure containing the nanonucleotide sequence TAATATTAC. Loop structures with the same nanonucleotide sequence also occur in the non-coding regions of MSV, WDV and BCTV. In view of the similarities of loop structure with the primosome assembly site of the ss bacteriophage ØX 174 (Arai and Kornberg, 1981), it has been widely proposed that it serves as the binding site for host encoded enzymes responsible for priming complementary strand synthesis. However, MSV is unique among geminiviruses so far examined in having an ~ 80 nucleotide DNA molecule which is complementary to virion DNA and can be used to prime the synthesis of a complementary strand *in vitro* (Donson *et al.*, 1984). This small DNA molecule maps not to the loop with the conserved nanonucleotide sequence in the large intergenic region, but in the small intergenic region which is on the opposite side of the circle (Fig. 2), close to two hairpin structures in the 17.8 k ORF.

Apart from the nanonucleotide sequence there is no significant homology between the common regions of different geminiviruses. This has led to the speculation that the common region could be the determinant of host range (Howarth *et al.*, 1985). However it is difficult to reconcile this interpretation with the finding that CLV, TGMV and BCTV all share a common host in *Nicotiana benthamiana*. Hopefully the question of host range determination will be resolved by the construction of chimeric viruses in which common regions have been exchanged.

III. Gene Expression

Genomic sequences of CLV, MSV and BCTV (Figures 1—3) show a number of consensus promoter sequences TATAA/TAA/T (Breathnach and Chambon, 1981) and polyadenylation signals AATAAA (Proudfoot and Brownlee, 1976) The position of these sequences relative to ORFs strongly suggests that transcription occurs bi-directionally from promoters lying outside or just within the common regions and converges on polyadenylation signals diametrically opposite the promoters. Such an arrangement implies that convergent overlapping transcription (antiparallel transcription) is restricted to short stretches downstream of the polyadenylation signals. In view of the obvious similarities to the genome organisation and transcription of SV40 and polyoma virus (reviewed by Griffin, 1982), expression of geminivirus genes might be divided into early and late phases corresponding to transcription from the virion on complementary sense template.

Bi-directional transcription has been demonstrated for CLV (Townsend *et al.*, 1985), MSV (Morris-Krsinich *et al.*, 1985) and BGMV (Kridl, 1986) by the hybridisation of polyadenylated viral RNAs to M13 ssDNA clones containing either virion (+) or complementary (−) sense full-length copies of viral DNAs.

The major CLV transcript is 1 Kb long and of + sense polarity. This RNA has been mapped to the 30,200 Da ORF of DNA 1 which codes for a product of similar size and composition to the viral coat protein (Townsend *et al.*, 1985). *In vitro* translation of poly(A)$^{+}$ selected RNAs yields a product of ~ 30,000 Da which is immunoprecipitated by anti-CLV coat protein serum. The synthesis of this product is arrested following hybridisation of RNA to an M13 ssDNA clone containing a − sense copy of CLV DNA 1 (nucleotides 485—804), complementary to about half the coding region of the putative coat protein gene. Synthesis of the ~ 30,000 Da product is restored upon melting of the heteroduplex indicating that the 1 Kb transcript is the mRNA encoding the coat protein. In view of its nature and location, the promoter region of the coat protein gene is of particular interest. High resolution S_1 mapping has located the 5′-terminus of the major 1 Kb RNA in the vicinity of nucleotide 278 (Townsend and Ward, personal communication). There is a consensus promoter sequence, TATATAAT, at position −27 relative to the transcript start site (+1) and a putative promoter modulator sequence, GGCCAATC, at position −75. However the ~ 160 nucleotide long leader contains two AUGs (286 and 305) upstream of the initiation codon of the coat protein gene at nucleotide 446, which are in phase with the small overlapping 13,100 Da ORF (Stanley and Davies, 1985). We have no evidence to suggest that the 1 Kb mRNA of CLV is functionally dicistronic. *In vitro* translation of poly(A)$^+$ RNA in rabbit reticulocyte lysates yielded only the coat protein and there was no evidence of a smaller product despite the relative abundance of the 1 Kb RNA. However, synthesis of a 13,100 Da product would require ribosomes to initiate on an extremely short leader of ~ 8 nucleotides and might be expected to result in a rather low level of expression (Kozak, 1983). Initiation at the second AUG would produce a product of ~ 11,000 Da. However, the sequence context about this AUG, and about two others at positions 326 and 340 which are succeeded almost immediately by termination codons, does not resemble the consensus eukaryotic initiation sequence and in particular lacks the preferred purine A in the −3 position (Kozak, 1984) which is found in eleven of the twelve ORFs which are conserved between DNA 1 of CLV and the corresponding components of TGMV and BGMV.

DNA 1 of CLV is slightly larger than the equivalent DNA (DNA A) of TGMV. Apart from the respective common regions, the DNAs are closely related except in the non-coding region 5′ of the coat protein gene of CLV. It is tempting to suggest that a small portion of CLV DNA 1 has been duplicated perhaps in response to a change in packaging constraints, so creating the 13,100 Da ORF, which is absent in TGMV, and effectively separating the functional region of the coat protein gene from its promoter. Intervening sequences may have been modified to prevent initiation at upstream sites. In this context it is interesting to note that the sequence GGUG is conserved in positions −2 to −5 of each of the non-functional AUGs, with the exception of that at the extreme 5′ end of the transcript; however the sequence immediately preceding the initiation codon of the coat protein is UAAUU.

Evidence for the duplication of a short piece of DNA may be provided by recent mapping experiments (Townsend and Ward, personal communication) which have revealed the presence of a very minor transcript, possibly the "vestigial" coat protein mRNA. This RNA is apparently 3′ co-terminal with the 1 Kb coat protein message but its 5′-terminus is located around nucleotide 378 and its 5′ untranslated leader contains no AUG sequences before the initiation codon of the coat protein gene. Significantly, the sequences about the two putative promoter regions show extensive homology and can be aligned relative to the transcript start sites (Fig. 4), but the minor transcript lacks a consensus promoter modulator sequence.

Two overlapping 3′ co-terminal transcripts (1.7 Kb and 0.7 Kb) of − sense polarity map to the other half of the CLV DNA 1 (Fig. 1). These two RNAs could encode the products of the three ORFs in the complementary sense of DNA 1, however, no evidence of splicing has yet been obtained to account for the expression of these overlapping ORFs. *In vitro* translation of complementary (−) polarity polyadenylated RNAs hybrid-selected with virion sense DNA in wheat germ extracts yielded products of ~ 40,000 Da and 15,000 Da (Davies and Robinson, unpublished). Site-directed mutagenesis studies are in progress to determine if both the

```
     250        260        270        280
        ┌───────┐
GGGCCACTATATAATTGCAGGCGAAGTTGTGGCTAGT
 ••• • •• ••• ••••••••      •     ●
TGGCAATTAAATATTTGCAGGCCTTAGAGGATACATA
        └───────┘
    350        360        370        380
```

Fig. 4. Comparison of the sequences about the two putative promoter regions on CLV DNA 1 (+) upstream of the coat protein and 13 K ORFs. The sequences are aligned to show the homology

15,000 Da and 15,800 Da ORFs are expressed *in vivo.* The presence of a possible promoter sequence in the proximity of the 5′-terminus of the small 0.7 Kb RNA, suggests that this might be transcribed independently of the large 1.7 Kb RNA, which appears to be transcribed from a promoter just inside the left hand border of the common region. Analysis of the sequence of BCTV reveals a similar arrangement of ORFs but the amino-terminus of the overlapping 20,100 Da ORF is located upstream of the sequence analogous to the CLV putative promoter suggesting this intact ORF is not expressed from the smaller transcript. Possible splice donor sequences occur within 80 nucleotides of the putative promoters in the intergenic regions of CLV and BCTV and there are possible acceptor sequences in the vicinity of the 5′-terminus of the small CLV transcript and 5′ of both the 20,100 Da and 16,100 Da ORFs of BCTV. Closer examination of the small CLV transcript may reveal the presence of a short spliced leader that would not have been detected by the methods used to map the RNAs.

Two polyadenylated RNAs of approximately 0.9 Kb and 1.1 Kb have been mapped to the 29,300 Da (+ sense) and 33,700 Da (− sense) ORFs respectively on CLV DNA 2 (Fig. 1). In the absence of a consensus promoter sequence in the proximity of the 5′-terminus of the − polarity transcript, it was suggested that a minor 1.35 Kb RNA species of the same polarity might represent a precursor RNA (Townsend *et al.,* 1985). However, preliminary mapping of this RNA (Townsend and Ward, personal communication) indicates that it is 5′ co-terminal with the 1.1 Kb molecule but may terminate at an alternative polyadenylation signal (AATAAT;1080) downstream of the 33,700 Da ORF. Similarly polyadenylation at multiple sites apparently gives rise to several minor virion (+) polarity transcripts of between 1.7 Kb and 2.2 Kb which are 5′ co-terminal with the 0.9 Kb RNA.

A preliminary report (Kridl, 1986) suggests that the expression of the BGMV genome is generally similar to that of CLV.

MSV-infected plants contain at least three polyadenylated virus specific RNAs (Morris-Krsinich *et al.,* 1985; Table 1). Two of these RNAs map to the virion sense DNA and are 3′ co-terminal about nucleotide 1114, 57 bases downstream of a consensus polyadenylation signal and 9 bases from the 3′-terminus of the major species of the "primerlike" DNA molecules found on virion DNA (Donson *et al.,* 1984). The more abundant of these transcripts has a 5′-terminus near nucleotide 163 and, although there is no consensus promoter sequence in the area, it is preceded by a T-rich sequence. This RNA could encode both the product of the 27,000 Da ORF, which has been identified on the basis of amino acid sequence as the coat protein, and the overlapping 11,200 Da ORF. However, this latter ORF is not conserved in the sequence of WDV or BCTV and may be fortuitous. The second and less abundant virion (+) polarity transcript is rather longer and has a 5′-terminus near nucleotide 2682. This transcript has the potential to encode all the products of the + sense ORFs although expression of the 10,900 Da ORF would require ribosomes to initiate on a 4 nucleotide leader. The next possible initiation site is the AUG immediately preceding the coat protein gene. Translation studies coupled with immunoprecipitation confirmed that either or both of these RNAs encode the coat protein gene but revealed no evidence of other gene products (Morris-Krsinich *et al.*, 1985). However it should be noted that ^{35}S-methionine was used to label *in vitro* synthesised products while the putative product of the 10,900 Da ORF contains only the amino-terminal methionine.

The single complementary (−) polarity MSV transcript maps between the two intergenic regions having a 5′-terminus near nucleotide 2360 and a 3′-terminus in the vicinity of nucleotide 1160. This places the 5′ end of the transcript within the 31,400 Da ORF. However, there is a second in-frame ATG sequence 99 bases downstream. Initiation at this site would give a product of 27,700 Da. The region of sequence encoding the overlapping 17,800 Da ORF is the most highly conserved between MSV and WDV (McDowell *et al.*, 1985). The corresponding 17,300 Da ORF in WDV lacks an ATG triplet; instead the sequence GTG occurs. This has led to the proposal that the products of the two overlapping reading frames in WDV could be expressed as a composite protein either as a consequence of RNA splicing or translational frameshift. In the absence of a unique mRNA mapping to the 17,800 Da ORF of MSV or evidence of RNA splicing, it seems probable that a similar mechanism might operate to express the − sense products of MSV. In this context it is interesting to note that the sequence of a Kenyan (K) isolate of MSV (Howell, 1984; 1985) is more than 99% homologous to the Nigerian (N) strain of MSV (Mullineaux *et al.*, 1985) but shows heterogeneity at the amino-terminus of the ORF (P16), corresponding to the 17,800 Da ORF of MSV-N. This is reflected in a shortening of the MSV-K ORF by 43 amino acids which gives rise to a tandem arrangement of the two − sense ORFs in the same reading frame. The presence of a potential amber termination codon at the carboxy-terminus coding region of the larger ORF suggests that a read-through product could be produced (Howell, 1984). The possibility that the two ORFs are expressed to give a single fused protein is particularly attractive in view of the extensive homology which exists between the 17,800 and 31,400 Da products of MSV and the 40,300 Da product of the complementary sense DNA 1 of CLV (Mullineaux *et al.*, 1985).

IV. Gene Functions

The only geminiviral gene product that has so far been observed in infected tissue is the coat protein. Considering that it is probably the most abundant viral gene product, the fact that its detection necessitates immunolocalisation (Stanley and Townsend, 1986) reflects the problem of detection of other less well represented virus-coded proteins. It has not yet been determined if the nuclear cytopathological structures (fibrillar rings), characteristic of infection by the whitefly-transmitted geminiviruses, comprise a viral-encoded product(s) or are a host response to virus infection.

In the majority of cases, only a single protein is resolved by SDS-PAGE of purified virus particles although occasionally minor species are apparent (Goodman *et al.,* 1980; Adejare and Coutts, 1982). Whether or not these minor species represent unique viral-coded proteins or are derived from the coat protein by post-translational modification or degradation remains to be seen. As discussed in previous sections, the coat protein gene has been correlated with the major + sense ORF of both the leafhopper-transmitted geminivirus DNA and DNA 1 of those that are whitefly-transmitted. Coat protein characteristics might be expected to play a major role in the interaction of the virus with the insect vector responsible for its transmission in the field. The fact that the coat protein ORFs are the most highly conserved coding regions within the genomes of BGMV, CLV and TGMV (Hamilton *et al.,* 1984; Howarth *et al.,* 1985), all of which are transmitted by the whitefly *Bemisia tabaci,* is consistent with this idea. In contrast, the coat protein ORFs of BCTV, MSV and WDV, each of which has a unique leafhopper vector, show significantly less homology with each other (McDowell *et al.,* 1985; Stanley *et al.,* 1986) than do those of whitefly-transmitted geminiviruses. However, the BCTV coat protein bears a greater resemblance to those of other leafhopper-transmitted geminiviruses, yet is located in a genome that is otherwise organised strictly on the lines of DNA 1 of the whitefly-transmitted geminiviruses (Stanley *et al.,* 1986).

Viral-coded insect transmission functions have been demonstrated for some potyviruses (Govier *et al.,* 1977; Sako and Ogata, 1981; Hellmann *et al.,* 1983) and caulimoviruses (Woolston *et al.,* 1983; Armour *et al.,* 1983; Markham and Hull, 1985). With the exception of the coat protein ORFs, the high degree of homology between the analogous ORFs of BCTV and the whitefly-transmitted geminiviruses suggests that they are not involved in encoding such a transmission factor. However, this does not rule out the possibility that a DNA 2-encoded product might be involved in transmission by *B.tabaci* although no transmission or acquisition factor dependence has yet been demonstrated. In view of the fact that a gene product required solely for insect transmission would be effectively non-essential when virus transmission is by mechanical means, its presence could be tested for by *in vitro* mutagenesis of each putative gene product.

While other viral gene products have yet to be detected in infected tissues, evidence is recently available to suggest additional gene functions and their location within the genome. The ability to infect dividing protoplasts from *Nicotiana plumbaginifolia* using either CLV or cloned copies of the CLV genome has allowed the demonstration of the independent replication of DNA 1 (Townsend *et al.*, 1986). A similar conclusion was subsequently reached following the release of integrated copies of the TGMV genome within transgenic petunia plants (Rogers *et al.*, 1986). Of the four CLV DNA 1 ORFs that are conserved between the whitefly-transmitted geminiviruses (Fig. 1), the + sense 30.2 K ORF has been implicated in encoding the coat protein, which implies that one or more of the – sense ORFs are concerned with DNA replication, either encoding a replicase or possibly a subunit or modifying protein thereof with which to exploit the existing cellular synthetic machinery. A CLV mutant containing a defective 15.8 K ORF, produced by *in vitro* mutagenesis, has retained the capacity for systemic infection suggesting that this – sense ORF is not an essential part of the replicative function (Etessami and Stanley, personal communication). The amino acid homologies that exist between the CLV DNA 1 40.3 K – sense ORF (Fig. 1) and the 31.4 K and 17.8 K – sense ORFs of MSV (Fig. 2) (Mullineaux *et al.*, 1985) are perhaps indicative of the conservation of a functional domain, suggesting that the replication function is contained within one or both of these MSV ORFs.

The ability to infect plants by mechanical inoculation of deproteinised geminivirus DNA (Goodman, 1977 a, b; Hamilton *et al.*, 1981; Ikegami *et al.*, 1984) indicates that viral-encoded products are not required for the production of transcriptionally active dsDNA from the encapsidated ssDNA form. In at least MSV, RNA-primed second strand synthesis probably occurs at the position of the small complementary fragment found associated with ssDNA within virus particles (Donson *et al.*, 1984). It has recently been suggested (Rogers *et al.*, 1986) that TGMV DNA A (equivalent to CLV DNA 1) encodes a product analogous to ∅X 174 gene A protein. This multifunctional protein acts at a specific sequence of RF I DNA (Heidekamp *et al.*, 1982) that shows significant homology to the above mentioned common nanonucleotide to produce circular ssDNA by a rolling circle mechanism (Brown *et al.*, 1982 and references therein). Computer-assisted comparison of ∅X 174 gene A protein with CLV – sense putative products has failed to reveal homologies between these proteins (Stanley, unpublished).

Previously, both DNA components of the CLV genome were shown to be necessary for the production of a systemic infection in host plants (Stanley, 1983). The self-replicative activity of DNA 1 in protoplasts implies that DNA 2 must contribute a function essential for systemic spread of the virus. The product of the 30 K ORF of tobacco mosaic virus, which has been implicated in cell-to-cell spread (Taliansky *et al.*, 1982 a, b), shows no significant homology with the CLV 29.3 K ORF (Stanley, unpublished). Furthermore, the isolation of a mutant of CLV defective in coat protein and particle production suggests that encapsidation is not prerequisite for systemic spread of the virus (Stanley and Townsend, 1986). Therefore, it would appear that the inability of DNA 1 to spread throughout the plant in the absence of DNA 2 is not due to a DNA 2-mediated virus assembly function, although of course its participation in such a function cannot yet be ruled out. This possibility can be tested by searching for intact particles in DNA 1-infected protoplasts.

The BCTV infectious unit closely resembles CLV DNA 1 yet can cause a systemic infection in a common host such as *Nicotiana benthamiana* in the absence of a genomic component equivalent to DNA 2 (Stanley *et al.*, 1986) and therefore without the requirement for the presupposed transport function. However, while CLV is readily mechanically transmissible to *N. benthamiana,* all attempts to transmit BCTV in this way have been unsuccessful (Stanley, unpublished). BCTV has only been successfully mechanically transmitted with difficulty into *Beta vulgaris* by pin-pricking or high pressure injection into seedling crowns (Severin, 1924; Mumford, 1972), which possibly serves to introduce the inoculum directly into the vascular system in which the virus proliferates. Both BCTV and CLV are predominantly phloem-limited although CLV has been observed less frequently in other cell types (Sequeira and Harrison, 1982). The possibility exists that a barrier to virus transport from peripheral leaf cells into the vascular tissue is circumvented to a limited extent by a DNA 2 gene product. Replication of MSV and WDV, which in common with BCTV are considered to have only a single genomic component, might be confined to incompletely differentiated meristematic cells and consequently may not spread from cell-to-cell in the manner envisaged for CLV.

V. The Potential of Geminiviruses as Gene Vectors

Plant viruses are infectious, non-integrative agents which replicate to high copy number in infected cells. As such they represent vehicles which are potentially capable of carrying "foreign" genes directly into plants and amplifying them in plant tissues as well as being sources of episomal replicons for the transfection of plant protoplasts which might eventually be modified to serve as vectors for the stable transformation of plant crops.

The finding that CaMV had a genome of circular dsDNA (Hull and Shepherd, 1977) prompted a wave of speculation as to the potential of DNA plant viruses as gene vectors (e. g. Hull, 1978). Unfortunately, attempts to subjugate CaMV to this end have largely been frustrated by certain inherent properties of the virus (Hull, 1985). Will geminiviruses prove to be more amenable to manipulation? Certainly some of the problems which beset CaMV will also prove major obstacles to the development of vectors based on geminiviruses but, in the light of recent findings, we can be more optimistic about the prospects for the development of short term expression vectors based on geminiviruses although their possible contribution to the long term transformation of crop plants remains largely a matter for conjecture.

The necessity to encapsidate recombinant CaMV genomes in an isometric particle proved a severe limitation on the amount of extra DNA which could be accommodated. Although small viral genomes may be convenient from the standpoint of *in vitro* manipulation, they are highly restrictive in terms of deletable sequences. The largest single non-essential region in the CaMV genome is ORF II which encodes an aphid transmission factor (Woolston *et al.,* 1983) and the largest piece of DNA which can be inserted and stably maintained in the CaMV genome is only about 250 bp in length (Gronenborn *et al.,* 1981). Although CLV must be encapsidated to be transmitted by its whitefly vector (Townsend and Markham, personal communication) encapsidation is not a prerequisite for successful mechanical inoculation of plants (Stanley, 1983) or systemic spread of the virus within its host (Stanley and Townsend, 1986). Consequently there may be no size restraint on infectious recombinant molecules of CLV and possibly other geminiviruses. However, chimeric molecules comprising CLV DNAs cloned in M 13 were found to undergo recombination *in vivo* such that virtually all the vector DNA was eliminated. The progeny molecules varied in size by no more than 2—3 % relative to virion DNA. Hence, while encapsidation may not be essential for replication, molecules of a size suitable for encapsidation do have a selective advantage in establishing infections.

Restrictions on the size of integrated sequences or disruption of essential viral gene functions can be circumvented by the use of helper viruses to complement defective chimeric molecules (Goff and Berry, 1976). However, pairs of defective CaMV molecules have proved to be highly recombinationally active and combine to give wild type virus (Walden and Howell, 1982). It is clear that extensive recombination also occurs between CLV DNAs. At least four distinct mutants were isolated following a single passage of a chimeric clone of DNA 1 (Stanley and Townsend, 1986). In some instances the deletion of the vector sequences had clearly occurred via intermolecular recombination between homologous sequences in the two components of CLV, but in other instances there were no identifiable primary or secondary structures to account for the elimination of the deleted sequences. In most cases it is the parental genome which will outcompete other recombinants. However, one group of recombinant molecules is maintained in the population of CLV DNAs present in infected tissue. These are approximately half molecules of component 2 which can be isolated from preparations of purified virus (Stanley and Townsend, 1986) and which may be encapsidated in isometric particles (Sequeira, 1982). Since these molecules carry both a functional origin of replication and signals necessary for encapsidation, they may be able to tolerate the introduction of at least 1,500 nucleotides of additional sequences, possibly more if other parts of the defective molecule can be deleted. The finding that all the viral-encoded functions necessary to replicate the CLV genome are encoded within the three overlapping ORFs in the complementary sense of DNA 1 (Townsend *et al.,* 1986; Stanley and Townsend, 1986) suggests the possibility of constructing a chimeric molecule in which the coat protein gene is replaced by the putative cell-to-cell spread factor which is presumed to be encoded on DNA 2 (Townsend *et al.,* 1986; Rogers *et al.,* 1986). The resulting molecule should be capable of replicating autonomously and establishing a systemic infection in plants but might not be subject to packaging constraints in inserted sequences. In practice such molecules may be difficult to construct without disrupting viral replicative processes. The expression of any viral genome is likely to be highly regulated and interrelated as exemplified by the necessity to maintain the relative positions of translation initiation and termination codons of CaMV where it appears that several contiguous ORFs are translated from a polycistronic mRNA (Dixon and Hohn, 1984). Although transcript mapping suggests that expression of the CLV genome may be relatively simple (Townsend *et al.,* 1985) this may not be the case for geminiviruses with monopartite genomes (Morris-Krsinich *et al.,* 1985).

The high rate of recombination observed between deleted CaMV molecules is probably in part a consequence of the unusual replicative strategy of CaMV which raises a fundamental question as to its usefulness as a vector. It has become apparent that CaMV has several features in common with retroviruses including a replication mechanism that includes reverse transcription (Hull and Covey, 1983). This involvement of enzymes which lack proofreading may lead to the accumulation of errors in non-viral sequence for which there is no selective pressure (Hull, 1985). Although little is known about the replication cycle of geminiviruses they show none of the characteristic features of retroviruses (Hull and Covey 1986) and it would be very surprising if their replication did not involve a DNA-dependent DNA polymerase with proofreading.

The extent to which the genomes of geminiviruses which infect monocotyledonous plants could be manipulated remains a matter for speculation. The failure to infect monocotyledonous hosts mechanically with virus, or more significantly, cloned DNA places severe constraints on their development as vectors at the present time. Even if the infectivity problem can be overcome, there are several fundamental aspects of geminivirus biology which may be difficult or impossible to circumvent. Replication of viral DNA will be associated with the expression of genes which lead to the production of disease symptoms which, in the case of geminiviruses are certainly deleterious and probably cytocidal. It may not be possible to dissect essential replicative functions from pathogenic determinants. In this respect, preliminary information gained from experiments with CLV is not encouraging. Pseudorecombinant experiments clearly show that major determinants of symptom production reside on DNA 1 (Stanley *et al.,* 1986). Disruption of most of the coding region of the coat protein gene has almost no effect on symptom expression (Stanley and Townsend, 1986) hence, symptom expression is probably localised within the overlapping ORFs in the complementary sense of DNA 1 which are also implicated in genome replication.

Another problem concerns the distribution of virus in infected plants. Evidence from thin sectioning of infected tissues suggests that most geminiviruses are largely restricted to phloem-associated cells. This could prove a major limitation on their use as gene vectors. The finding that this tissue specificity can be relaxed by coinfection of BGMV with TMV, resulting in a more generalised infection of the host (Carr and Kim, 1983) suggests a solution to the problem might be provided by the incorporation of a cDNA copy of that part of the TMV genome which encodes the 30 K cell-to-cell spread protein (Taliansky *et al.*, 1982 a, b). The implication of this strategy in restricting the distribution of viral-based vectors has been discussed elsewhere (Hull, 1985). It is clearly impracticable to transform most seed-propagated crops on a field scale by mechanical inoculation of each generation. A commercially viable vector for long term transformation of cereal groups must be one which is efficiently transmitted from one generation to the next via the seed or possibly via pollen. None of the members of the geminiviruses group is seed transmitted. Although about 10 % of known viruses are transmitted through the seed (Matthews, 1981) most are only transmitted with efficiencies of a few percent and as yet we have no understanding of the molecular or biological determinants of seed transmission which could be exploited in vector construction.

Episomal replication systems based on animal viruses have been widely exploited to study fundamental aspects of gene structure and expression in cells including the regulation of cloned genes in homologous genetic backgrounds and high level expression of genes to facilitate studies of RNA and protein processing (Hamer, 1980). By introducing origins of replication from *E. coli* plasmids, hybrid shuttle vectors have been constructed which are capable of replicating in both animal and bacterial cells enabling cDNAs to be screened in animal cells and recovered by cloning in *E. coli* (Sarner *et al.*, 1984). Recent developments including the infection of protoplasts with cloned DNAs (Townsend *et al.*, 1986) and the production of transgenic plants containing integrated TGMV genomes (Rogers *et al.*, 1986) suggest that geminivirus-based episomal vectors analogous to both papilloma virus and polyomavirus vectors (Rigby, 1983) are realistic possibilities. Furthermore, the potential of an increasing number of plant species, including some monocots, to be regenerated from protoplasts, has stimulated interest in the development of stably inherited minichromosome vectors based on geminivirus replicons (Buck and Coutts, 1983).

Our preliminary experiments showed that up to 5 % of *N. plumbaginifolia* protoplasts could be infected with CLV as demonstrated by the synthesis of coat protein antigen and its accumulation in the nucleus detected by fluorescent antibody staining (Townsend and Watts, personal communication). Although this efficiency of infection is low, control experiments using highly infectious cowpea mosaic virus RNA suggest that there is room for improvement in the protoplast isolation procedure and that modified inoculation procedures or inoculation via electroporation (Fromm *et al.*, 1985) might increase the rate of transfection substantially. A disturbing possibility is that the tissue specificity exhibited by geminiviruses in plants may extend to protoplasts and that the low transfection rate reflects the proportion of protoplastable leaf cells which are competent to replicate the virus. Much may depend on the outcome of experiments, currently in progress, to investigate the replication of CLV in protoplasts derived from suspension cultures of *N. plumbaginifolia* callus tissue. Furthermore, since it appears that replication of CLV is dependent upon host DNA synthesis (Townsend *et al.,* 1986), it may be possible to improve transfection frequencies by synchronising the division cycle of cells cultured in suspension.

Since we have shown that CLV component 1 is capable of independent replication in protoplasts and that the coat protein gene on component 1 is non-essential it should be relatively simple to replace the coat protein gene with an assayable reporter gene. Chloramphenicol acetyl transferase (CAT) (Gorman *et al.,* 1982) is particularly suitable for such experiments since it can be isolated on an 800 bp DNA fragment so that the recombinant molecule will maintain any replicative advantage conferred by the size of the wild type genome (see above). Autonomous replication of the chimeric molecule might be expected to produce levels of CAT expression in transfected protoplasts that were substantially higher than those of comparable molecules which had been rendered replication defective; preliminary results (Stanley and Etessami, personal communication) suggest that this could be achieved by mutagenesis of the 40.3 ORF which appears to be essential for replication. The demonstration of enhanced CAT activity would pave the way for the development of transient expression vectors suitable for the analysis of plant genes in homologous backgrounds.

An alternative approach to the construction of episomal vector systems is exemplified by COS monkey cells which have been transformed with large T-antigen and can replicate early region defective SV40 viral DNA (Gluzman, 1981) or plasmids based on pBR322 that include the 85 bp SV40 replication origin fragment (Myers and Tjian, 1980) but from which the inhibitory pBR322 "poison sequences" have been removed (Lusky and Botchan, 1981). The COS-SV40 system is capable of synthesising up to 4×10^5 recombinant molecules per infected cell (Mellan *et al.,* 1981) which can compensate for low efficiencies of transfection.

The bipartite nature of the geminivirus genome is eminently suited to investigating the potential of a plant cell equivalent of the COS cell system. In its basic form, this might comprise plant cells in which the presumptive viral replication functions encoded in the complementary sense of DNA 1 were integrated in the chromosomes and would supply *in trans* the functions necessary to replicate a plasmid containing the origin of replication of the second component. The preliminary experiments of Rogers *et al.* (1986) suggest that this approach is feasible. Using Ti-plasmid vectors, tandem copies of either component A or B of TGMV (equivalent to CLV components 1 and 2 respectively) were introduced into plants which remained phenotypically normal even though free ss and ds DNA A forms were detected in plants carrying tandem inserts of that component. This is consistent with the capacity of CLV DNA 1 for autonomous replication. When the plants were crossed, one quarter of the progeny showed disease symptoms and contained free forms of DNAs A and B. Transgenic plants containing a single integrated copy of component A crossed with plants containing tandem copies of component B remained symptomless inferring that virus becomes free by homologous recombination within the tandem copy or that the tandem copy serves as template for the sythesis of a full length ss DNA copy of the viral genome. Although the absence of symptoms in such crosses suggests that complementation of DNA B does not occur, the single copy of DNA A was cloned at the unique EcoRI site (2251) which disrupts the long ORF in the complementary sense of DNA A which is equivalent to the 40.3 ORF of CLV. These experiments might also suggest that, in contrast to CLV (Stanley *et al.,* 1985), TGMV component A functions do not induce symptoms (Rogers *et al.,* 1986). This is obviously an important consideration in the development of a COS cell type system. However, no evidence was presented to show that any of the integrated viral genes were expressed and, although symptomless plants containing tandem inserts of component A also contained free ds DNA A forms, the release of this free DNA occurred in relatively few cells. In the absence of a cell-to-cell spread factor to amplify the infection, it is perhaps unlikely that visible symptoms would be produced.

To progress from these promising beginnings to the development of episomal vectors capable of stably transforming plant tissue will require a much more extensive knowledge of the biology of geminiviruses as well as advances in other areas such as resistance markers for transformed cells and developments in plant tissue culture including the isolation of auxotrophic mutants which can be complemented by dominant selectable markers. Such systems have been fundamental in the development of yeast vectors (Beggs, 1978; Struhl *et al.,* 1979) and may offer a mechanism for maintaining continued selection to prevent plasmid loss during plant regeneration. Taking yeast replicating plasmids as their model (Hsiao and Carbon, 1981), Buck and Coutts (1982) speculated that the potential problem of unequal plasmid segregation during disjunction might be overcome by incorporation of plant centromeric DNA into a geminivirus-based replicon so creating a minichromosome which would be stable during mitosis and meiosis. While this remains an attractive concept, it carries with it the implication that such a minichromosome would be maintained at low copy number.

Clearly, the geminiviruses are intriguing models for gene expression studies and pathogen-host interactions, but they also offer considerable potential for designing constructs that could be useful tools in plant molecular biology (see also Grimsley, Hohn, Davies and Hohn, 1987, and Grimsley, Chapter 4 this volume) and which might lead to practical gene vectors.

Acknowledgements

We wish to thank our colleagues for their contributions to much of the work reviewed here: P. Mullineaux, P. Markham, M. Boulton, A. Ward, P. Etessami, J. Donson, B. Morris-Krsinich and others.

VI. References

Adejare, G. O., Coutts, R. H. A., 1982: The isolation and characterisation of a virus from Nigerian cassava plants affected by the cassava mosaic disease, and attempted transmission of the disease. Phytopathol. Z. **103,** 198—210.

Arai, K-I., Kornberg, A., 1981: Unique primed start of ∅X174 DNA replication and mobility of the primosome in a direction opposite chain synthesis. Proc. Natl. Acad. Sci. U.S.A. **78,** 69—73.

Armour, S. L., Melcher, U., Pirone, T. P., Lyttle, D. J., Essenberg, R. C., 1983: Helper component for aphid transmission encoded by region II of cauliflower mosaic virus DNA. Virology **192,** 25—30.

Beggs, J. D., 1978: Transformation of yeast by a replicating hybrid plasmid. Nature **275,** 104—109.

Breathnach, R., Chambon, P., 1981: Organisation and expression of eukaryotic split genes coding proteins. Ann. Rev. Biochem. **50,** 349—383.

Brown, D. R., Rainberg, D., Schmidt-Glenewinkel, T., Roth, M., Zipursky, S. L., Hurwitz, J., 1982: DNA structures required for ∅X174 A-protein-directed initiation and termination of DNA replication. Cold Spring Harbor Symp. Quant. Biol. **42,** 701—715.

Buck, K. W., Coutts, R. H. A., 1983: The potential of geminiviruses in the construction of plant host vector systems. Plant Mol. Biol. **2,** 351—357.

Carr, R. J., Kim, K. S., 1983: Evidence that bean golden mosaic virus invades non phloem tissue in double infections with tobacco mosaic virus. J. Gen. Virol. **64,** 2489—2492.

Dixon, L. K., Hohn, T., 1984: Initiation of translation of the cauliflower mosaic virus genome from a polycistronic mRNA: evidence from deletion mutagenesis. EMBO J. **3,** 2731—2736.

Donson, J., Morris-Krsinich, B. A. M., Mullineaux, P. M., Boulton, M. I., Davies, J. W., 1984: A putative primer for second strand DNA synthesis of maize streak virus is virion associated. EMBO J. **3,** 3069—3073.

Fromm, M., Taylor, L., Walbot, V., 1985: Expression of genes electroporated into monocot and dicot plant cells. Proc. Natl. Acad. Sci. U.S.A. **82,** 5284.

Gluzman, Y., 1981: SV40-transformed simian cells support the replication of early SV40 mutants. Cell **23,** 175—182.

Goff, S. O., Berry, P., 1976: Construction of hybrid virus containing SV40 and λ phage DNA segments and their propagation in cultured monkey cells. Cell **9,** 695—705.

Goodmann, R. M., 1977a: Infectious DNA from a whitefly-transmitted virus of *Phaseolus vulgaris*. Nature **266,** 54—55.

Goodmann, R. M., 1977b: Infectious DNA from a whitefly-transmitted virus. Virology **83,** 171—179.

Goodman, R. M., Shock, T. L., Haber, S., Browning, K. S., Bowers, G. R., 1980: The composition of bean golden mosaic virus and its single stranded DNA genome. Virology **106,** 168—172.

Gorman, C. M., Moffatt, L. F., Howard, B. H., 1982: Recombinant genomes which express chloramphenicol acetyltransferase in mammalian cells. Mol. and Cell. Biol. **2,** 1044—1051.

Govier, D. A., Kassanis, B., Pirone, T. P., 1977: Partial purification and characterisation of the potato virus Y helper component. Virology **78,** 306—314

Griffin, B. E., 1982: Structure and genomic organisation of SV40 and polyoma virus. In: J. Tooze (ed.), DNA Tumor Viruses. Molecular Biology of Tumor Viruses. New York: Cold Spring Harbor Laboratories.

Grimsley, N. Hohn, T., Davies, J. W., Hohn, B., 1987: Agrobacterium-mediated delivery of infectious maize streak virus into maize plants. Nature **325,** 177—179.

Gronenborn, B., Gardner, R. C., Schaefer, S., Shepherd, R. J., 1981: Propagation of foreign DNA in plants using cauliflower mosaic virus as vector. Nature **272,** 375—377.

Hamer, D. H., 1980: DNA cloning in mammalian cells with SV40 vectors. In: J. K. Setlow and A. Hollaender (eds.), Genetics Engineering, Principles and Methods. Vol. 2, pp. 83—401. New York: Plenum.

Hamilton, W. D. O., Saunders, R. C., Coutts, R. H. A., Buck, K. W., 1981: Characterisation of tomato golden mosaic virus as a geminivirus. FEMS microbiol. Lett. **11,** 263—267.

Hamilton, W. D. O., Bisaro, D. M., Coutts, R. H. A., Buck, K. W., 1983: Demonstration of the bipartite nature of the genome of a single-stranded DNA plant virus by infection with the cloned DNA components. Nucl. Acids Res. **11,** 7387—7396.

Hamilton, W. D. O., Stein, V. E., Coutts, R. H. A., Buck, K. W., 1984: complete nucleotide sequence of the infectious cloned DNA components of tomato golden mosaic virus: potential coding regions and regulatory sequences. EMBO J. **3,** 2197—2205.

Harrison, B. D., 1985: Advances in geminivirus research. Ann. Rev. Phytopathol. **23,** 55—82.

Heidekamp, F., Baas, P. D., Jansz, H. S., 1982: Nucleotide sequences at the ØX gene A protein cleavage site in replicative form I DNAs of bacteriophages U3, G14 and 3. J. Virology **42,** 91—99.

Hellman, G. M., Thornbury, D. W., Hiebert, E., Shaw, J. G., Pirone, T. P., Rhoads, R. E., 1983: Cell-free translation of tobacco vein mottling virus RNA. II. Immunoprecipitation of products by antisera to cylindrical inclusion, nuclear inclusion and helper component proteins. Virology **124,** 434—444.

Howarth, A. J., Caton, J., Bossert, M., Goodman, R. M., 1985: Nucleotide sequence of bean golden mosaic virus and a model for gene regulation in geminiviruses. Proc. Natl. Acad. Sci. U.S.A. **82,** 3572—3576.

Howell, S. H., 1984: Physical structure and genetic organisation of the genome of maize streak virus (Kenyan isolate). Nucl. Acids Res. **12,** 7359—7375.

Howell, S. H., 1985: Physical structure and genetic organisation of the genome of maize streak virus (Kenyan isolate). Nucl. Acids Res. **13,** 3018—3019.Hsiao, C. L., Carbon, J., 1981: Characterisation of a yeast replication origin (ars 2) and construction of stable minichromosomes containing cloned yeast centromeric DNA (CEN3). Gene **15,** 157—166.

Hull, R., 1978: The possible use of plant virus DNAs in genetic manipulation in plants. Trends Biochem. Sci. **3,** 254—256.

Hull, R., 1985: Viruses as vectors for plant genes. In: J. H. Dodds (ed.), Plant Genetic Engineering, pp. 95—113. Cambridge: Cambridge University Press.Hull, R., Covey, S. N., 1983: Does cauliflower mosaic virus replicate by reverse transcription? Trends Biochem. Sci **8,** 119—121.

Hull, R., Covey, S. N., 1986: Genome organisation and expression of reverse transcribing elements: variation and a theme. J. Gen. Virol. **67,** 1751—1758.

Hull, R., Shepherd, R. J., 1977: Structure of cauliflower mosaic virus genome. Virology **79,** 216—230.

Ikegami, M., Morinaga, T., Muira, K., 1984: Infectivity of virus-specific double-stranded DNA from tissue infected by mungbean yellow mosaic virus. Virus Res. **1,** 507—512.

Kozak, M., 1983: Comparison of initiation of protein synthesis in procaryotes, eucaryotes and organelles. Microbiological Reviews **47,** 1—45.

Kozak, M., 1984: Compilation and analysis of sequences upstream from the translational start site in eukaryotic mRNAs. Nucl. Acids Res. **12,** 857—872.

Kridl, J. C., 1986: Expression of bean golden mosaic virus and use of promoter sequences in chimeric constructs. 15th Ann. Meet. UCLA Symposium on Molecular and Cellular Biology, March/April 1986, Abstr. J107. J. Cell. Biochem. Supp. 10C, 40.

Lusky, M., Botchan, M., 1981: Inhibitory effect of specific pBR322 DNA sequences upon SV40 replication in simian cells. Nature **293,** 79—81.

Markham, P. G., Hull, R., 1985: Cauliflower mosaic virus aphid transmission factors from other caulimoviruses. J. Gen. Virol. **66,** 921—923.

Matthews, R. E. F., 1981: Plant Virology, 2nd ed. New York: Academic Press.

Matzeit, V., Lindsten, K. Lörz, H., Schaefer, S., Gronenborn, B., 1985: Cloned WDV DNA replicates in protoplast derived cells of *Triticum monococcum.* 1st International Congress of Plant Molecular Biolaogy, Savannah 1985. (G. A. Galu, ed.), Abstr. OR-04-03. Athens, Georgia: University of Georgia.

McDowell, S. W., MacDonald, H., Hamilton, W. D. O., Coutts, R. H. A., Buck, K. W., 1985: The nucleotide sequence of cloned wheat dwarf virus DNA. EMBO J. 4, 2173—2180.

Mellan, P., Parker, V., Gluzman, Y, Maniatis, T., 1981: Identification of sequences required for transcription of the human-globin gene in a new SV40 host-vector system. Cell **27,** 279—288.

Morris-Krsinich, B. A. M., Mullineaux, P. M., Donson, J., Boulton, M. I., Markham, P. G., Short, M. N., Davies, J. W., 1985: Bidirectional transcription of maize streak virus DNA and identification of the coat protein gene. Nucl. Acids Res. **13,** 7237—7256.

Mullineaux, P. M., Donson, J., Morris-Krsinich, B. A. M., Boulton, M. I., Davies, J. W. 1984: The nucleotide sequence of maize streak virus DNA. EMBO J. 3, 3063—3068.

Mullineaux, P. M., Donson, J., Stanley, J., Boulton, M. I., Morris-Krsinich, B. A. M., Markham, P. G., Davies, J. W., 1985: Computer analysis identifies sequence homologies among potential gene products of maize streak virus and those of cassava latent virus and tomato golden mosaic virus. Plant Mol. Biol. **5,** 125—131.

Mumford, D. L., 1972: A new method of mechanically transmitting curly top virus. Phytopathol. **62,** 1217—1218.

Myers, R. M., Tjian, R., 1980: Construction analysis of simian virus 40 origins defective in tumor antigen binding and DNA replication. Proc. Natl. Acad. Sci. U.S.A. **77,** 6491—6495.

Proudfoot, N. J., Brownlee, G. G., 1976: 3′ non-coding region sequences in eukaryotic messenger RNA. Nature **263,** 211—214.

Rigby, P. W. J., 1983: Cloning vectors derived from animal viruses. J. Gen. Virol. **64,** 255—266.

Rogers, S. G., Bisaro, D. M., Horsch, R. B., Fraley, R. T., Hoffmann, N. L., Brand, L., Elmer, J. S., Lloyd, A. M., 1986: Tomato golden mosaic virus A component DNA replicates autonomously in transgenic plants. Cell **45,** 593—600.

Sako, N., Ogata, K., 1981: Different helper factors associated with aphid transmission of some potyviruses. Virology **112,** 762—765.

Sarner, N., Mitrani-Rosenbaum, S., Law, M-F., McAllister, W. T., Byrne, J. C., Howley, P. M., 1984: Bovine papillomavirus shuttle vectors. In: J. K. Setlow, A.

Hollaender (eds.), Genetic Engineering: Principles and Methods, Vol, 5, pp. 173—190. New York: Plenum.

Sequeira, J. C., 1982: Purification, properties and relationships of cassava latent virus. PhD Thesis, University of Dundee, U. K.

Sequeira, J. C., Harrison, B. D., 1982: Seriological studies on cassava latent virus. Ann. Appl. Biol. **101,** 33—42.

Severin, H. H. P., 1924: Curly leaf transmission experiments. Phytopathol. **14,** 80—93.

Stanley, J., 1983: Infectivity of the cloned geminivirus genome requires sequences from both DNAs. Nature **305,** 643—645.

Stanley, J., 1985: The molecular biology of geminiviruses. Adv. Virus Res. **30,** 139—177.

Stanley, J., Davies, J. W., 1985: Structure and function of the DNA genome of geminiviruses. In: Davies, J. W. (ed.), Plant Molecular Virology, Vol. II, pp. 191—218. Boca Raton, Florida: CRC Press.

Stanley, J., Gay, M. R., 1983: Nucleotide sequence of cassava latent virus DNA. Nature **301,** 260—262.

Stanley, J., Markham, P. G., Callis, R. J., Pinner, M. S., 1986: The nucleotide sequence of an infectious clone of the geminivirus beet curly top virus. EMBO J. (in press).

Stanley, J., Townsend, R., Curson, S. J., 1985: Pseudorecombinants between cloned DNAs of two isolates of cassava latent virus. J. Gen. Virol. **66,** 1055—1061.

Stanley, J., Townsend R., 1986: Infections mutants of cassava latent virus generated *in vivo* from single copies of the genome. Nucl. Acids Res. **14,** 5981—5998.

Struhl, K., Stinchcomb, D. T., Schener, S., Davis, R. W., 1979: High frequency transformation of yeast: autonomous replication of hybrid DNA molecules. Proc. Natl. Acad. Sci. U.S.A. **76,** 1035—1039.

Taliansky, M. E., Malyshenko, S. I., Pshennikova, E. S., Atabekov, J. G., 1982a: Plant virus-specific transport function. II. A factor controlling virus host range. Virology **122,** 327—331.

Taliansky, M. E., Malyshenko, S. I., Pshennikova, E. S., Kaplan, I. B., Ulanova, E. F., Atabekov, J. G., 1982b: Plant virus-specific transport function. I. Virus genetic control required for systemic spread. Virology **122,** 318—326.

Townsend, R., Stanley, J., Curson, S. J., Short, M. N., 1985: Major polyadenylated transcripts of cassava latent virus and location of the gene encoding coat protein. EMBO J. **4,** 33—38.

Townsend, R., Watts, J., Stanley, J., 1986: Synthesis of viral DNA forms in *Nicotiana plumbaginifolia* protoplasts inoculated with cassava latent virus (CLV); evidence for the independent replication of one component of the CLV genome. Nucl. Acids. Res. **14,** 1253—1265.

Walden, R. M., Howell, S. H:, 1982: Intergenomic recombination events among pairs of defective cauliflower mosaic virus genomes in plants. J. Mol. Appl. Genet. **1,** 447—456.

Woolston, C. J., Covey, S. N., Penswick, J. R., Davies, J. W., 1983: Aphid transmission and a polypeptide are specified by a defined region of the cauliflower mosaic virus genome. Gene **23,** 15—23.

Chapter 3

cDNA Cloning of Plant RNA Viruses and Viroids

Pieter Vos

Department of Molecular Biology
Agricultural University
De Dreijen 11, NL-6703 BC Wageningen
The Netherlands

With 9 Figures

Contents

I. General Introduction

After the development of recombinant DNA technology and convenient techniques for rapid determination of nucleotide sequences, the construction of well-defined cloned DNA copies of plant viral RNA genomes and viroids is becoming more and more an integral part of the research on these pathogens. In this chapter I will attempt to show how the availability of such clones indeed has added a new dimension to plant virology.

In the first part of this chapter (section II) the technical aspects of synthesising cDNA will be described while, in addition, various cloning strategies will be compared and discussed. The following part (section III) will deal with applications of cDNA clones in experiments designed to unravel the multiplication cycle of plant RNA viruses. Hence, the use of these clones to study RNA replication, RNA recombination, gene expression and functions of viral proteins will be discussed. Section IV, in addition, will focus on the use of cDNA clones in viroid and satellite research whereas, finally, in section V some attention will be paid to diagnosis of plant virus diseases using cDNA clones.

II. Construction of Full-Length cDNA Clones

A. Introduction

Following the development of modern molecular cloning techniques, an ever growing number of plant virologists have been trying to construct full-length cDNA clones of plant RNA viruses and viroids, aiming to obtain specific probes for studying plant viruses, viroids and their diseases. While for several viroids full-size cDNA clones have been readily obtained (see e. g. Van Wezenbeek *et al.*, 1982; Cress *et al.*, 1983; Ohno *et al.*, 1983; Sano *et al.*, 1984; Tabler and Sanger, 1984), relatively few reports have been published on the construction of full-length cDNA clones of plant RNA viruses. This already indicates that synthesis of full-length double-stranded cDNA from relatively large RNA molecules, often containing extensive secondary structures, is far from easy. When writing this chapter full-size cDNA clones have been reported for the RNA of satellite tobacco necrosis virus (STNV) (Van Emmelo *et al.*, 1984), tobacco mosaic virus (TMV) (Meshi *et al.*, 1986; Dawson *et al.*, 1986), the three genomic RNAs of bromo mosaic virus (BMV) (Ahlquist and Janda, 1984) and the two genomic RNAs of cowpea mosaic virus CPMV) (Vos *et al.*, 1984; Verver *et al.*, submitted). In the following paragraphs I will discuss in some detail how such clones can be prepared.

B. Synthesis of Double-Stranded cDNA

For first strand synthesis the reverse transcriptase of avian myeloblastosis virus (AMV) is routinely used. This enzyme, which is commercially available since 1978, synthesises complementary DNA using RNA as template and requires like DNA polymerases a primer. Essential for successful first strand synthesis are both the quality of the reverse transcriptase used and the absolute purity and intactness of the template RNA. Potent RNase inhibitors are often included in the reaction mixture to prevent degradation of template RNA during the transcription reaction by ribonucleases present in most commercial reverse transcriptase preparations. Suitable reaction conditions have been described by several authors (see e. g. Efstratiadis *et al.*, 1976; Retzel *et al.*, 1980; Fields and Winters, 1982) and usually differ only in the monovalent cation concentration used. Actually, the optimal conditions for reverse transcription vary for different template RNAs and should therefore be determined independently for each RNA species. First strand synthesis can be followed by addition of labelled dNTPs to the reaction mixture and analysis of the reaction products on denaturing agarose gels (see Fig. 1).

The next step in the procedure is to synthesise the second strand using again reverse transcriptase or, as mostly preferred, *E. coli* DNA polymerase I. If the latter enzyme is used in combination with RNase H the first strand material synthesised can directly be used for second strand synthesis. The holo-enzyme of DNA polymerase I will perform "repair synthesis" on the RNA-DNA hybrid, replacing thereby the RNA strand

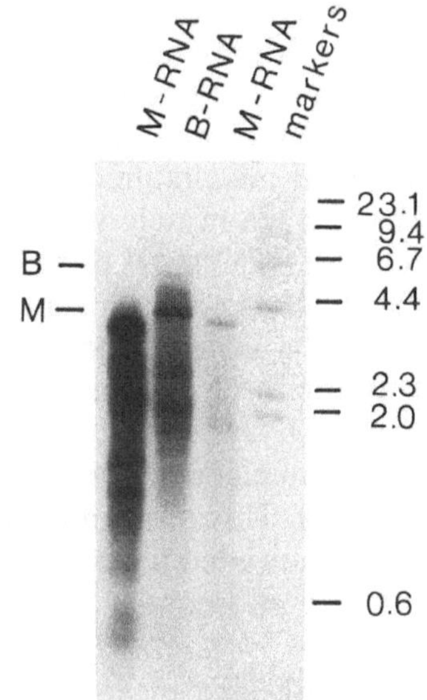

Fig. 1. Reaction products of first strand cDNA synthesis of cowpea mosaic virus M and B RNA analysed on a 1.2% alkaline agarose gel (McDonnell *et al.*, 1977). On the left hand site the positions of full-size transcripts of M (3.5 kb) and B RNA (5.9 kb) are indicated. On the right hand site the positions of marker DNA fragments are shown

attacked by the RNase H (Okayama and Berg, 1982). As a result a complementary strand will be obtained, containing several nicks that can be easily repaired by ligation. Although this approach is very efficient for synthesising the complementary strand and has the advantage that no primer is needed for second strand synthesis, it does not result in full-length double-stranded cDNA molecules comprising the ultimate 5′-terminal sequence of the RNA. To achieve the latter, second strand synthesis should be primed with an oligonucleotide homologous to the 5′-end of this RNA and extended using the Klenow fragment of DNA polymerase I. Since the Klenow enzyme cannot replace RNA hybridised to the template DNA, this RNA should be removed previously by alkali treatment or RNase treatment after heat denaturation. To improve the yield of full-length double-stranded cDNA it is advisable to seperate full-size single-stranded cDNA from smaller products prior to second strand synthesis. This can be accomplished in several ways, e. g. alkaline sucrose density gradient centrifugation (Meshi *et al.,* 1983), elution from denaturing polyacrylamide gels (Maxam and Gilbert, 1980) or elution from alkaline agarose gels (P. Vos, unpublished results).

C. Cloning Strategies

Once the double-stranded cDNA has been obtained, insertion in a suitable acceptor plasmid should follow. Blunt-end cloning of the double-stranded cDNA is most simple, but generally not very efficient. Furthermore, DNA fragments can be orientated in either direction, which in some cases may lead to a considerable number of unsuitable clones if one orientation, the wrong one, is preferred. For cowpea mosaic virus (CPMV) a strong orientation preference for full-length DNA copies of M RNA has been observed in some vectors (P. Vos, unpublished results). That, on the other hand, this approach can be very successful was demonstrated by Ahlquist and Janda (1984) by cloning in this way the three genomic RNAs of BMV.

Addition of linkers to the cDNA certainly enhances the efficiency of cloning, but knowledge of the nucleotide sequence of the RNA is then required to select suitable linkers. This disadvantage may be circumvented by methylation of the double-stranded cDNA (by specific methylases) prior to addition of linkers and subsequent digestion to generate sticky ends.

Alternatively first strand synthesis can be primed with an oligonucleotide extended at its 5′-end with the recognition sequence of a restriction enzyme for which no cleavage sites are present in the cDNA sequence. Upon second strand synthesis, initiated with an oligonucleotide homologous to the 5′-end of the RNA sequence, the double-stranded cDNA is digested with the restriction enzyme in question and ligated into a suitable vector using one sticky (3′-)end and one blunt (5′-)end. This protocol ensures both efficient ligation and insertion in the desired orientation. For plant RNA viruses having a 3′-poly(A) tail (e. g. comoviruses, potyviruses, nepoviruses) first strand synthesis may be primed with a plasmid provided with an oligo(dT) tail as described by Okayama and Berg (1982). Second strand synthesis can next be primed with an oligonucleotide homologous to the 5′-end of the RNA and cloning involves only circularisation of the synthesised double-strand cDNA linked to the plasmid. This protócol serves again the important goal of efficient cloning in the desired orientation while, furthermore, the minimal length of the poly-d(A-T) track in the cDNA clone to be constructed can be previously determined by the length of the oligo (dT) tail linked to the plasmid primer.

Finally, when the RNA sequence is known, double-stranded cDNA may be cleaved with restriction enzymes yielding a number of fragments, which can be cloned separately and joined afterwards to a full-size copy (Meshi *et al.*, 1986; Dawson *et al.*, 1986; Vos *et al.*, 1984; Verver *et al.*, submitted). This approach is especially attractive for cloning genomic RNAs of considerable length.

D. Cloning in Transcription Vectors

Recently several *in vitro* transcription systems have been developed, which allow the synthesis of large quantities of RNA from cloned cDNA (Contreras *et al.*, 1982; Green *et al.*, 1983; Davanloo *et al.*, 1984). These systems are based on a homologous combination of a specific RNA polymerase and its promoter. RNA transcripts generated from such systems have been proven to be efficient messenger RNAs *in vitro* (Melton *et al.*, 1984; Ahlquist and Janda, 1984; Vos *et al.*, 1984), which in the case of transcripts from viral DNA copies may express infectivity (Ahlquist *et al.*, 1984a; Mizutani and Colonno, 1985; Van der Werf *et al.*, 1986). Since DNA copies of most plant RNA viruses are not directly infectious (exception STNV, Van Emmelo, submitted), cloning downstream of a promoter sequence of an RNA polymerase, which allows efficient transcription *in vitro*, may be an attractive possibility to generate infectious RNA from cDNA clones (see for further discussion on this point section III. 2.). Efficient *in vitro* transcription systems have been described based on *E. coli* RNA polymerase on one hand or on the RNA polymerases of bacteriophages like SP6 or T7 on the other hand.

The *E. coli in vitro* transcription system is not as efficient as the transcription systems based on RNA polymerases of bacteriophages, but has the advantage that sequences downstream from the transcription initiation site are dispensable for efficient transcription *in vitro*. The *E. coli* promoter is defined by the "-35" and "-10" region, allowing cDNA inserts to be positioned directly at the transcription initiation site of a strong *E. coli* or *E. coli* phage promoter without affecting transcriptional efficiency. Hence Ahlquist and Janda (1984) designed a transcription vector, pPM1, based on the phage lambda PR promoter (see Fig. 2). Introduction of any DNA fragment with a 5′-proximal purine into the unique *SmaI* site of this vector will result in transcriptional start by *E. coli* RNA polymerase at the first nucleotide of the DNA fragment inserted. Moreover, addition of the cap analog m^{7}GpppN to the transcription mixture leads to the generation of capped transcripts (Contreras *et al.*, 1982; Ahlquist and Janda, 1984). Using this *E. coli* cloning/transcription system RNA transcripts from full-length cDNA clones of the three genomic RNAs of BMV and of TMV RNA have been synthesised, which perfectly resemble the natural viral RNAs with respect to their capped 5′-ends (Ahlquist *et al.*, 1984a; Meshi *et al.*, 1986; Dawson *et al.*, 1986).

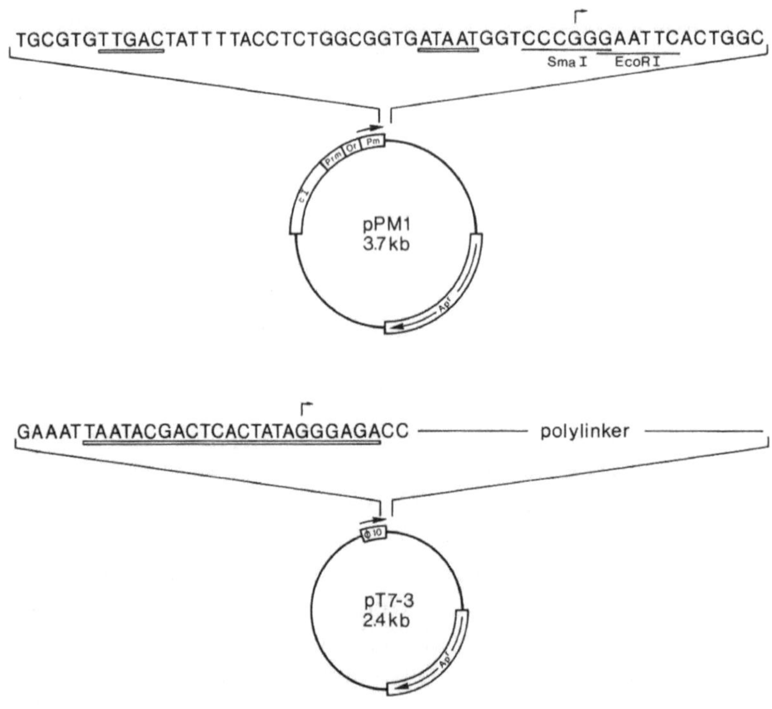

Fig. 2. Schematic representation of transcription vectors pPM1 and pT7-3 (see text) with in detail the sequences surrounding the transcription initiation sites (indicated by arrows). Sequences important for efficient transcription by either the *E. coli* or T7 RNA polymerase are indicated by double lines. In the plasmids the β-lactamase gene for ampicillin resistance is shown, pBR322 derived sequences are indicated by a single line and open boxes refer to phage or T7 derived sequences (Ahlquist and Janda, 1984; Queen, 1983; Dunn and Studier, 1983: Tabor and Richardson, 1985)

Similarly the RNA polymerase of bacteriophages like SP6 and T7 are very suitable to generate RNA transcripts *in vitro* from any DNA fragment positioned downstream of an SP6 or T7 promoter respectively (Green *et al.*, 1983; Vos *et al.*, 1984; Verver *et al.*, submitted, and Fig. 2). These RNA polymerases have the advantage over the *E. coli* RNA polymerase that they are very specific for their own promoter sequence and that they are at least ten times more active *in vitro* compared to the *E. coli* enzyme. Yields of up to 20 μg of RNA synthesised from 1 μg of template DNA have been described (Tabler and Sänger, 1985; Verver *et al.*, submitted). Similar to *E. coli* RNA polymerase these phage-encoded polymerases are capable of incorporating the cap precursor m^7GpppN generating capped transcripts (Konarska *et al.*, 1984; P. Vos, unpublished results). A general disadvantage of the phage polymerases is, however, that sequences downstream of the transcription initiation site are important for efficient transcription and thus cannot be removed. With both SP6 and T7 promoters the first five nucleotides downstream of the transcription initiation site form part of the promoter sequence (Dunn and Studier, 1983; P. Vos, unpublished results). Nucleotide changes in this region decreases the efficiency of transcription, although considerable amounts of RNA transcripts may still be produced. Investigation of this issue has indicated that as long as the transcription initiation site and the first downstream nucleotide remain unchanged, transcription efficiency is still 10—20% compared to normal (Van der Werf *et al.*, 1986; P. Vos, unpublished results), which, in most cases, still gives sufficient yield. Vectors for the SP6 or T7 *in vitro* transcription system are commercially available and have multiple cloning sites downstream of the promoter sequence facilitating insertion of DNA fragments. To date construction of cDNA clones downstream of bacteriophage promoters has been reported for several RNA viruses like human rhino virus (Mizutani and Colonno, 1985), poliovirus (Van der Werf *et al.*, 1986), black beetle virus (Dasmahapatra *et al.*, 1986) and cowpea mosaic virus (CPMV) (Vos *et al.*, 1984; Verver *et al.*, submitted). Infectious transcripts from these cDNA clones were obtained in spite of non-viral extra sequences at the ends of the transcripts (see also III. 2).

III. DNA Copies as Tools to Study the Molecular Biology of Plant RNA Viruses

A. Introduction

This section will deal with the application of cDNA clones in studies on the molecular biology of plant RNA viruses. Full-length cDNA clones are currently available only for a few plant RNA viruses, although their number will certainly grow in the near future. When writing this chapter, full-size cDNA clones have been described only for the genomic RNAs of cowpea mosaic virus (CPMV), bromo mosaic virus (BMV) and tobacco mosaic virus (TMV). As a consequence the following pages will mainly focus on these viruses. Whenever necessary, background information on these viruses will be given throughout the different paragraphs.

B. Infectivity of cDNA Clones

i) Infectivity of DNA Copies

Since plant RNA viruses have RNA as the source of genetic information it is interesting to know if a double-strand DNA copy of such a genomic RNA is infectious upon inoculation of host cells. This phenomenon has indeed been found for some RNA viruses, like bacteriophage Qβ (Taniguchi *et al.*, 1978), and for human poliovirus (Racaniello and Baltimore, 1981). In all these cases the infectivity of the DNA copy was very low compared to that of viral RNA. This is not surprising, especially not for RNA viruses of eukaryotes, if one tries to imagine the pathway that such a copy must follow after introduction into the host cell to eventually produce an infectious RNA molecule. To achieve this the DNA copy, after entering the cytoplasm, has to reach the nucleus where transcription should take place. Transcripts having probably extra non-viral sequences at both ends should then be transported to the cytoplasm, where the multiplication process of the virus takes place. To get multiplication of these transcripts, they should be translated (to obtain viral replication proteins) and probably should be processed to RNA molecules that can be recognised by the viral replicase. One or more of these processes can be expected to be very inefficient and therefore may lead to a very low level of infectivity. While for viroids direct infectivity of double-stranded DNA copies has been demonstrated (see section IV. 2 for a more detailed discussion), for plant RNA viruses infectious DNA copies have so far not been reported. It could very well be that the infectivity of such copies is very low, as in the case of poliovirus, and therefore escaped detection. For CPMV, direct infectivity of double-stranded DNA copies has been tested extensively, but never resulted in a detectable level of infection (unpublished results). Even the use of (strong) eukaryotic promoter sequences, like the SV40 early promoter, the nopaline synthase promoter from *Agrobacterium tumefaciens* or the cauliflower mosaic virus (CaMV) 35 S promoter to enhance transient transcription after introduction in cowpea cells, was not successful.

An alternative way to express double-stranded DNA copies to give infectious particles again may be transfer of the cDNA to plants using *Agrobacterium tumefaciens,* which can also result in stable integration of the DNA copy into the plant chromosome (see chapter on agroinfection). Thus far this approach has only been successfully applied for plant DNA viruses, i. e. cauliflower mosaic virus (Grimsley *et al.,* 1986) and tomato golden mosaic virus (Rogers *et al.,* 1986).

With respect to RNA viruses, partial DNA copies of TMV RNA have been integrated into tobacco (Bevan *et al.,* 1985; Powell Abel *et al.,* 1986). It was demonstrated that these DNA copies were transcribed in transformed plants and furthermore that also TMV-specific protein, in casu coat protein, could be detected. For CPMV cowpea cells have been transformed with a full-size DNA copy of the smaller of the two genomic RNAs, the M-RNA, provided either with the nopaline synthase (nos) promoter or the CaMV 35 S promoter, using *Agrobacterium*-mediated gene transfer. This

DNA copy was efficiently expressed in transformed callus tissue obtained, yielding transcripts of the expected length of more than 3500 nucleotides long (Garcia *et al.,* 1986). These transcripts were remarkably stable, although synthesis of CPMV RNAs occurs normally in the cytoplasm. Since the M-RNA of CPMV is dependent for its replication on the B-RNA encoded RNA dependent RNA polymerase, it will be of interest to determine if infection with B-RNA of transformed cowpea cells will result in replication of the M-RNA-like transcripts.

ii) Infectivity of in vitro Transcripts from DNA Copies

The development of efficient *in vitro* transcription systems has offered an attractive alternative for the approaches described above to achieve infectivity from plant viral DNA copies. DNA copies preceded by a strong promoter specific for a certain RNA polymerase can be transcribed *in vitro* into their RNA counterparts. Using this approach infectious transcripts have been described for human rhinovirus and poliovirus (Mizutani and Collono, 1985; Van der Werf *et al.,* 1986), the insect virus black beetle virus (Dasmahapatra *et al.,* 1986) and the plant RNA viruses BMV (Ahlquist *et al.,* 1984 a) and TMV (Dawson *et al.,* 1986; Meshi *et al.,* 1986).

The problem when synthesising RNA *in vitro* is to generate transcripts which resemble the original viral RNA with respect to its 5′ and 3′ end. It is obvious that extra terminal sequences may interfere with viral RNA replication, since this is generally a very specific process. From experiments done so far it is clear that the more the termini of *in vitro* RNA transcripts resemble those of the natural viral RNA the more infectious they are. For both the picornaviruses and viroids it was found that RNA transcripts were some orders of magnitude more infectious than the DNA copies themselves (Van der Werf *et al.,* 1986; Tabler and Sänger, 1985; Mizutani and Collono, 1985). However the way in which extra sequences affect the infectivity of *in vitro* transcripts may differ from virus to virus depending on their replication mechanism.

As mentioned above for plant RNA viruses infectious transcripts have thus far only been obtained for BMV and TMV (Ahlquist *et al.,* 1984 a; Dawson *et al.,* 1986; Meshi *et al.,* 1986). Both viruses have RNAs with a cap structure at their 5′ end, and for both capping of the *in vitro* transcripts was essential for infectivity. Extra nucleotides at the 5′ ends and/or the absence of a cap (almost) abolished the infectivity of the *in vitro* transcripts. On the other hand extra sequences at the 3′-end seem not to have such drastic effects. Such extra sequences have never been found in the progeny of *in vitro* transcripts, indicating that they are removed upon replication. Furthermore, for TMV reconstitution *in vitro* of the RNA transcripts with coat protein have been shown to further increase their infectivity (Meshi *et al.,* 1986). It is obvious that transcripts are protected in this way from nucleases and that the infection process occurs much more efficiently.

In various cases it has been observed that, without any obvious reason, some cDNA clones did not produce infectious transcripts. Although none of these clones have been analysed in detail, changes at the nucleotide level are probably the reason for this lack of infectivity. Indeed it has been demonstrated that the RNA population of a virus may be rather heterogenous in sequence (Goelet *et al.,* 1982; Lomonossoff and Shanks, 1983), probably due to miscopying during RNA replication. The error rate of RNA-dependent RNA polymerases in general is 10^{-3} to 10^{-4}. Non-infectious clones can therefore have originated from RNA molecules which were not infectious. Alternatively, mistakes may be introduced by the reverse transcriptase during first strand synthesis. In this view the use of several independently obtained clones for infectivity assays is advisable. However, sequence analysis of non-infectious cDNA clones could still be very important to gain more insight in the process of virus infection and multiplication.

C. RNA Replication

Thus far little is known about the replication of eukaryotic RNA viruses, and this is certainly true for RNA viruses of plants. Only recently the idea is more and more emerging, that most, if not all, eukaryotic RNA viruses encode their own RNA-dependent RNA polymerase (replicase) which, as far as known, is very specific for its own RNA (for a review see Van Kammen, 1985). Most probably this specificity must have its origin in the primary and/or secondary structure at the termini of the RNA molecule, where RNA replication starts. Determination of the sequence signals involved in the initiation of replication would be greatly facilitated by the availability of appropriate systems to study this process *in vitro.* Such a system has thus far only been described for BMV (Bujarski *et al.,* 1982; Miller and Hall, 1983). From BMV-infected barley leaves viral RNA-synthesising extracts can be prepared, which are RNA dependent and capable of synthesising full-length complementary strands from added BMV RNA. Hence, since for BMV full-length cDNA clones have been constructed, a powerful system is available to study sequences involved in the initiation of replication of BMV RNA, by addition of RNA transcripts prepared from modified DNA templates to BMV-infected leaf extracts. Before discussing the results obtained so far in this system it is necessary to give a short description of the virus.

BMV is a virus of which the genetic information is divided among three positive-stranded RNA molecules, designated RNA 1, 2 and 3 respectively, with sizes of 3.2, 2.9 and 2.1 kilobases (kb) (Lane and Kaesberg, 1971; Lane, 1981; Ahlquist *et al.,* 1984 b). In infected plants an additional RNA is abundantly present, RNA 4, which is a subgenomic mRNA of RNA 3 and codes for the capsid protein. The viral RNAs all have a cap structure at their 5′-end and possess a tRNA-like structure at their 3′-end, including a CCA-terminus which can be aminoacetylated with tyrosine (Kiberstis and Hall, 1983; Loesch-Fries and Hall, 1982) (see Fig. 3). The sequence of the 3′-terminal 200 nucleotides is virtually identical among the three genomic BMV RNAs, indicating that this region may play an important

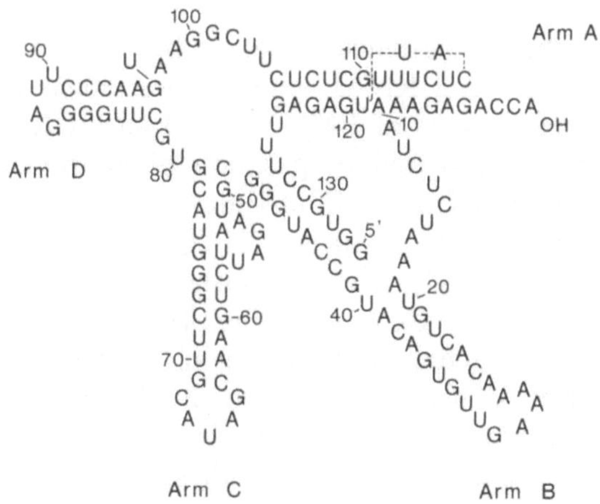

Fig. 3. t-RNA-like structure at the 3′ end of the bromo mosaic virus RNAs as proposed by Rietveld *et al.* (1983). The stem structures of arms A, B, C and D are indicated in the figure. The CCA-terminus is located in arm A, the anticodon AUA in arm C

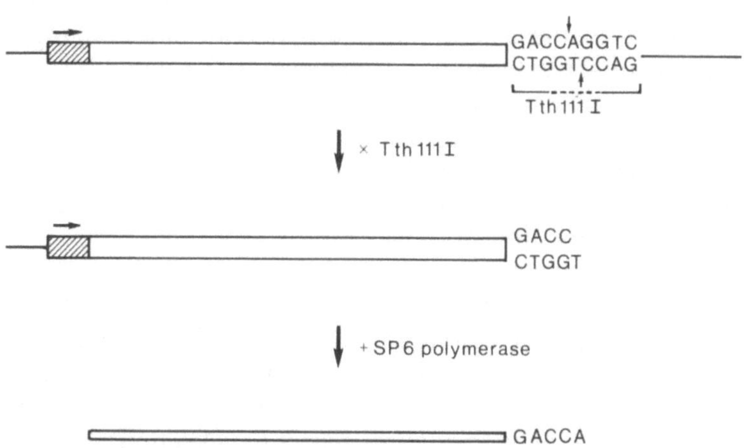

Fig. 4. Schematic representation of the cDNA clone containing the ultimate 201 nucleotides of RNA3 of bromo mosaic virus downstream of the SP6 promoter. Single lines indicate vector sequences, the SP6 promoter is indicated as a hatched bar, BMV sequences are indicated as an open bar. Note that the central three nucleotides in the nonameric recognition sequence of the restriction enzyme Tth111I are not important for cleavage, allowing site-directed mutagenesis at these three nucleotides without affecting restriction enzyme digestion

role in initiation of (–) strand synthesis and aminoacylation (Ahlquist *et al.*, 1984b). To investigate sequences involved in (–) strand synthesis and aminoacylation, mutant viral RNAs were synthesised *in vitro* and offered as templates to either extracts of BMV-infected barley leaves (Miller and Hall, 1983) or partially purified preparation of aminoacetyl-tRNA synthetase from wheat germ (Kiberstis and Hall, 1983).

Since for aminoacylation a correct 3′ CCA-terminus is an absolute requirement (Deutscher, 1983), it was important to generate *in vitro* BMV RNAs containing a correct CCA terminus. To meet this requirement a cDNA clone was constructed, downstream of an SP 6 promoter, which contained the 3′-terminal 201 nucleotides of BMV RNA3 terminated at a unique Tth111I restriction site. Upon transcription of Tth111I-linearised template DNA, RNA transcripts were synthesised ending with a correct CCA-terminus devoid of additional nucleotides (see Fig. 4). Furthermore the recognition sequence of Tth111I (GACNNNGTC) permitted changes in one or more of the three centrally located nucleotides of the nonameric sequence, allowing study of the effect of mutations at this end (see Fig. 4). The effect of several mutations in the RNA sequence on both template activity and aminoacylation were analysed, a. o. point mutations at the CCA terminus and anticodon loop (Dreher *et al.*, 1984), deletions in the stem and loop structures of the tRNA-like region (Bujarski *et al.*, 1985) and 5′-deletions and 3′-extensions of the terminal 201 nucleotides present in the cDNA clone (Miller *et al.*, 1986). The results obtained with the mutagenised transcripts are summarised below.

As expected both point mutations and extensions at the 3′-CCA-terminus dramatically affected aminoacylation, but the effect on the template activity of the RNA was variable. Replication was not affected by limited extensions to the 3′ CCA terminus or nucleotide changes at the ultimate A, but longer extensions or changes of the penultimate C abolished replicase template activity.

Besides, mutations in the anticodon sequence had no effect on efficiency of aminoacylation, nor on the type of amino acid linked, but caused a drastic decrease of replicase template activity, at a rate independent of the nature of point mutation in the anticodon loop.

Finally, deletions in several regions of the stem and loop structures of the tRNA-like region had differential effects on both aminoacylation and template activity. Complete deletion of arm D (see Fig. 3) had no effect on aminoacylation or template activity and even seemed to stimulate replication. On the other hand, deletions in arm C abolished replicase template activity and inhibited aminoacylation partially. Deletion in arm B also affected both tyrosylation and (–) strand synthesis depending on the size of the deleted fragment. From deletion mutants at the 5′-end of the tRNA-like structure it became clear that all transcripts having the last 134 nucleotides of BMV RNA3 undergo normal aminoacylation and initiation of (–) strand synthesis.

One of the most important conclusions from these experiments is that the tRNA-like structure itself contains all signals for both aminoacylation and initiation of (−) strand synthesis. At least *in vitro* the entire t-RNA-like region appears to be essential for replication, except for loop D. In this context it is noticeable that this latter loop is missing in the closely related broad bean mottle virus (BBMV). Furthermore initiation of (−) strand synthesis seems to start opposite the penultimate C of the CCA-terminus, since point mutations in the ultimate A or 3′ extensions did not affect replication. This observation has been confirmed by other experiments on BMV replication (Miller *et al.,* 1986).

The experiments furthermore confirm that for aminoacylation a correct CCA-terminus is a prerequisite. Nucleotide substitutions in the anticodon sequence, on the other hand, do not affect this process, nor the identity of the linked amino acid. Besides, it is interesting to note that RNA templates that cannot be aminoacylated are still very efficient templates *in vitro,* which makes any model demanding the use of esterified tyrosine for priming (−) strand synthesis very unlikely. The role aminoacylation plays in the life cycle of BMV therefore remains obscure and still awaits to be elucidated.

The results described above indicate the attractive possibilities of *in vitro* transcripts in studying viral RNA sequences involved in replication. Using *in vitro* transcripts of RNA3 of (−) strand polarity it has recently been demonstrated that synthesis of the subgenomic RNA4 occurs by internal initiation on (−) strand RNA3 (Miller *et al.,* 1985). The results obtained with BMV are probably representative for other RNA viruses having genomic RNAs with similar structural features (cap, tRNA-like end). For other plant RNA viruses having a different genome structure, e. g. viruses with 3′-polyadenylated RNAs, a similar approach to study their RNA replication is indicated. Although full-length clones suitable for the preparation of transcripts with well defined modifications are available for some of these viruses, the development of template-dependent *in vitro* replication systems, however, seems to impede this kind of experiment at the moment.

D. RNA Recombination

Until recently for plant RNA viruses evidence for RNA recombination was completely lacking. On the other hand, it has been known for years that for some animal RNA viruses RNA recombination is an important feature in their multiplication cycle. Synthesis of mRNAs of influenza virus for instance (Plotch *et al.,* 1981) is dependent upon certain RNA recombination events. RNA recombination has also been found and is regarded to be important in evolutionary variation of picornaviruses (King *et al.,* 1982; Emini *et al.,* 1984) The availability of cDNA clones, which are biologically active, creates the opportunity to generate well-defined, non-lethal RNA-mutants. Infection of plants with such mutants allows us to investigate if new RNA molecules are generated through certain recombination events. The first example of studying recombination with a plant RNA virus, e. g. BMV, has been described by Bujarski and Kaesberg (1986) and will be discussed here. In the previous section (III. 3) it was demonstrated that BMV RNA3-transcripts derived from DNA-templates lacking arm D in the tRNA-like structure are still efficient templates for aminoacylation and replication *in vitro*. Since this arm D seems to be dispensable *in vitro*, Bujarski and Kaesberg inoculated barley plants with a mixture of wild type RNA1 and 2 and a mutant RNA3, lacking this loop, and wild type RNA1 and 2, and followed the fate of the altered RNA3 *in vivo*.

Although barley plants became infected with this combination of RNAs, less RNA3 and subgenomic RNA4 were found to accumulate than normal. The progeny RNA3 and RNA4 both possessed 3′ non-coding regions missing arm D in their tRNA-like structures and thus originated from the modified RNA3 inoculated. Since the deletion in the tRNA-like structure seemed to reduce the viability of RNA3, infected plants were incubated over a longer period to test if RNA variants more similar to wild type RNA3 would arise by recombination events. Indeed after prolonged infection revertant forms of RNA3 were observed in several plants, which eventually outcompeted the original RNA3 mutant and which had regained the deleted stem and loop region. These revertants were different from wild type RNA3 in having the t-RNA-like structure from RNA1 or 2. Different plants had different revertants and sometimes more than one type of pseudo wild type RNA3 could be detected within the same plant. Analysis of the different RNA 3 variants showed that in most cases recombination had occurred between RNA3 and an homologous region in RNA1 or 2, while recombination outside homologous regions was also demonstrated. Clearly the D-loop in the tRNA-like structure of BMV RNAs is not essential for infectivity, but its absence is somehow disadvantageous. As a result RNAs which have regained an intact t-RNA-like region by some recombination event have a selective advantage over the RNA3 mutant with the deleted stem and loop structure.

In summary it can be concluded that the experiments with the modified BMV RNA3 very elegantly demonstrate that recombination between plant viral RNAs is possible. It is probable that such a mechanism plays an important role in the evolution of plant RNA viruses. It certainly makes RNA viruses more flexible in adapting quickly to new environments or hosts.

E. Genetic Organisation and Gene Expression

Another major application of cDNA clones in plant virology has been to study the genetic organisation by expression of cDNA clones modified in specific regions. Since no methods for site-specific mutagenesis on RNA are available, the only approach is to generate RNA mutants indirectly by *in vitro* transcription of modified DNA clones. Relatively few reports have been published on the use of cDNA clones in identifying gene functions of RNA viruses *in vivo* for the simple reason that most mutations result in non-infectious genotypes. This means that in a number of cases the only outcome of a mutation in a gene of which the function is to be determined will be that this gene is essential for infectivity. It is clear that there is a need for alternative ways to study functions of genes, for instance expression of these genes *in vitro*. However, this will only be possible in those cases where suitable assays for probing gene functions are available. An example of such a gene function that lends itself well for analysis *in vitro* is proteolytic processing, which is employed by various RNA viruses in their genetic expression, e. g. picornaviruses (Putnak and Phillips, 1981), comoviruses (Goldbach and Van Kammen, 1985) and potyviruses (Vance and Beachy, 1985; Dougherty *et al.*, 1985; Dougherty and Hiebert, 1985). This processing occurs faithfully upon translation *in vitro* of the viral RNAs (Pelham, 1978; Franssen *et al.*, 1984; Jackson, 1986). The *in vitro* translation and processing of mutant RNA transcripts of cowpea mosaic virus (CPMV) has resulted in important information on the proteolytic processing of this virus. Some of the most interesting results obtained will be discussed here.

The genome of cowpea mosaic virus (CPMV) is divided among two, separately encapsidated, single-stranded RNA molecules designated M and B RNA (for a recent review see Goldbach and Van Kammen, 1985). The viral RNAs are characterised by a small protein covalently linked to the 5′-end called VPg (Stanley *et al.*, 1978; Daubert *et al.*, 1978) and a poly(A) tail at their 3′-end (El Manna and Bruening, 1973). The translation of both RNAs into polyproteins and their subsequent proteolytic processing to yield the functional proteins have been described extensively (for a review see Goldbach and Van Kammen, 1985) (see Fig. 5). From nucleotide sequence analysis of the genomic RNAs (Van Wezenbeek *et al.*, 1983; Lomonossoff and Shanks, 1983) and from amino acid sequences of the N-terminal ends of various viral proteins (Wellink *et al.*, 1986) it is known that three types of cleavage sites are employed in proteolytic processing: glutamine-glycine, glutamine-serine and glutamine-methionine dipeptide sequences. There are several interesting aspects about this proteolytic processing which demand further analysis. What is the nature of the proteolytic activities; are these activities host or virus encoded and if they are virus encoded, which viral protein(s) is/are involved? What determines a cleavage site; is it only the dipeptide sequence or are surrounding amino acids or the three dimensional folding of the protein also important? In the following pages it will be demonstrated how the answers to these questions can be obtained by *in vitro* expression of cDNA clones modified by site-directed mutagenesis.

For this purpose full-length cDNA clones of both CPMV M and B RNA have been constructed downstream of T7 or SP6 promoters, which gave upon *in vitro* transcription and subsequent *in vitro* translation in rabbit reticulocyte lysate translation products indistinguishable from the *in vitro* translation products obtained with natural viral RNA (Vos *et al.*, 1984; Verver *et al.*, submitted). Based on time course translation studies and sequence homology to the picornaviral protease P-3c, the 24 K polypeptide encoded by B RNA (Fig. 5) has previously been proposed to be a viral protease involved in the proteolytic cleavage of the CPMV polyproteins

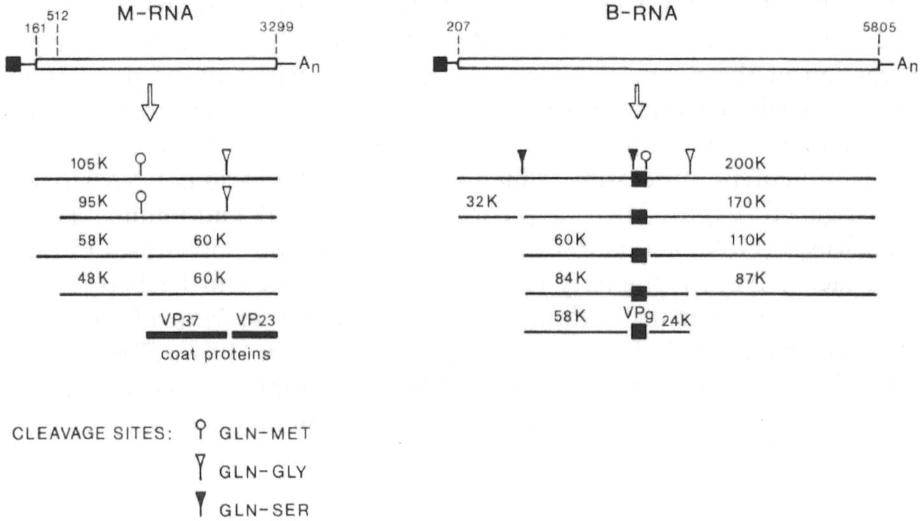

Fig. 5. Translation and processing of the genomic RNAs of cowpea mosaic virus. On the RNAs the long open reading is indicated by a double lined bar and the positions of start and stop condons are shown. Black squares refer to the genome linked protein VPg; other proteins are indicated as single lines with their estimated sizes in kilodaltons (K) presented above the protein

(Franssen *et al.*, 1984a and 1984c). To test this a small deletion was introduced into the coding region for the 24K polypeptide within a full-length B cDNA clone (Verver *et al.*, submitted). This deletion did not disturb the reading frame of the B RNA-encoded polyprotein in this clone, but resulted only in the loss of the coding sequence for 29 amino acids centrally located in the 24K polypeptide (Fig. 5). Whereas translation of viral B RNA or *in vitro* transcripts of full-length cDNA clones of B RNA resulted in a 200K polyprotein, which is rapidly processed into polypeptides of 170K and 32K (see Fig. 5), translation of the transcript from the deletion-containing B cDNA clone resulted in a slightly smaller primary translation product ($\approx$ 197K), which was not proteolytically processed. This result demonstrates that cleavage at the glutamine-serine dipeptide between the 170K and 32K polypeptides is blocked by the small deletion in the 24K polypeptide representing therefore the protease involved in this cleavage.

The action of the 24K polypeptide was further investigated by constructing a hybrid cDNA clone containing sequences derived from both M and B RNA. In this hybrid clone a DNA fragment of a full-size B clone (pSPB21) containing the coding region for the 24K polypeptide plus surrounding sequences, was inserted into a full-length M cDNA clone (pSPM5 Δ Bgl II; see Fig. 6). The result of this construction is clone pMB210,

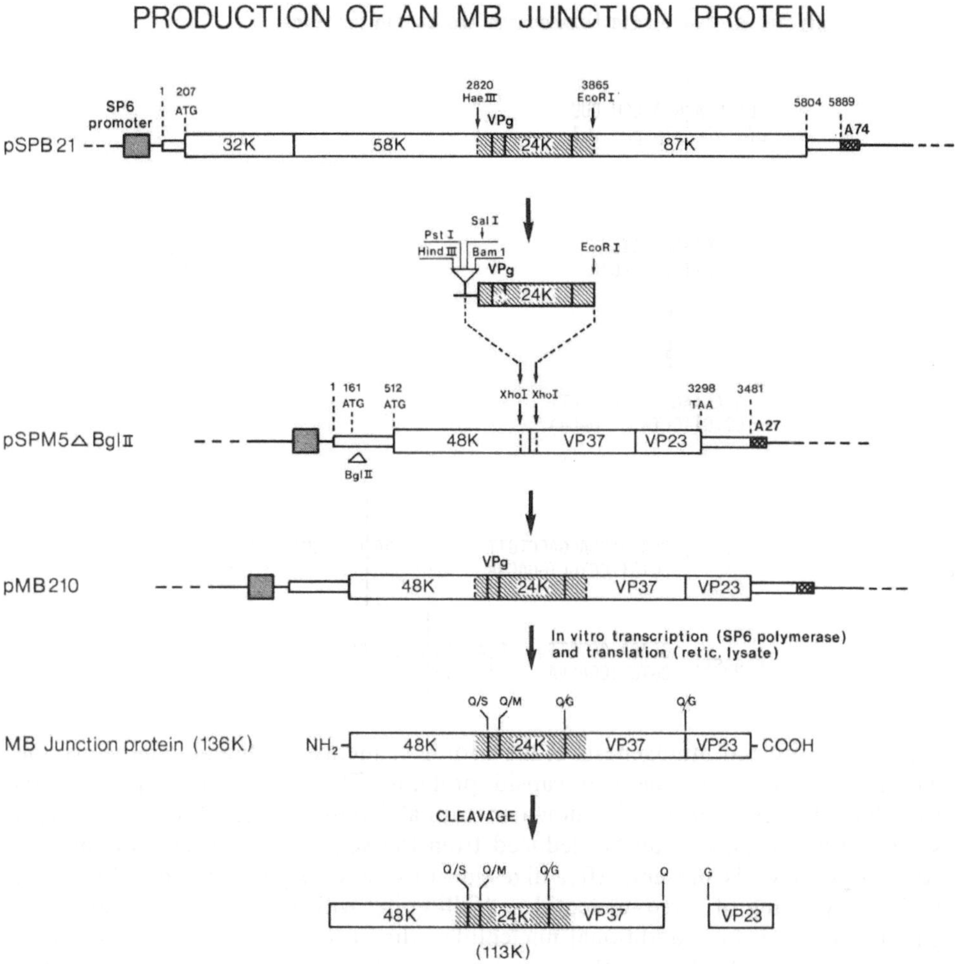

Fig. 6. Construction and schematic representation of a hybrid cDNA clone, containing sequences derived from M and B RNA of cowpea mosaic virus. A HaeIII-EcoRI fragment containing the coding region for the 24K polypeptide plus surrounding sequences was inserted into pUC9 to introduce a number of restriction sites at the left hand site of this DNA fragment. Subsequently this DNA fragment was released with *SalI* and *EcoRI* and inserted into the M cDNA clone pSPM5 △ Bg1II (Vos *et al.*, 1984) digested with *XhoI*. The resulting clone pMB210 contained one contiguous open reading frame which could code for a protein of 136kD. After *in vitro* transcription with SP6 polymerase and subsequent *in vitro* translation in rabbit reticulocyte lysates indeed a 136K polypeptide is produced, which is rapidly cleaved at the right most glutamine-glycine site (see text). Open bars represent the open reading frame of polypeptides originating from M RNA (open region) or B RNA (hatched region). Small double-lined bars indicate the non-translated regions. The positions of the cleavage sites on the translation products are indicated by Q/S (glutamine-serine), Q/G (glutamine-glycine) and Q/M (glutamine-methionine)

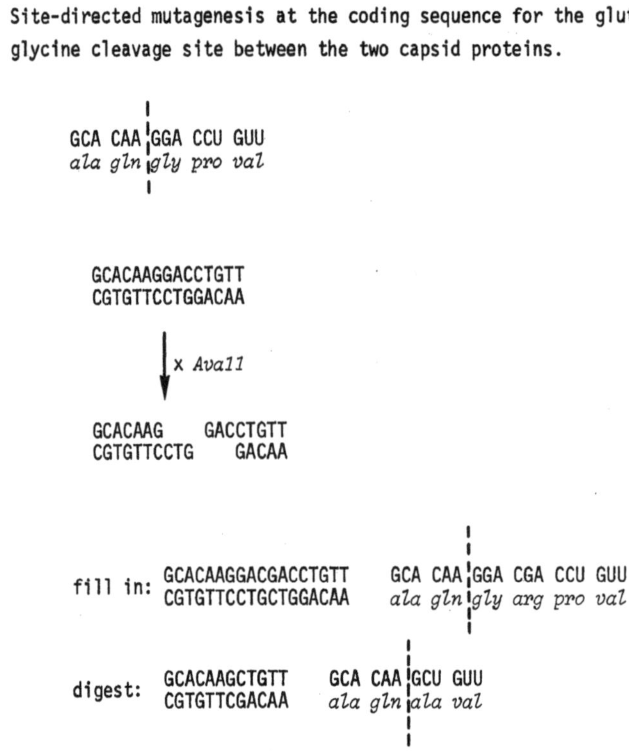

Fig. 7. Site-directed mutagenesis at the coding sequence for the glutamine-glycine cleavage site between the two capsid proteins. The nucleotide sequence surrounding the cleavage site is shown in capital letters, the resulting amino acid sequence in italics. As can be deduced from the sequence a recognition site for AvaII (GGACC) is present. After digestion this enzyme gives 5′ protruding ends with an overhang of three nucleotides. "Filling in" with Klenow enzyme and religation results in three additional nucleotides, digestion sith S1-nuclease and religation results in the loss of three nucleotides. Mutagenesis is therefore possible without interrupting the open reading frame

containing a long open reading frame of approximately 3800 nucleotides, and which produced upon *in vitro* transcription and translation a protein with a molecular weight of 136K (Fig. 6). In this protein, comprising sequences of both the B-RNA- and M-RNA-encoded polyproteins, four different proteolytic cleavage sites were present, one of which, the glutamine-glycine cleavage site between the two capsid protein (VP37 and VP23) sequences, was rapidly cleaved. This cleavage was not observed when the small deletion of 87 basepairs (see above) was introduced in the 24K protein-coding region. This result demonstrates that the 24K protein not only has proteolytic activity involved in cleavage of glutamine-serine sites (see above) but also recognizes glutamine-glycine cleavage sites.

Additionally, to analyse preliminary the properties of the proteolytic cleavage sites two derivatives were constructed from the hybrid cDNA clone pMB210, which were modified at the coding sequence for the glutamine-glycine site between the two capsid proteins (Vos *et al.*, manuscript in preparation, and Fig. 7). the result of the first mutation was a deletion of one amino acid, changing the glutamine-glycine dipeptide sequence into a glutamine-alanine sequence. The result of the second mutation was an insertion of one amino acid residue at the right hand site of the glutamine-glycine dipeptide sequence, changing the sequence glutamine-glycine-proline into glutamine-glycine-arginine. Upon subsequent *in vitro* transcription and translation of these modified cDNA clones it was shown that removal of the glycine in the cleavage site abolished proteolytic processing completely, while introduction of the extra amino acid next to the glycine resulted in a considerable decrease of proteolytic cleavages at this site.

These and the other results discussed above nicely demonstrate how RNA transcripts derived from modified DNA templates can be utilised to identify and analyse the function and enzymatic activities of viral proteins. Using this approach for CPMV the 24 K polypeptide encoded by CPMV B RNA was definitely identified as a viral protease, recognizing both glutamine-serine and glutamine-glycine dipeptide sequences. Furthermore it was shown that not only is a correct dipeptide sequence an absolute requirement for proteolytic cleavage, but that neighbouring amino acids also affect the recognition of the cleavage site by viral proteases. As this approach has resulted in an almost complete understanding of the proteolytic processing of the CPMV virus, this may also be expected for other viruses, which employ proteolytic processing for their genetic expression (e. g. nepoviruses, potyviruses).

IV. Viroids and Satellites

A. Introduction

This section will deal with the applications of cDNA clones in studies on viroids and satellites. Viroids and satellites are groups of small RNA molecules which differ from plant RNA viruses in many aspects. One of the major differences is that they are totally dependent for their replication on host enzymes (viroids) or the polymerase of a "helper virus" (satellites). In fact viroids and most satellites do not code for any protein at all, the former occuring as naked RNA molecules. An exception are the so-called "satellite viruses", i. e. satellite RNAs which encode their own capsid protein. The best studied example of these is satellite tobacco necrosis virus (STNV), a satellite virus of tobacco necrosis virus (TNV), and part of this section will concern cDNA studies on this satellite virus. Most attention here will however be paid to viroids, since these pathogens have been extensively studied for many years and for several of them full-length cDNA clones have been constructed (Van Wezenbeek *et al.*, 1982; Cress *et al.*, 1983; Ohno *et al.*, 1983; Sano *et al.*, 1984; Tabler and Sänger, 1984).

Viroids are the smallest known agents of infectious diseases. They consist of a small circular RNA molecule of 270—360 nucleotides that is extensively basepaired to form a rod-like structure (for reviews see Diener, 1983 or Sänger, 1984) (see Fig. 8). Thus far, viroids have exclusively been found to be associated with diseases of higher plants and it has still remained obscure how such small molecules can cause diseases with sometimes dramatic consequences. Dickson *et al.* (1979) demonstrated that sequence heterogeneity among different isolates of the same viroid results in different virulence. Nucleotide sequencing of various isolates of potato spindle tuber viroid (PSTV) (Schölzer *et al.*, 1985) has indicated that mutations affecting pathogenicity are all centered in one region, denoted virulence modulation region. Recombinant DNA technology offers the opportunity to analyse these observations in more detail.

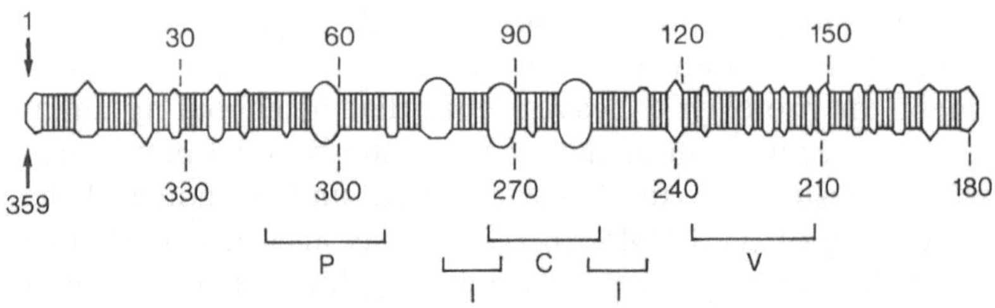

Fig. 8. Schematic representation of the secondary structure of PSTV, a similar structure can be drawn for other viroids as well. V refers to variable region, a region with relatively little sequence homology among the different viroids. P indicates pathogenicity region, a region in which most point mutations affecting virulence are centered. C refers to central conserved region, a region with a very high homology among the different viroids (Keese and Symons, 1985). I indicates the so-called hairpin I, one of the three hairpins, not present in the native configuration but formed transiently during thermal denaturation (Riesner *et al.*, 1979; Henco *et al.*, 1979; Diener, 1986)

Another interesting aspect of viroids concerns the replication mechanism. Replication of viroids takes place via oligomeric RNA intermediates of both positive and negative strand polarity, suggesting a rolling circle-type replication mechanism (Branch *et al.*, 1981; Owens and Diener, 1982; Branch and Robertson, 1984). The existence of oligomeric RNA intermediates requires specific cleavage and subsequent ligation to obtain covalently closed circular molecules (Branch and Robertson, 1984). Since isolation of oligomeric RNA intermediates from infected plants is very difficult, it was until recently not possible to follow the cleavage/ligation reaction in detail. However the availability of cloned multimeric viroid DNA copies has opened the way to study this process as will be demonstrated in this section.

In the first part of this section I will focus on the molecular cloning of viroids. For the construction of full-size cDNA clones of satellites and satellite viruses I would like to refer to paragraph 2, since cloning of satellite nucleic acid generally meets the same difficulties as cloning of the "helper" virus genome. In fact molecular cloning of satellites mostly is considerably less difficult than cloning of RNA viruses as their size is usually very small. On the other hand molecular cloning of viroids is different from RNA viruses, due to the circularity and high level of secondary structure of the RNA. Additionally, in the second part of this section the use of DNA copies to improve our understanding of the molecular basis of replication and biological activities of viroids and satellite viruses will be discussed.

B. Molecular Cloning of Viroids

If the nucleotide sequence of the viroid in question is known, first strand synthesis may be primed with one or more specific oligonucleotides, complementary to sequences in the viroid RNA. When this approach is followed two aspects are very important. Firstly, since viroid preparations always contain circular and linear RNA molecules, the circular form of the RNA should be purified from preparative denaturing gels (Maxam and Gilbert, 1980; Van Wezenbeek *et al.,* 1982). Only in this way will a high yield of full-length transcripts be obtained, since linear template RNAs will not result in such transcripts. Secondly, the oligonucleotide primers used should be complementary to regions in the RNA sequence, in which internal basepairing of the viroid RNA is relatively weak, assuring efficient hybridisation of the primer to the RNA. To obtain of full-length double-stranded cDNA second strand synthesis must also be primed with an oligonucleotide.

An elegant approach for this has been developed by Haseloff and Zimmern (personal communication) in which a primer is employed that is partially complementary to the primer used for first strand synthesis (see Fig. 9). Upon annealing of such a primer to full-size single-stranded cDNA a partially double-stranded circle will be formed, which can be converted to a completely double-stranded molecule using the Klenow enzyme. The remaining single-stranded nicks are repaired by DNA ligase and finally the circular ds cDNA is linearised at a unique site and inserted into an appropriate vector.

An alternative approach is to use two primers for first strand synthesis that are complementary to different regions of the viroid RNA. Upon second strand synthesis by hairpin priming or by using the RNaseH/DNA-polymerase I method of Okayama and Berg (1982), cDNA clones will be obtained which together comprise the whole RNA sequence. After characterisation of the cDNA clones a full-size DNA copy can be constructed by joining of selected restriction fragments from these clones.

In addition, molecular cloning of viroid RNA can be achieved by polyadenylation of (linear) viroid molecules followed by priming with oligo(dT) (Sippel, 1973, Ohno *et al.,* 1983). Prior knowledge of the nucleotide sequence is not necessary in this case. Another advantage of this method is that generally cDNA clones will be obtained starting at a dif-

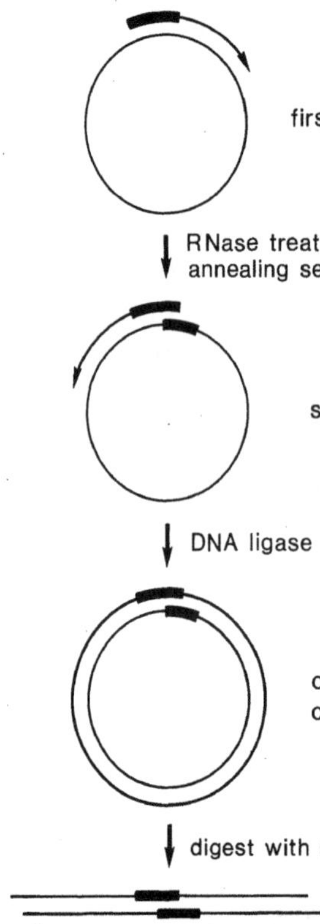

Fig. 9. Cloning strategy for circular RNA molecules, i. e. viroids, for which the nucleotide sequence is known, using an oligonucleotide primer for first strand synthesis and a primer for second strand synthesis partially complementary to the first strand primer. The result is a double-stranded circular DNA copy with a nick in both strands, which can be repaired by DNA ligase. Upon digestion with a unique restriction enzyme a linear molecule is formed, which can be cloned in monomeric of multimeric forms in a suitable vector molecule

ferent positions of the viroid sequence, resulting in molecular cloning of the entire RNA sequence. This is due to the fact that in general viroid RNA will be linearised and thus polyadenylated at different positions of the sequence. Finally, after polyadenylation, all cloning protocols originally developed for cellular mRNAs, like the highly efficient Okayama and Berg (1982) method, can be applied for viroid RNAs.

C. Application of cDNA Clones

i) Viroids

In contrast to the DNA copies of plant viruses available to date all DNA copies from viroids tested so far have proven to be infectious. A prerequisite for this infectivity however is that the viroid-specific cDNA has to be cloned in a multimeric form. Monomeric DNA copies have only very low infectivities or are not infectious at all (Diener, 1986). The same holds for RNA transcripts of viroid-specific cDNA clones. The rate of infectivity of multimeric transcripts of positive polarity is indistinguishable from that of natural viroid RNA. On the other hand, neither monomeric transcripts are infectious, nor multimeric transcripts of negative polarity (Diener, 1986). These observations suggest that multimeric transcripts of positive polarity represent true intermediates in viroid replication and that they are faithfully cleaved and ligated to circular monomers *in vivo* and therefore confer a high level of infectivity. It should be possible to identify sequences involved in this specific cleavage ligation reaction using recombinant DNA techniques.

Meshi *et al.*, (1985) have been trying to locate the region involved in cleavage of viroid multimers by testing the infectivity of cDNA clones containing more than one but less than two units of hop stunt viroid (HSV) sequence. They found that duplication of the upper portion of the highly conserved central region and some neighbouring sequences (see Fig. 9) was sufficient for infectivity of monomeric DNA copies. Site-directed mutagenesis of cDNA clones of citrus exocortis viroid (CEV) (Visvader *et al.*, 1985) have also identified this conserved central region as the possible processing site for viroid oligomers. Based on these and other observations Diener (1986) has recently proposed a model for the cleavage-ligation process of viroid multimers in which these multimers should form a highly basepaired, thermodynamically stable configuration to achieve the generation of monomers. In this model the upper portions of the central conserved region and hairpin I (see Fig. 8) of every two neighbouring monomeric sequences can form an extremely stable G-C-rich double-stranded structure exceeding the stability of native monomeric viroid RNA. Clearly additional experiments with modified cDNA clones will soon reveal the exact location of the cleavage ligation site.

Another promising application of cDNA clones in viroid research is to identify sequences involved in the virulence of the pathogen. These sequences, involved in symptom formation, are of great interest, because they might shed light on the interaction between viroid and host plant. Different isolates of the same viroid have been known for many years now, which replicate at a similar level but differ markedly in pathogenicity, varying from lethality to symptomless expression. Sequence analysis of various field isolates of both potato spindle tuber viroid (PSTV) and citrus exocortis viroid (CEV) have revealed a region in which most point mutations affecting virulence are centered (Schölzer *et al.*, 1985; Visvader and Symons, 1985). This region is called the virulence modulation region (Fig. 8). When writing this review no data were yet available on site-directed mutagenesis of DNA copies with the purpose of affecting virulence. However, their availability in the near future may contribute to a better understanding of how viroids cause diseases.

ii) Satellite Viruses

Satellite viruses can be defined as encapsidated nucleic acids that encode their own capsid protein but are unable to multiply in cells without the assistance of a specific "helper virus" and that do not contribute in any way to the replication or infection of this helper virus nor have any sequence homology with their helper virus genome (for a review see Murant and Mayo, 1982). The best studied satellite virus is the satellite of tobacco necrosis virus (STNV). It was the first plant RNA "virus" of which the total nucleotide sequence was elucidated (Ysebaert *et al.,* 1980). The genome of STNV is 1239 residues long, the 5′ half of the RNA containing the coding region for the capsid protein and the 3′-terminal 622 nucleotides being untranslated. A full-sized cDNA clone of STNV has been constructed (Van Emmelo *et al.,* 1980; Van Emmelo *et al.,* submitted), which surprisingly turned out to be infectious. A plasmid containing a full-size DNA copy of STNV inserted by G-C-tailing in the PstI-site of a pBR322 derivative resulted in the production of STNV-particles when inoculated together with the TNV helper virus. A plasmid lacking the 5′-terminal 23 nucleotides of the STNV DNA copy was not infectious and a full-length cDNA clone lacking G-C-tails was not infectious either. This indicates that the G-C-tail introduced for cloning could be of some importance for the generation of eukaryotic viral RNA from the prokaryotic plasmid DNA, although the mechanism remains obscure (Van Emmelo *et al.,* submitted).

Linker insertion at various places in the STNV DNA copy generally did not affect the infectivity, while the mutation in the STNV RNA was stably maintained in the progeny satellite virus. Insertions in the coat protein region, affecting the reading frame of this protein, did not influence the infectivity either, although the ratio of free single-stranded STNV compared to the double-stranded replicative form was highly reduced (Van Emmelo *et al.,* submitted). Probably the STNV RNA is rapidly degraded when it is not encapsidated. Analysis of several deletion mutants revealed that in some cases short deletions were allowed without affecting infectivity.

A very interesting observation was the occurrence of a kind of cross-protection by inoculating leaves with the STNV cDNA clone. It appeared that in leaves that were previously inoculated with TNV and a STNV DNA copy, modified in the coat protein region, secondary infection with TNV/STNV did not result in the appearance of STNV particles. The availability of infectious cDNA clones of satellite viruses offers therefore a good opportunity to study these viruses and their relation to the helper virus in great detail. Furthermore, since some satellite viruses probably replicate by the same mechanism as the helper virus, they could shed more light on the replication of the helper virus genome. Sequences involved in initiation of replication *in vivo* could be analysed by introducing specific mutations in the DNA copy of the satellite, without affecting the multiplication and expression of the helper virus itself.

V. Diagnosis of Plant Diseases Using DNA Copies of Plant Viruses and Viroids

A. Introduction

Plant viruses and viroids cause diseases in several economically important crops, e. g. potatoes, tomatoes, cucumbers and tobacco. Infection of plants often results in significant decrease in yield and in some cases total crop loss is possible. It is therefore very important to obtain seeds and seed plants absolutely free of viruses or viroids.

To test plant tissue for the presence of viruses or viroids, methods should be applied which allow rapid but sensitive screening of large amounts of samples routinely. For most plant viruses such methods exist based on immunochemical techniques. These methods, however, cannot be applied to viroids. Several years ago, a method was developed based on hybridisation of radioactively labelled cDNA to viroid RNA bound to nitrocellulose membranes to detect viroids in infected plants (Owens and Diener, 1981). This method, called "spot hybridisation", can also be applied to plant viruses for which immunochemical methods give poor results. In this section I will focus on the characteristics of this method in comparison with other methods used for diagnosis and detection of viruses and viroids.

B. Spot Hybridisation

Since viroids lack coat proteins, immunochemical methods cannot be applied for their detection in infected tissue. Until recently the only alternative detection methods were bioassays on suitable test plants or gel electrophoresis of extracted nucleic acids. In most cases the two methods were combined, since multiplication of viroids in intermediate hosts was necessary for detection on poly-acrylamide gels. Although gel electrophoretic detection techniques have improved a lot, recently culminating in the development of a bidirectional electrophoresis approach in combination with a highly sensitive silver staining technique (Schumacher *et al.*, 1983), they still are laborious and expensive and therefore not very suitable for routine indexing of large amounts of samples. Bioassays often show slow symptom development (Allen and Dale, 1981; Palukaitis *et al.*, 1981) and are not reliable since some viroids can replicate in test plants to normal levels without symptoms (Schölzner *et al.*, 1985).

Molecular hybridisation is a technique routinely used for detection of nucleic acids. Using radioactively labelled probes of high specific activity, amounts as low as 0.1 pg of purified nucleic acid bound to a nitrocellulose membrane have been reported (Melton *et al.*, 1984). Therefore hybridisation should be a sensitive technique to detect viroid RNA in infected tissue and because of its simplicity should allow simultaneous screening of large numbers of samples. During the last few years this method indeed has been applied successfully to a number of viroids (Owens and Diener,

1981 and 1984; Palukaitis *et al.*, 1985; Symons, 1985). Since methods for molecular hybridisation were already well established, molecular cloning of viroid sequences and methods for preparations of samples from infected plants were the only problems to be solved. Molecular cloning techniques for viroids have been described in the previous section (IV.2). To date for most viroids cDNA clones are available which can be used for preparation of radioactively labelled cDNA. Furthermore methods have been developed for rapid and simple preparation of extracts of plant tissue, which can be performed in a single Eppendorf tube, giving samples suitable for direct "spotting" on membranes, that can bind nucleic acids (Owens and Diener, 1984; Palukaitis *et al.*, 1985). At this moment spot hybridisation is routinely used in several laboratories for the detection of potato spindle tuber viroid, cirtus exocortis viroid, coconut cadang cadang viroid and avocado sunblotch viroid.

Because of its simplicity and sensitivity the enzyme-linked immunosorbent assay (ELISA) will remain the principal diagnostic assay for detection of plant viruses. However, in some cases when ELISA is not satisfactory, molecular hybridisation might provide a solution. When viruses are present in a very low concentration in infected plants dot hybridisation, because of its higher sensitivity, is to be preferred. Other viruses are difficult to purify, are poor antigens or have variable antigenic properties. In these cases problems with ELISA can be expected and should be replaced by the spot hybridisation technique. Likewise, detection of non-encapsidated viruses, like the particle-deficient isolates of tobacco rattle virus (Harrison and Robinson, 1982) cannot be performed by ELISA and should give no problems using spot hybridisation.

Spot hybridisation is an attractive alternative for detection of plant viruses especially when immunochemical techniques fail to succeed. To date, for a number of plant RNA and DNA viruses this technique has been applied successfully and shown to be more sensitive than immunochemical techniques (Maule *et al.*, 1983; Sela *et al.*, 1984). The major disadvantage of the spot hybridisation is the use of radioactivity in the procedure, especially for diagnostic laboratories which are not well equipped for this kind of work. However, the development of non-radioactive probes (Leary *et al.*, 1983) can circumvent this problem and will certainly make this procedure accessible for more laboratories.

VI. Conclusions and Future Aspects

In this chapter we have seen how recombinant DNA technology has enabled experiments with plant RNA viruses and viroids, which until recently were considered to be impossible. Of course it is very difficult to predict how plant virology will develop in the near future, but we certainly may anticipate that the application of cDNA clones will become more and more crucial in experiments which are intended to unravel the properties of plant viruses and viroids. It is obvious that the number of plant RNA viruses for which full-length DNA copies are available will grow steadily.

Apart from studying viral replication and gene expression such clones may be used to identify viral genes involved in specific virus-host plant interactions, hence determining pathogenicity, host range and symptom development. Another application of plant RNA viruses may be the use as gene vectors. Since generally viral genes are expressed very efficiently (e. g. genes for coat and inclusion body proteins), viruses may be altered in suitable gene vectors. Recently French *et al.* (1986) indeed detected chloramphenicol acetyltransferase (CAT) activity in barley protoplasts when infected with BMV RNA 1 and 2 and a modified BMV RNA 3 containing a bacterial CAT gene. A possible advantage of plant viruses as gene vectors is that difficult transformation procedures can be circumvented since viral RNAs containing desired genes can simply be inoculated on host plants. Viruses have, on the other hand, various serious disadvantages and limitations as gene vectors. Important disadvantages could be the limited host range of most viruses and the limitations of the size of the genes to be introduced. Furthermore, the error rate of RNA polymerases is known to be relatively high, which may cause rapid introduction of non-desired mutations in inserted genes. Finally viral genes involved in pathogenicity should be altered in such a way that the virus is tolerated by the plant and does not cause crop losses. In summary, it may therefore be concluded that the application of plant viruses as gene vectors is only a distant prospect.

Another application of plant viral cDNAs (and this may sound contradictory) is their use as antiviral agents. As early as 1929 it was described that infection of a plant with a (mild) virus strain may result in resistance of this plant to other (severe) strains of the same virus (McKinney, 1929). This phenomenon, called "cross-protection", also occurs with viroids (Fernow, 1967). This property has for some plant viruses and viroids been used in practice. There are, however, several disadvantages about this procedure. For example, a mild strain may change into a highly pathogenic strain by only one or a few point mutations (Palukaitis and Zaitlin, 1984), and a virus strain causing mild symptoms in one host plant can be very virulent in others. These problems could be circumvented if the viral genes responsible for cross protection could be transferred to host plants instead of using whole viruses. Powell Abel *et al.* (1986) recently demonstrated that tobacco plants transformed with a partial DNA copy of TMV, encoding and expressing the coat protein, indeed showed symptom delay upon infection with the corresponding whole virus. Although the molecular basis of this form of induced resistance is still obscure, these first experiments promise good possibilities for protecting economically important crops against virus infections.

VII. References

Ahlquist, P., Janda, H., 1984: cDNA cloning and *in vitro* transcription of the complete bromo mosaic virus genome. Mol. Cell. Biol., **4**, 2876—2882.

Ahlquist, P., French, R., Janda, M., Loesch-Fries, L. S., 1984a: Multicomponent

RNA plant virus infection derived from cloned viral cDNA. Proc. Natl. Acad. Sci. U.S.A., **81,** 7066—7070.

Ahlquist, P., Dasgupta, R., Kaesberg, P., 1984b: Nucleotide sequence of the bromo mosaic virus genome and its implications for viral replication. J. Mol. Biol., **172,** 369—383.

Allen, R. N., Dale, J. L., 1981: Application of rapid biochemical methods for detecting avocado sunblotch disease. Ann. Appl. Biol., **98,** 451—461.

Bevan, M. W., Mason, S. E., Goelet, P., 1985: Expression of tobacco mosaic virus coat protein by a cauliflower mosaic virus promoter in plants transformed by *Agrobacterium.* EMBO J., **4,** 1921—1926.

Branch, A. D., Robertson, H. D., Dickson, E., 1981: Longer-than- unit-length viroid minus strands are present in RNA from infected plants. Proc. Natl. Acad. Sci. U.S.A., **78,** 6381—6385.

Branch, A. D., Robertson, H. D., 1984: A replication cycle for viroids and other small infectious RNAs. Science, **223,** 450—455.

Bujarski, J. J., Hardy, S. F., Miller, W. A., Hall, T. C., 1982: Use of dodecyl-B-D-maltoside in the purification and stabilisation of RNA polymerase from bromo mosaic virus-infected barley. Virology, **11,** 465—473.

Bujarski, J. J., Dreher, T. W., Hall, T. C., 1985: Deletions in the 3′-terminal tRNA-like structure of bromo mosaic virus RNA differentially affect aminoacylation and replication *in vitro.* Proc. Natl. Acad. Sci. U.S.A., **82,** 5636—5640.

Bujarski, J. J., Kaesberg, P., 1986: Genetic recombination between RNA components of a multipartite plant virus. Nature, **321,** 528—531.

Contreras, R., Cheroute, H., Degrave, W., Fiers, W., 1982: Simple, efficient *in vitro* synthesis of capped RNA useful for direct expression of cloned eukaryotic genes. Nucleic Acids Res., **10,** 3133—3142.

Cress, D. E., Kiefer, M. C., Owens, R. A., 1983: Construction of infectious potato spindle tuber viroid cDNA clones. Nucleic Acids Res., **11,** 6821—6835.

Dasmahapatra, B., Dasgupta, R., Saunders, K., Selling, B., Gallagher, T., Kaesberg, P., 1986: Infectious RNA derived by transcription from cloned cDNA copies of the genomic RNA of an insect virus. Proc. Natl. Acad. Sci. U.S.A., **83,** 63—66.

Daubert, S. D., Bruening, G., Najarian, R. C., 1978: Protein bound to the genome RNAs of cowpea mosaic virus. Eur. J. Biochem., **92,** 45—51.

Davanloo, P., Rosenberg, A. H., Dunn, J. J., Studier, F. W., 1984: Cloning and expression of the gene for bacteriophage T7 RNA polymerase. Proc. Natl. Acad. Sci. U.S.A., **81,** 2035—2039.

Dawson, W. O., Beck, D. L., Knorr, D. A., Grantham, G. L., 1986: cDNA cloning of the complete genome of tobacco mosaic virus and production of infectious transcripts. Proc. Natl. Acad. Sci. U.S.A., **83,** 1832—1836.

Deutscher, H. P., 1983: Enzymes of nucleic acid synthesis and modification, **vol. 2** (Jacob, S. T., ed.) 159—183. CRC Press, Boca Raton, Florida.

Dickson, E., Robertson, H. D., Niblett, C. L., Horst, R. K., Zaitlin, M., 1979: Minor differences between nucleotide sequences of mild and severe strains of potato spindle tuber viroid. Nature, **277,** 60—62.

Diener, T. O., 1983: Viroids. Adv. Virus. Res., **28,** 214—283.

Diener, T. O., 1986: Viroid processing: A model involving the central conserved region and hairpin I. Proc. Natl. Acad. Sci. U.S.A., **83,** 58—62.

Dougherty, W. G., Hiebert, E., 1985: Molecular Plant Virology, **Vol. II** (Davies, J. W., ed.) 39—42. CRC Press, Boca Raton, Florida.

Dougherty, W. G., Allison, R. F., Parks, T. D., Johnston, R., E., Feild, M. J., Armstrong, F. B., 1985: Nucleotide sequence at the 3′ terminus of pepper mottle virus genomic RNA: evidence for an alternative mode of potyvirus capsid protein gene organization. Virology, **146,** 282—291.

Dreher, T. W., Bujarski, J. J., Hall, T. C., 1984: Mutant viral RNAs synthesised *in vitro* show altered aminoacylation and replicase template activities. Nature, **311,** 171—175.

Dunn, J. J., Studier, F. W., 1983: Complete nucleotide sequence of bacteriophage T7 DNA and the locations of T7 genetic elements. J. Mol. Biol., **166,** 477—435.

Efstratiadis, A., Kafatos, F. C., Maxam, A. M., Maniatis, T., 1976: Enzymatic *in vitro* synthesis of globin genes. Cell, **7,** 279—288.

El Manna, H., Bruening, G., 1973: Polyadenylate sequences in the ribonucleic acids of cowpea mosaic virus. Virology, **56,** 198—206.

Emini, E. A., Leibowitz, J., Diamond, D. C., Bonin, Wimmer, E., 1984: Recombinants of mahoney and sabin strain poliovirus type 1: analysis of *in vitro* phenotypic markers and evidence that resistance to guadine maps in the non-structural proteins. Virology, **137,** 74—85.

Van Emmelo, H., Devos, R., Ysebaert, H., Fiers, W., 1982: Construction and characterisation of a plasmid containing a nearly full-size DNA copy of satellite tobacco necrosis virus RNA. J. Mol. Biol., **143,** 259—270.

Van Emmelo, J., Ameloot, P., Plaetinck, G., Fiers, W., 1984: Controlled synthesis of the coat protein of satellite tobacco necrosis virus in *Escherichia coli.* Virology, **136,** 32—40.

Van Emmelo, J., Ameloot, P., Fiers, W. (submitted for publication): Linker insertion mutagenesis of satellite tobacco necrosis virus genome and its expression in plants.

Fernow, K. H., 1967: Tomato as a test plant for detecting mild strains of potato spindle tuber viroid. Phytopathology, **57,** 1347—1352.

Fields, S., Winter, G., 1982: Nucleotide sequences of influenza virus segments 1 and 3 reveal mosaic structure of a small viral RNA segment. Cell, **28,** 303—313.

Franssen, H., Goldbach, R., Broekhuijsen, H., Moerman, M., Van Kammen, A., 1982: Expression of middle-component RNA of cowpea mosaic virus: *in vitro* generation of a precursor to both capsid proteins by a bottom-component RNA-encoded protease from infected cells. J. Virol., **41,** 8—17.

Franssen, H., Goldbach, R., Van Kammen, A., 1984a: Translation of bottom-component RNA of cowpea mosaic virus in reticulocyte lysate: faithful proteolytic processing of the primary translation product. Virus Research, **1,** 39—49.

Franssen, H. J., Moerman, M., Rezelman, G., Goldbach, R., 1984b: Evidence that the 32,000-dalton protein encoded by the bottom-component RNA of cowpea mosaic virus is a proteolytic processing enzyme. J. Virol., **50,** 183—190.

Franssen, H., Leunissen, J., Goldbach, R., Lomonossoff, G., Zimmern, D., 1984c: Homologous sequences in non-structural proteins from cowpea mosaic virus and picorna viruses. EMBO J., **3,** 855—861.

French, R., Janda, M., Ahlquist, P., 1986: Bacterial gene inserted in an engineered RNA virus: Efficient expression in monocotyledonous plant cells. Science, **231,** 1294—1297.

Garcia, J. A., Hille, J., Vos, P., Goldbach, R., 1986: Transformation of cowpea *Vigna unguiculata* cells with a full-length DNA copy of cowpea mosaic virus M RNA. Plant Science, in press.

Goelet, P., Lomonossoff, G. P., Butler, P. J. G., Akam, M. E., Gait, N. J., Karn, J., 1982: Nucleotide sequence of tobacco mosaic virus RNA. Proc. Natl. Acad. Sci. U.S.A., **79,** 5818—5823.

Goldbach, R. W., Schilthuis, J. G., Van Kammen, A., 1981: Comparison of *in vitro* translation of cowpea mosaic virus RNAs. Biochem. Biophys. Res. Commun. **99**, 89—94.

Goldbach, R. W., Van Kammen, A., 1985: Molecular Plant Virology, Vol. **II** (Davies, J. W., ed.) 83—120. CRC Press, Boca Raton, Florida.

Green, M., Maniatis, T., and Melton, D., 1983: Human B-globin pre-mRNA synthesised *in vitro* is accurately spliced in Xenopus oocyte nuclei. Cell, **32**, 681—694.

Grimsley, N., Hohn, B., Hohn, Th., Walden, R., 1986: "Agroinfection", an alternative route for viral infection of plants by using the Ti-plasmid. Proc. Natl. Acad. Sci. U.S.A., **83**, 3282—3286.

Harrison, B. D., Robinson, D. J., 1982: Genomic reconstitution and nucleic acid hybridisation as methods of identifying particle-deficient isolates of tobacco rattle virus in potato plan with stem-mottle disease. J. of Virol. Methods, **5**, 255—266.

Henco, K., Sänger, H. L., Riesner, D., 1979: Fine structure melting of viroids as studied by kinetic methods. Nucleic Acids Res., **6**, 3041—3059.

Jackson, R. J., 1986: A detailed kinetic analysis of the *in vitro* synthesis and processing of encephalomyocarditis virus products. Virology, **149**, 114—127.

Van Kammen, A., 1985: The replication of plant virus RNA. Microbiological Sciences, **2**, 170—174.

Keese, P., Symons, R. H., 1985: Domains in viroids: Evidence of intermolecular RNA rearrangements and their contribution to viroid evolution. Proc. Natl. Acad. Sci. U.S.A., **82**, 4582—4586.

Kiberstis, P. A., Hall, T. C., 1983: N-acetyl-tyrosine at the 3′ end of bromo mosaic virus RNA has little effect on infectivity. J. Gen. Virol., **64**, 2073—2077.

King, A. M. Q., McCahon, D., Slade, W. R., Newman, J. W. I., 1982: Biochemical evidence of recombination within the unsegmented RNA genome of aphtovirus. J. Virol., **41**, 66—77.

Lane, L., Kaesberg, P., 1971: Biophysically homogenous bromegrass mosaic virus contains multiple genetic components. Nature, **232**, 40—43.

Lane, L., 1981: Handbook of plant virus infections and comparative diagnosis (Kurstak, E., ed.) 333—376. Elsevier North Holland Biomedical Press, Amsterdam.

Leary, J. L., Brigati D. J., Word, D. C., 1981: Rapid and sensitive colorimetric method for visualising biotin-labeled DNA probes hybridised to DNA or RNA immobilised on nitrocellulose: Bio-blots. Proc. Natl. Acad. Sci. U.S.A., **80**, 4045—4059.

Loesch-Fries, L. S., Hall, T. C., 1982: *In vivo* aminoacylation of bromo mosaic and barley strip mosaic virus RNAs. Nature, **298**, 771—773.

Lomonossoff, G., Shanks, M., 1983: The nucleotide sequence of cowpea mosaic virus B RNA. EMBO J., **2**, 2253—2258.

Maule, A. J., Hull, R., Donson, J., 1983: The application of spot hybridisation to the detection of DNA and RNA viruses in plant tissues. J. of Virol. Methods, **6**, 215—224.

Maxam, A. M., Gilbert, W., 1980: In: Methods in Enzymology, **65**, 499—560 (Grossman, L., Moldave, K., eds.). Academic Press, Inc., New York.

McDonnell, M. W., Simon, H. N., Studier, F. W., 1977: Analysis of restriction fragments of T7 DNA and determination of molecular weights in neutral and alkaline gels. J. Mol. Biol., **110**, 119—146.

McKinney, H. H., 1929: Mosaic diseases in the Canary Islands, West Africa and Gibraltar. J. Agric. Res., **39**, 557—566.

Melton, D. A., Krieg, P. A., Rebagliati, M. R., Maniatis, T., Zinn, K., Green, M. R., 1984: Efficient *in vitro* synthesis of biologically active RNA and RNA hybridisation probes from plasmids containing a bacteriophage SP_6 promoter. Nucleic Acids Res., **12,** 7035—7056.

Meshi, T., Kiyanna, R., Ohno, T., Okada, Y., 1983: Nucleotide sequence of the coat protein cistron and the 3′ non-coding region of cucumber green mottle mosaic virus (watermelon strain) RNA. Virology, **127,** 54—64.

Meshi, T., Ishikama, H., Watamabe, Y., Yamaya, H., Okada, Y., Sano, T., Shikata, E., 1985: The sequence necessary for the infectivity of hop stunt viroid cDNA clones. Mol. Gen. Genet., **200,** 199—206.

Meshi, T., Ishikawa, M., Motoyoshi, F., Semba, K., Okada, Y., 1986: *In vitro* transcription of infectious RNAs from full-length cDNAs of tobacco mosaic virus. Proc. Natl. Acad. Sci. U.S.A., **83,** 5043—5047.

Miller, W. A., Bujarski, J. J., Dreher, T. W., Hall, T. C., 1986: Minusstrand initiation by bromo mosaic virus replicase within the 3′ tRNA-like structure of native and modified RNA templates. J. Mol. Biol., **187,** 537—546.

Miller, W. A., Hall, T. C., 1983: Use of micrococcal nuclease in the purification of highly template dependent RNA-dependent RNA polymerase from brome mosaic virus-infected barley. Virology, **125,** 236—241.

Miller, W. A., Dreher, T. A., Hall, T. C., (1985): Synthesis of bromo mosaic virus subgenomic RNA *in vitro* initiation on (−) sense genomic RNA. Nature, **313,** 68—70.

Mizutani, S., Colonno, R. J., 1985: *In vitro* synthesis of an infectious RNA from cDNA clones of human rhinovirus type 14. J. Virol., **56,** 628—632.

Murant, A. F., Mayo, M. A., 1982: Satellites of plant viruses. Ann. Rev. Phytopathol., **20,** 49—70.

Okayama, H., Berg, P., 1982: High-efficiency cloning of full-length cDNA. Mol. Cell. Biol., **2,** 161—170.

Ohno, T., Takamatsu, N., Meshi, T., Okado, Y., 1982: Hop shunt viroid: Molecular cloning and nucleotide sequence of the complementary cDNA copy. Nucleic Acids Res., **11,** 6180—6197.

Owens, R. A., Diener, T. O., 1981: Sensitive and rapid diagnosis of potato spindle tuber viroid disease by nucleic acid hybridization. Science, **213,** 670—672.

Owens, R. A., Diener, T. O., 1982: RNA intermediates in potato spindle tuber in viroid replication. Proc. Natl. Acad. Sci. U.S.A., **79,** 113—117.

Owens, R. A., Diener, T. O., 1984: Spot hybridisation for detection of viroids and viruses. In: Methods in Virology, vol. VII (Maramorosch, K., Koprowski, H., eds.). Academic Press Inc., New York.

Palukaitis, P., Rakowski, A. G., Alexander, D., McE, Symons, R. H., 1981: Rapid indexing of the sunblotch disease of avocados using a complementary DNA probe to avocado sunblotch viroid. Ann. Appl. Biol., **98,** 439—449.

Palukaitis, P., Cotts, S., Zaitlin, M., 1985: Detection and identification of viroids and viral nucleic acids by "dot-blot" hybridisation. Acta Horticulturae, **164,** 109—118.

Pelham, H. R. B., 1978: Translation of encephalomyocarditis virus RNA *in vitro* yields an active proteolytic processing enzyme. Eur. J. Biochem., **85,** 457—462.

Plotch, S., J., Bouloy, M., Ulmanen, I., Krug, R. H., 1981: A unique cap (m^7GpppXm)-dependent influenza virion endonuclease cleaves capped RNAs to generate the primer that initiate viral RNA transcription. Cell, **23,** 847—858.

Putnak, J. R., Philips, B. A., 1981: Picornaviral structure and assembly. Microbiol. Rev., **45,** 287—315.

Queen, C., 1983: A vector that uses phage signals for efficient synthesis of proteins in *Escherichia coli.* J. Mol. Appl. Genet., **2,** 1—10.

Racaniello, V. R., Baltimore, D., 1981: Cloned poliovirus complementary DNA is infectious in mammalian cells. Science, **214**, 916—919.

Retzel, E. F., Collet, M. S., Faras, A. J., 1980: Enzymatic synthesis of deoxyribonucleic acid by the avian retrovirus reverse transcriptase *in vitro*: optimum conditions required for transcription of large ribonucleic acid templates. Biochemistry, **19**, 513—518.

Rezelman, G., Goldbach, R., Van Kammen, A., 1980: Expression of bottom component RNA of cowpea mosaic virus in cowpea protoplasts. J. Virol., **36**, 366—373.

Rogers, S. G., Bisaro, D. H., Horsch, R. B., Fraley, R. T., Hoffmann, N. L., Brand, L., Elmer, J. S., Lloyd, A. H., 1986: Tomato golden mosaic virus A component DNA replicates autonomously in transgenic plants. Cell, **45**, 593—600.

Sano, T., Uyeda, I., Shikata, E., Ohno, T., Okada, Y., 1984: Nucleotide sequence of cucumber pale fruit viroid: homology to hop stunt viroid. Nucleic Acids Res., **12**, 3427—3434.

Riesner, D., Henco, K., Rokol, U., Klotz, G., Kleinschmidt, A. K., Gross, H. J., Domdey, H., Sänger, H. L., 1979: Structure and structure formation of viroids. J. Mol. Biol., **133**, 85—15.

Sänger, H. L., 1984: In: The Microbe 1984: Part I Viruses, 281—334 (Mahy, B. W. J., Pattison, J. R., eds.). Cambridge University Press, Cambridge, U. K.

Schuhmacher, H., Randles, J. W., Riesner, d., 1983: a two-dimensional electrophoretic technique for the detection of circular viroids and virusoids. Anal. Biochem., **135**, 288—295.

Schnölzer, M., Haas, B., Ramm, K., Hoffman, H., Sänger, H. L., 1985: Correlation between structure and pathogenicity of potato spindle tuber viroid (PSTV). EMBO J., **3**, 2181—2190.

Sela, I., Riechman, H., Weissbach, A., 1982: Comparison of dot molecular hybridisation and enzyme linked immunosorbent assay for detecting tobacco mosaic virus in plant tissue and protoplasts. Phytopathology, **74**, 385—389.

Sippel, A. E., 1973: Pruification and characterisation of adenosine triphosphate: ribonucleic acid adenyltransferase from *Escherichia coli*. Eur. J. Biochem., **37**, 31.

Stanley, J., Rottier, R., Davies, J. W., Zabel, P., Van Kammen, A., 1978: A protein linked to the 5′-termini of both RNA components of the cowpea mosaic virus genome. Nucleic Acids Res., **5**, 4505—4522.

Symons, R. H., 1985: Diagnostic approaches for the rapid and specific detection of plant viruses and viroids. In: Plant Microbe Interactions, Molecular and Genetic Perspectives, vol. I, 93—124 (Kosuge, T. and Nester, E. W., eds.). Macmillan Publishing Co., New York.

Tabler, M., Sänger, H. L., 1984: Cloned single- and double-stranded DNA copies of potato spindle tuber viroid (PSTV) RNA and co-inoculated subgenomic DNA fragments are infectious. EMBO J., **3**, 3055—3062.

Tabor, S., Richardson, C. C., 1985, A bacteriophage T7 RNA polymerase/promotor system for controlled exclusive expression of specific genes. Proc. Natl. Acad. Sci. U.S.A., **82**, 1074—1078.

Taniguchi, T., Palmieri, M., Weissmann, C., 1978: QB DNA-containing hybrid plasmids giving rise to QB phage formation in the bacterial host. Nature, **274**, 223—228.

Vance, V. B., Beachy, R. N., 1984: Translation of soybean mosaic virus RNA *in vitro:* evidence of protein processing. Virology, **132**, 271—281.

Verver, J., Goldbach, R., Garcia, J., Vos, P., submitted for publication. *In vitro* expression of a full-length DNA copy of cowpea mosaic virus B RNA: Identification of the B RNA-encoded 24-kilodalton protein as a viral protease.

Visvader, J. E., Symons, R. N., 1985: Eleven new sequence variants of citrus exocortis viroid and the correlation of sequence with pathogenicity. Nucleic Acids Res., **13**, 2907—2920.

Vos, P., Verver, J., Van Wezenbeek, P., van Kammen, A., Goldbach, R., 1984: Study of the genetic organisation of a plant viral genome by *in vitro* expression of a full-length DNA copy. EMBO J., **3**, 3049—3053.

Wellink, J., Rezelman, G., Goldbach, R., Beyreuther, K., 1986: Determination of the proteolytic processing sites in the polyprotein encoded by the bottom-component RNA of cowpea mosaic virus. J. Virol., **60**, 50—58.

Van der Werf, S., Bradley, J., Wimmer, E., Studier, F. W., Dunn, J. J., 1986: Synthesis of infectious poliovirus RNA by purified T7 RNA polymerase. Proc. Natl. Acad. Sci. U.S.A., **83**, 2330—2334.

Van Wezenbeek, P., Vos, P., Van Boom, J., Van Kammen, A., 1982: Molecular cloning and characterization of a complete DNA copy of potato spindle tuber viroid RNA. Nucleic Acids Res., **10**, 7947—7957.

Van Wezenbeek, P., Verver, J., Harmsen, J., Vos, P., Van Kammen, A., 1983: Primary structure and gene organization of the middle component RNA of cowpea mosaic virus. EMBO J., **2**, 941—946.

Ysebaert, N., Van Emmelo, J., Fiers, W., 1980: Total nucleotide sequence of a nearly full-size DNA copy of satellite tobacco necrosis virus RNA. J. Mol. Biol., **143**, 273—287.

Chapter 4

Agroinfection

Nigel Grimsley and David Bisaro*

Friedrich Miescher-Institut, P. O. Box 2543, CH-4002 Basle, Switzerland

* Auburn University, Department of Botany and Microbiology, 129 Funchess Hall, Auburn, AL 36849, U.S.A.

With 4 Figures

Contents

I. Introduction

Most plant viruses are transmitted in nature by invertebrate vectors, commonly either insects or nematodes, which have previously fed on infected plants. While it is sometimes desirable to use natural routes of infection in studies in which an understanding of virus ecology or epidemiology is the aim, these methods may be inconvenient for other types of experiments. Vector stocks may be difficult to maintain, and their introduction to plants may be difficult to perform in a controlled manner. It is also impossible to determine the number of virus particles which constitutes the inoculum. Mechanical inoculation of plants with purified virus preparations or isolated viral nucleic acid is often an effective alternative and has the advantage of being quantitative. However, some viruses are intractable in this regard and are mechanically transmitted to their host plants either poorly or not at all. The same is true for the isolated nucleic acid of certain viruses, whether in native or cloned form. The reasons for this are perhaps many, and could include such factors as the inability of the virus or nucleic acid to enter cells, or an inability to replicate in or move through those cell types which are most likely to receive infectious virus particles or nucleic acid during mechanical abrasion of the leaf.

The recent development of *Agrobacterium tumefaciens* Ti plasmid vectors permits the delivery of heritable foreign DNA into plant cells and regeneration of transformed cells to intact plants (Horsch *et al.*, 1984; DeBlock *et al.*, 1984). This technology suggests a new method for the experimental transmission of viral DNA, or cDNA, for which the term "agroinfection" has been proposed (Grimsley *et al.*, 1986a). Agroinfection can be broadly defined as the use of *Agrobacterium* to introduce viral (or viroidal) genetic information into plants. Agroinfection may be performed in two ways (see Fig. 1, parts A and B). Route A is by mechanical inoculation of intact plants with bacterial suspensions and does not require the regeneration of transformed plants, whereas route B relates to the incubation of cultured plant cells with bacteria. In the latter case the use of a non-oncogenic *Agrobacterium* vector allows, in plant species which are amenable to regeneration, plants that contain copies of the viral genome in every cell to be produced. The possibility of obtaining plants transgenic for viral sequences opens the way to many approaches, such as the study of mutant strains of viruses which are normally lethal, the investigation of host-pathogen interactions by superinfection of transformed plants, and the development of meiotically transmissible viral vectors. In this review the potential applications of this new technique will be discussed and examples of the use of agroinfection will be presented.

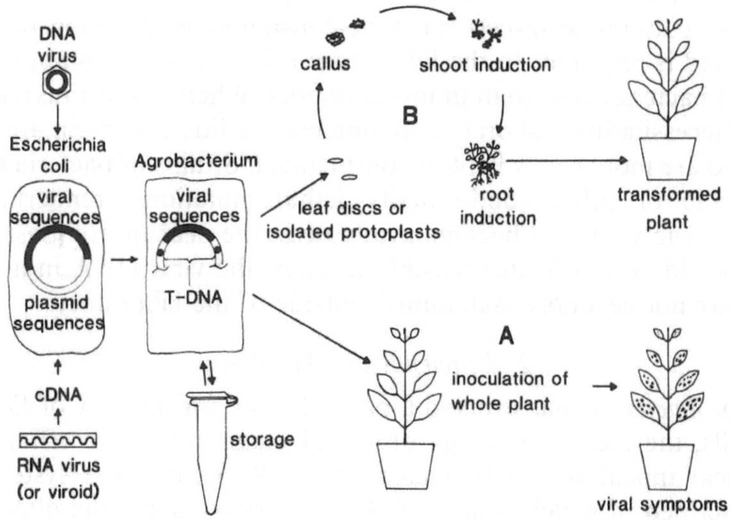

Fig. 1. Agroinfection: transfer or viral or viroidal to plants *via Agrobacterium*. *Agrobacterium* can conveniently be stored at −70 °C, and used for inoculation of whole plants (A) or used to obtain plants transgenic for viral sequences (B)

II. Potential Applications of Agroinfection

A. *Agrobacterium* as an Organism for the Experimental Storage and Transmission of Plant Viruses

Agroinfection offers an alternative to the propagation of viruses by natural vectors or as isolated virus of nucleic acid preparations. The procedure usually requires an oligomer (more than one complete unit of viral genetic information) to be placed into the T-DNA of a strain of *Agrobacterium* in such a way that infectious viral molecules can be produced by recombination, replication or transcription during or after the process of plant transformation. Preparation of the viral DNA oligomer is effected by routine cloning techniques, using a combination of *in vitro* nucleic acid reactions and *in vivo* bacterial manipulations. Use of an RNA virus demands the synthesis of cDNA molecules from viral RNA prior to cloning. Suspensions of bacteria carrying T-DNA with inserted viral DNA are applied to parts of intact plants by wounding or abrasion in much the same way as mechanical inoculation of plants with virus preparations or viral nucleic acids is performed. The advantages of agroinfection for storage and transmission of plant viruses can be summarised as follows:

1. *Storage and Safety*

Once a suitable strain of *Agrobacterium* containing the viral DNA has been made, it can be stored at −70 °C until required. One or two days prior to plant inoculation, a bacterial culture is set up from the stock. Neither maintenance of cultures of insects nor time-consuming preparation of virus of nucleic acid is required. In the laboratory it is safer to maintain strains of viruses in bacterial cells than in insect vectors. Whereas most insect vectors can fly, necessitating elaborate containment facilities, bacteria are not airborne and are more easily kept in confinement. Strains of bacteria carrying auxotrophic or other conditionally lethal mutations, rendering them inviable in the wild, will become available in the near future (Sastry *et al.*, 1986). In addition, it is also feasible to place the viral DNA in a plasmid which may not be stably maintained outside of the laboratory.

2. *Efficiency and Flexibility*

Since *Agrobacterium* has evolved a natural route for delivery of DNA into plant cells, the method is many orders of magnitude more efficient than mechanical inoculation with naked DNA. Recently three systems have been described in which strains of *Agrobacterium* containing oligomers of complete viral or viroidal genomes in the T-DNA have been used to inoculate whole plants; (1) cauliflower mosaic virus (CaMV) (Grimsley *et al.*, 1986a, Hille *et al.*, 1986); (2) potato spindle tuber viroid (PSTV) (Gardner *et al.*, 1986) and (3) maize streak virus (MSV) (Grimsley *et al.*, 1987).

Cauliflower Mosaic Virus is an 8 kb long double-stranded DNA virus which is transmitted in nature by aphids. It is thought to replicate in a manner analogous to animal retroviruses, namely *via* reverse transcription of an RNA molecule which spans the entire viral genome and bears a terminal direct repeat (Hohn *et al.*, 1985; Covey and Hull, 1985; Howell, 1985).

Although exactly unit length genomes of the viral DNA cloned in plasmid vectors are not infectious when inoculated as naked DNA onto host plant leaves, oligomers of viral genomes cloned in the same way do reproducibly give rise to systemic infection when 5—10 μg DNA per plant is inoculated (Lebeurier *et al.*, 1982; Walden and Howell, 1983; Grimsley *et al.*, 1986a). Thus relatively large amounts of plasmid DNA must be prepared for inoculation following growth of the isolate of interest in *Escherichia coli.*

In order to test the use of *Agrobacterium* for the introduction of viral DNA into host plants an oligomer comprising 1.4 genomes of CaMV, known to be infectious as naked DNA, was placed in the T-DNA of a non-oncogenic Ti-plasmid and the resulting agrobacterial strain was used for inoculation of turnip plants *Brassica campestris* var. *rapa.* Since naked DNA is usually inoculated by rubbing leaves with a mixture of a solution of DNA and an abrasive powder, and *Agrobacterium* is usually inoculated at the crown of a plant, agroinfection was used for both routes of inoculation. Under the experimental conditions employed, the first signs of systemic infection appeared within two weeks of inoculation, about one week earlier than in similar plants inoculated simultaneously with naked DNA. Plants agroinfected on the leaves also showed more pronounced symptoms on the primary inoculated leaves, as judged by the extent of chlorosis (unpublished observations). Agroinfection by leaf inoculation was extremely efficient, since inoculation of only 10^4 *Agrobacterium* cells, containing approximately 0.1 pg of CaMV DNA, were sufficient to produce systemic infection (Table 1). As a control, the same viral oligomer was placed in a broad host range plasmid lacking T-DNA borders, and plants were inoculated following introduction of this construct into *Agrobac-*

Table 1. The Sensitivity of Agroinfection as a Test for T-DNA Transfer

Type of inoculum	Number of bacteria inoculated per plant	Weight of CaMV DNA in the inoculum	Whole plants showing symptoms/total plants inoculated, no./no.
plasmid DNA	none	3 μg	8/12
plasmid DNA	none	2 μg	5/12
A. tumefaciens:			
C 58 (pEA1)	4×10^{10}	3 μg	0/24
C 58 (pGV3850::pCa305)	4×10^{10}	400 ng	4/4
C 58 (pGV3850::pCa305)	4×10^{9}	40 ng	4/4
C 58 (pGV3850::pCa305)	4×10^{8}	4 ng	4/4
C 58 (pGV3850::pCa305)	4×10^{7}	400 pg	4/4
C 58 (pGV3850::pCa305)	4×10^{6}	40 pg	4/4
C 58 (pGV3850::pCa305)	4×10^{5}	4 pg	4/4
C 58 (pGV3850::pCa305)	4×10^{4}	400 fg	2/4
C 58 (pGV3850::pCa305)	4×10^{3}	40 fg	4/4
C 58 (pGV3850::pCa305)	4×10^{2}	4 fg	0/4

Table 1. Symptoms were assessed six weeks after inoculation. Whole turnip plants were inoculated on the leaves with naked plasmid DNA or with an appropriately diluted suspension of *Agrobacterium* cells (Grimsley *et al.*, 1986a). The plasmid pCa305 is 17.5 kb long, containing about 10 kb of CaMV DNA, and pEA1 is about 24 kb long, containing the same 10 kb of CaMV sequences. The weight of CaMV DNA in the inoculum is calculated in relation to the proportion of CaMV DNA in the plasmid used for inoculation, and in the case of bacterial inoculation assumes that the copy number of pEA1 is 10 per cell (mini-Sa origin of replication) and the copy number of pGV3850::pCa305 is 1 (Ti plasmid origin of replication). pEA1 serves as a negative control, showing that infection does not arise by lysis of bacterial cells

terium. No viral symptoms developed in recipient plants, showing that infection did not occur by lysis of the bacterial cells and release of DNA. Thus, in terms of the amount of DNA required to produce infection, agroinfection (DNA in *Agrobacterium*) is many orders of magnitude more efficient than inoculation of leaves with naked DNA. In this comparison no account is taken of possible multiplication of bacterial cells at the wound site, but this is unlikely since the inoculated leaf dries out shortly after application of the bacterial suspension.

In an independent series of experiments using the same system, *Agrobacterium* cells containing the viral oligomer in the T-DNA were mixed with a similar strain of *Agrobacterium* which did not contain viral DNA, and the mixture was used to inoculate tubers of growing turnips just below the crown of the plant (Hille *et al.*, 1986). Using this competition test, agroinfection was judged to be at least a hundred times more efficient in detection of T-DNA transfer than production of a crown gall tumour. Avirulent mutants of *Agrobacterium* containing the oligomer between the T-DNA borders did not produce symptoms on inoculated plants, again demonstrating the requirement for intact *Agrobacterium* plant transformation functions.

Viroids are small (250—370 nucleotides), circular RNA molecules which can replicate autonomously in host plants. Their replication cycle probably involves both tandem linear and circular RNA intermediates (Branch and Robertson, 1984; Ishikawa *et al.*, 1984; Hutchins *et al.*, 1985). Full length cDNA copies of PSTV have been cloned in plasmid vectors (Cress *et al.*, 1983; Tabler and Sanger, 1984), and whereas monomeric PSTV cDNA has little or no infectivity when inoculated as naked DNA (unless it is cut out from the cloning vehicle at the cloning site), tandem linear cDNA copies are infectious. A variety of PSTV constructions were made, comprising monomeric or multimeric genomes, either not adjacent to a plant promoter or downstream of a plant promoter, and introduced to whole tomato plants by agroinfection (Gardner *et al.*, 1986). In agreement with experiments done with naked DNA, the efficiency of agroinfection was shown to increase with increasing numbers of tandem PSTV units. In controls, avirulent strains of *Agrobacterium* were shown not to mediate agroinfection with PSTV.

In some cases it has not been possible to infect plants with cloned viral nucleic acids, and agroinfection may be an effective way of circumventing this problem. A good example of this is provided by maize streak virus (MSV), for which no way has yet been found for the introduction of isolated viral DNA or cloned DNA (Mullineaux *et al.*, 1984) into whole maize plants. There is a report on the introduction of viral particles by eletroendosmosis (Polson and Von Wechmar, 1980), but this technique cannot be used to show that a cloned viral DNA is infectious, and consequently does not allow studies of *in vitro* manipulated viral DNA. A recent report indicates that it is nevertheless possible to infect whole maize plants with viral DNA *via Agrobacterium* (Grimsley *et al.*, 1987); agroinfection should now permit the analysis of *in vitro* produced mutant strains of MSV and allow the potential of MSV as a vector for foreign DNA to be evaluated.

3. *Release of Viral Genomes from the T-DNA*

There are several possible ways by which viral sequences may be released from the T-DNA vector after it has entered a plant cell. If a single-stranded intermediate is involved in T-DNA transfer (Stachel *et al.*, 1986), this could either be used directly as a substrate for the production of an infectious viral molecule (for example if it is the (+) strand of a single-stranded DNA virus), or synthesis of the complementary DNA strand could occur first. Whether or not the transferred nucleic acid is single- or double-stranded, escape could occur by recombination between tandemly repeated copies of the genome inserted into the vector, or by a replication or transcription mechanism whereby one strand of the genome is initially copied. These mechanisms need not be mutually exclusive. This question has been investigated using the systems which have already been mentioned.

The CaMV oligomer referred to above contained one genome of CaMV isolate "CM4-184" (CaMV.CM4-184) and 0.4 genome of CaMV isolate "S" (CaMV.S). Since these two isolates show differences in their DNA sequences and in certain restriction enzyme sites genetic analysis of progeny viruses produced by agroinfection allowed deductions to be made about the mechanisms by which infectious viruses escape from the transforming DNA (Grimsley *et al.*, 1986b). The 1.4 viral genomes were arranged in such a way that the long transcript thought to be an intermediate in viral replication by reverse transcription could be produced, with its 5′ end starting in one viral strain and its 3′ end finishing in the other. Escape by reverse transcription/replication would therefore give rise to one kind of progeny virus molecule. Homologous DNA recombination would, on the other hand, be expected to produce progeny viruses with a variety of genetic constitutions, depending upon where the individual events took place when circular DNA molecules recombined out from the transforming DNA. The authors concluded that about 90 % of the progeny viruses found in agroinfected plants had arisen by reverse transcription, and detailed genetic mapping by sequencing gave information about the position within the RNA terminal repeat at which the reverse transcriptase molecule normally switches from the 5′ end to the 3′ end of its template. The remaining 10 % of progeny viruses probably arose due to homologous DNA recombination. Escape of infectious viruses could have occurred either before or after integration of the DNA into the host chromosome, but no information about this could be gained from the available data.

The mechanism by which PSTV excapes from the transforming DNA is less clear, but may involve an RNA intermediate which is processed in some way to yield an infectious PSTV genome (Gardner *et al.*, 1986). When a monomer of PSTV cDNA is placed adjacent to a CaMV 35 S promoter agroinfection was successful if the PSTV genome was in the orientation required to produce a (+) strand of RNA, but not if the CaMV promoter directed synthesis of a PSTV (−) strand. Thus PSTV might also escape from the transforming DNA by an RNA polymerase II derived transcript, which is subsequently processed to produce an RNA replicative intermediate, and this then undergoes RNA - RNA replication. On the other hand, all constructs which contained multimers of the PSTV genome were infectious, and infectivity was shown to increase with increasing numbers of PSTV units. This was true even of constructs in which a plant promoter was absent, as well as those in which PSTV (−) strand RNA would be produced. It is suggested in these cases that infectious RNA results from (+) transcripts produced by "background" transcription from adjacentT-DNA promoters in the plant. However, this does not explain the non-infectious nature of the (−) strand monomer construct discussed above. An alternative explanation for the escape of PSTV from oligomers of PSTV cDNA, which would be independent of any adjacent plant promoter, is that a homologous DNA recombination event first occurs to produce a circular DNA molecule, which is then transcribed. In view of the observation that PSTV cDNA monomers are infectious when excised from the cloning vehicle, this possibility should not be overlooked; indeed, it would be one explanation for the observation that trimers are more infectious than monomers.

4. Analysis of T-DNA Transfer

Replication of viral molecules in recipient host plant cells can be used as a sensitive assay for the transfer of the *Agrobacterium* T-DNA. Unlike other assays for T-DNA transfer, such as proliferation of crown gall tumour tissue or growth of antibiotic resistant cells, integration of the T-DNA into the recipient's nuclear genome may not be required for escape of replicating viral progeny (Fig. 2). Early events in the T-DNA transfer process can thus be studied using replicating viral sequences as marker, and this approach may also permit the rescue and study of transfer intermediates. Agroinfection with appropriate viral genomes could also form the basis of screening experiments to determine the "host range" of *Agrobacterium* among monocot species. In the monocots especially, virus replication may be a more suitable assay than either tumour production or opine synthesis.

The role of T-DNA border sequences and Ti-plasmid encoded *vir* genes, which are required for normal tumour formation and are known to play important roles in the transfer of the T-DNA from *Agrobacterium* cells to plant cells (Gheysen *et al.*, 1985, Koukolíková *et al.*, this volume), and have been the first aspect of this process to be analysed by agroinfection.

A PSTV cDNA trimer was placed in *cis* with left and/or right T-DNA border sequences in a broad host range plasmid, and this plasmid was introduced into a variety of *Agrobacterium* strains (Gardner and Knauf, 1986). In *trans* with a Ti plasmid bearing intact *vir* genes, systemic infection of agroinfected tomato plants was found to occur when a left border or a right border or a left plus right border was present, but not in the absence of a border sequence. This is in agreement with previous findings about the essential nature of these borders. A right border was more efficient in agroinfection than a left border, as judged by the time of appearance of infection symptoms. This observation is in accord with the recent discovery

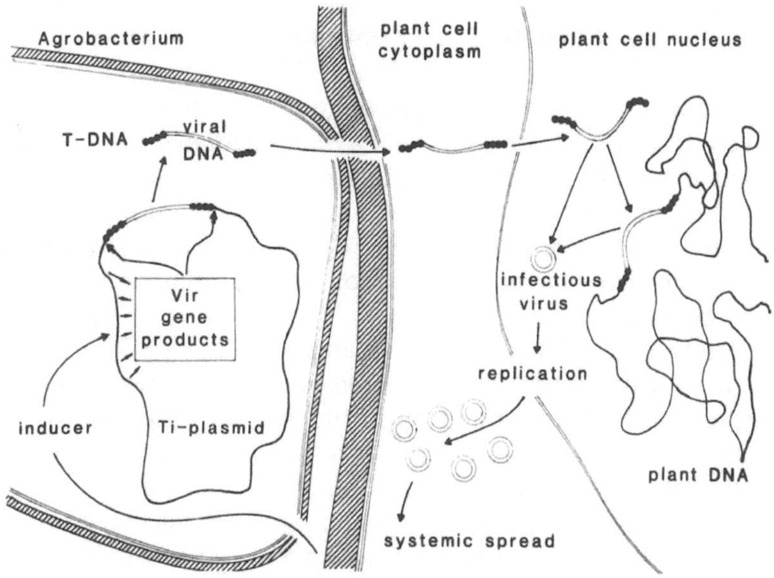

Fig. 2. Systemic viral infection as an assay for T-DNA transfer. The structure of the T-DNA intermediate is currently being studied (Koukolíková *et al.*, 1985; Koukolíková *et al.*, this volume; Stachel *et al.*, 1986), and little is known about the mechanism by which it moves from bacterium to plant, where it may eventually integrate into the nuclear DNA of the host. Agroinfection provides a sensitive assay, since theoretically only one infectious molecule is required, and may not require integration of the transferred DNA

of an "overdrive" sequence in close proximity to the right border, which was present in the constructions used. The "overdrive" sequence allows expeditious transfer of the T-DNA (Peralta *et al.*, 1986). Surprisingly, PSTV cDNA that was located outside of the T-DNA borders was also transferred to the plant.

The binary vector containing a right border and the PSTV trimer was then tested for agroinfection in pairwise combinations in *trans* with Ti plasmids bearing a variety of *vir* gene mutations. Whereas *Agrobacterium* strains with mutations in the virulence loci A, B, C, and D could not agroinfect tomatoes, those carrying a mutation in *vir*E could. A CaMV oligomer (Grimsley *et al.*, 1986 a) in a binary vector was also tested in *trans* in a variety of Ti-plasmid *vir* mutant strains (Hille *et al.*, 1986), and a result similar to that obtained with PSTV was found in this system using turnips as host plants. *Agrobacterium* strains carrying the *vir*E mutation could not induce tumours but were nevertheless effective in agroinfection, albeit at a reduced efficiency compared to wild-type strains. However, the results of the foregoing experiments with PSTV and CaMV must be interpreted with great caution, since agroinfection is orders of magnitude more sensitive than the production of a tumour as a T-DNA transfer marker (see above).

Mutant strains of *Agrobacterium* which have a much reduced efficiency of T-DNA transfer might therefore be mistaken for mutant strains which are transfer-proficient but integration-deficient. The observation that agroinfection is so much more efficient than tumour formation implies in itself that relatively few of the copies of T-DNA intermediates transferred from bacterium to plant are eventually integrated and expressed in the plant nucleus. In this context, it should also be mentioned that the agroinfection test most likely requires movement of the T-DNA intermediate to the plant nucleus, and this may be only a fraction of the number of molecules entering the cytoplasm.

The agroinfection of whole maize plants with MSV (Grimsley *et al.*, 1987) conclusively demonstrates that *Agrobacterium* can transfer DNA to a graminaceous monocot, providing an invaluable substitute for the insect vector. A sensitive test for T-DNA transfer to maize was thus developed, requiring the delivery of only one infectious MSV molecule to a recipient plant. In an attempt to use the system to analyse the mechanism of T-DNA transfer, a dimer of the MSV genome was tested in two constructs, differing only in the orientation of the MSV sequences with respect to the T-DNA borders. Either a (+) or a (−) strand of the virus would therefore have been transferred, but since both orientations were equally agroinfectious no information about the single- or double-stranded nature of the T-DNA intermediate could be gained.

B. *Transformation of Plant Cells with Viral Genetic Information*

The *Agrobacterium*-mediated transformation of cultured plant cells (see Rogers and Klee, this volume, for a review) with viral genomes may be used to obtain transgenic tissue, or plants in cases where regeneration is possible. This agroinfection approach has the potential to facilitate studies of viruses which may be limited to particular tissues, or to study the behaviour of viruses in non-host plants, where virus movement rather than replication is normally restricted. Further, this technique is of particular interest for the mutational analysis of viral genomes, since mutant viral sequences will be present in every cell independently of whether they can generate active virus particles capable of normal systemic spread throughout the various tissues of the host. It is of course also possible to use *Agrobacterium* vectors to deliver fragments of viral genomes, or more specifically selected viral genes, for expression in transgenic plants.

1. *Transient Expression*

The assay of transient expression of transforming DNA in isolated or regenerating plant protoplasts is an important tool for the elucidation of the mechanisms involved in plant gene expression. One limitation of experiments conducted until now is the relatively small amount of nucleic acid which is taken up and expressed in the recipient cells. Very sensitive tests for expression, such as the acetylation of radioactive chloramphenicol (Shaw, 1975), or the expression of luciferase (Ow *et al.*, 1986), are usually required. The use of agroinfection allows amplification of the incoming genetic information to occur by replication of viral sequences, thus increasing the quantity of assayable gene products. Since *Agrobacterium* transfers its T-DNA following attachment to a recipient cell wall it may not be necessary to prepare isolated protoplasts, allowing transient assays to be done, for example, *in vivo* on whole leaves, or in culture with cell suspensions which do not give a high plating efficiency after protoplasting.

2. *Expression of Viral Genes in Host and Non-Host Plants*

Since the host range of *Agrobacterium* is much broader than that of individual kinds of viruses, it is possible to agroinfect non-host plants with viral sequences and observe how individual viral genes are expressed in different environments. Viral genomes constitute one of the most readily available sources of plant gene regulatory sequences. The development of selectable marker genes for plants has been greatly facilitated by their use, since viral promoters are usually strong and constitutive. When such CaMV regulatory sequences are appropriately linked to the coding sequence of bacterial neomycin phosphotransferase, for example, chimaeric genes giving high levels of resistance to kanamycin in transgenic plants can be obtained (Paszkowski *et al.,* 1984; Bevan *et al.,* 1985; Fromm *et al.,* 1986). We will limit examples in this review to those cases where expression in agroinfected plants or plant cells of the viral sequences *per se* is of primary interest, but note that chimaeric genes containing viral regulatory sequences have also been used on other contexts *(op. cit.).*

In one study, a CaMV genome was placed in the T-DNA of a wild-type Ti-plasmid, and the resulting strain of *Agrobacterium* was used to induce galls on a variety of host or non-host plants (Shewmaker *et al.,* 1985). Northern blot analysis indicated that the two previously identified promoters (Covey *et al.,* 1981; Odell *et al.,* 1981; Guilley *et al.,* 1982) were active when integrated into the nuclear genome of a variety of plants, directing transcription, on the whole, more efficiently in host than in non-host plants. The authors did not observe excision and autonomous replication of the CaMV DNA, but this is not surprising since a monomer broken by non-viral DNA in two positions (at the cloning site and at another site by an introduced selective marker) was used. One drawback for these experiments was that gall tissue was used for the analysis, and it was therefore not possible to determine what proportion of the cells was actually transformed.

The DNA sequences required for the activity of the CaMV 35S RNA promoter in tobacco (a non-host plant) have been studied in transgenic plants following agroinfection (Odell *et al.,* 1985). Using a CaMV 35S RNA promoter — human growth hormone chimaeric construction, a series of deletions in the 5′ end of the promoter was made, and transferred into tobacco plants. Whereas sequences extending to −46 were sufficient for accurate transcription, sequences between −46 and −105 increased the level of transcription. No tissue-specificity of expression was shown by the promoter.

Two different laboratories report on the expression of tobacco mosaic virus (TMV) coat protein (cp) by a CaMV promoter in regenerated tobacco plants, (see also section B 4 below). Using a TMV cDNA clone, a chimaeric CaMV 35 S promoter-TMV cp gene was made and leaf pieces were agroinfected (Bevan *et al.*, 1985). Transgenic plants were selected on the basis of resistance to kanamycin, encoded by a CaMV 35 S promoter — Tn 5 hybrid gene. The level of coat protein synthesised from the nuclear gene was *ca.* 6000 times less than that observed in acute viral infection. However, somewhat higher levels (50—100 fold) were obtained in independent experiments with a similar construct (Abel *et al.*, 1986).

Brassica tissue (a non-host) was also agroinfected with the aforementioned oligomers of PSTV which were agroinfectious on tomato (Gardner *et al.*, 1986), but no PSTV RNA could be found in crown galls, suggesting that no replication of PSTV occurs in turnip when the viroid is introduced to the chromosome in the form of multimeric DNA.

3. *Transgenic Plants Containing Oligomers of Viral Genomes or Genome Components*

i) *Complementation Between Different Components of a Multi-Component System*

Plants transgenic for an oligomer of one component of a multi-component viral genome which is by itself unable to produce infectious viruses can be agroinfected by another component of the virus in any of the following ways; (1) by inoculation or agroinfection of the plant with another component, (2) by transformation, producing plants transgenic for both components, or (3) by crossing to plants transgenic for another component. In all cases, the assay for functional complementation would be the replication of a virus containing all of the components.

Tomato golden mosaic virus (TGMV) belongs to the geminivirus group, whose members are characterised by their distinctive twin icosahedral particle morphology and genomes of circular single stranded DNA (Stanley, 1985; Davies *et al.*, this volume). The genome of TGMV is divided between two such DNA molecules, designated A and B, both of which are about 2.5 Kb in length (Bisaro *et al.*, 1982; Hamilton *et al.*, 1984). Viral DNA replication is thought to occur through circular double stranded replicative form DNAs found in the nuclei of infected cells (Hamilton *et al.*, 1982; Coutts and Buck, 1985) and the cloned double stranded DNA components are infectious when both are released from the cloning vector and inoculated together on tobacco leaves (Hamilton *et al.*, 1983). Neither component alone produces symptoms or replicates to a detectable extent under these conditions.

Viral genome components were separately introduced into petunia cells using a Ti transformation-regeneration system (Rogers *et al.*, 1986). In these experiments both single and tandemly repeated inserts of each TGMV component in the Ti plasmid vector were constructed. These derivative plasmids were introduced into agrobacteria which were subsequently used in the leaf disc procedure (Horsch *et al.*, 1985) to obtain transformed regenerated petunia plants. All transformed plants were normal in morphology and most had only one or a few T-DNA insertions at a single chromosomal site.

The transformed plants flowered normally and were used in all possible combinations of crosses (Rogers *et al.*, 1986). All progeny seedlings were normal in appearance with the exception of those which resulted from crosses of tandem A by tandem B plants. One-quarter of these seedlings displayed symptoms of virus infection. Analysis of total unrestricted DNA from the symptomatic A+B progeny plants revealed the presence of free, replicating single- and double stranded viral DNA forms identical to those previously identified in TGMV infected tobacco (Hamilton *et al.*, 1982). Both the A and B components were represented in these DNA forms. Further, a similar analysis of non-symptomatic parents showed that free viral DNA was also present in plants containing tandem insertions of the A component alone. These DNAs hybridised only with A-specific probes. Free viral DNA was not detected in plants transformed with a monomeric A component, or in any plants transgenic only for the monomeric or tandemly repeated B component. These experiments allow several interesting conclusions and speculations concerning the organisation of the TGMV genome in particular and agroinfection in general. The results indicate that the TGMV A genome component encodes all viral functions necessary for the replication of viral DNA and, since plants which carry tandem A insertions do not show symptoms, that the B component must encode some function(s) essential to their production. Thus the B component, itself incapable of replication, was shown to be activated by the independently replicating A component in progeny that result from the crossing of transformed A and B parents. Systemic infection also results at high frequency when transformed B plants are inoculated with free A component DNA (S. Rogers and D. Bisaro, unpublished). Further, the results clearly demonstrate that viral DNA can escape from the T-DNA after chromosomal integration. Free viral DNA could be detected in the progeny of agroinfected plants, while during natural infection TGMV is not known to be seed transmitted. Information concerning the mechanism by which TGMV DNA is released from the chromosomally integrated T-DNA is not yet available, although it might occur either by single strand replication or by homologous recombination between tandemly repeated copies of the viral genome. The first alternative is analogous to the transcriptional mechanism observed in the case of CaMV (Grimsley *et al.*, 1986b). If this method of escape proves general, it would be reasonable to suppose that vectors derived from viral replicons in T-DNA could indefinitely generate replicating extrachromosomal DNA in the host cell since replication or transcription should not alter the chromosomal viral sequences. This could have important implications for the development of genetically transmissible high copy number plant expression vectors (see below).

Cucumber Mosaic Virus (CMV) particles contain three genomic single stranded RNA molecules and a subgenomic species which acts as an mRNA for the CMV coat protein. Some isolates also contain a satellite RNA (Kaper *et al.*, 1976). Recently, cDNA copies of the satellite have been made and, as monomers or tandem dimers, cloned in a plant expression vector (Baulcombe *et al.*, 1986). Tobacco leaf discs were agroinfected with such constructs and transgenic plants were found to express the RNA species predicted from the positions of the promoter and the polyadenylation sites present in the expression vector, producing an RNA molecule much longer than the natural satellite RNA. Upon inoculation of parts of these transformed plants with a satellite-free isolate of CMV, satellite-length RNA molecules were found in the tissue 4—7 days later. Hybridisation with cloned satellite sequences and passaging of the CMV with its satellite acquired through another host plant confirmed its identity and biological activity. Thus the CMV had acquired genetic information from the genome of its host plant. The authors suggest that it might be possible to protect crop plants from viral attack by transforming them with viral satellite sequences, since the presence of a satellite in a viral isolate sometimes leads to a reduction in severity of the viral symptoms (Waterworth *et al.*, 1979; Mossop and Francki, 1979).

ii) Analysis of in vitro *Produced Mutant Viral Strains*

A variety of mutations can be produced in viral genomes by *in vitro* manipulations, but the majority of these are lethal, precluding subsequent *in vivo* analysis (e. g. in CaMV Howell *et al.*, 1981; Daubert *et al.*, 1983; Dixon *et al.*, 1983). Agroinfection would allow integration of the mutated viral DNA, in oligomeric form, into the host plant nuclear DNA, as outlined schematically for caulimoviruses and geminiviruses, in Fig. 3. In this situation part of the viral life-cycle may occur within each plant cell, allowing deductions to be made about the role of particular viral genes or nucleic acid sequences. This method offers an important advantage over the use of cultured plant cells for mutant analysis since viral genes which affect symptom generation or some other aspect of the virus-host interaction can be studied. At present there are no reports on the use of this approach, although such work is in progress with TGMV (D. B., unpublished) and CaMV (B. Pisan, T. Hohn, B. Hohn and N. G., unpublished).

iii) Development of Proviral Vectors

A combination of the *Agrobacterium* based transformation system with viral vector systems can be made which draws together advantages of both types of system. A viral replicon that contains a cargo DNA of interest when integrated into the plant nuclear genome as a kind of "provirus", can provide a substrate for escape of free extra chromosomal DNAs carrying cargo DNA (Fig. 3). Suitable constructions should allow the expression of viral and cargo DNA gene products to be regulated. Genetic studies outlined above should yield information concerning the viral functions required to initiate and maintain replication of the free nucleic acid sequences, and this knowledge would be of use in the development of these vector systems. While the nuclear DNA copy provides for stability and

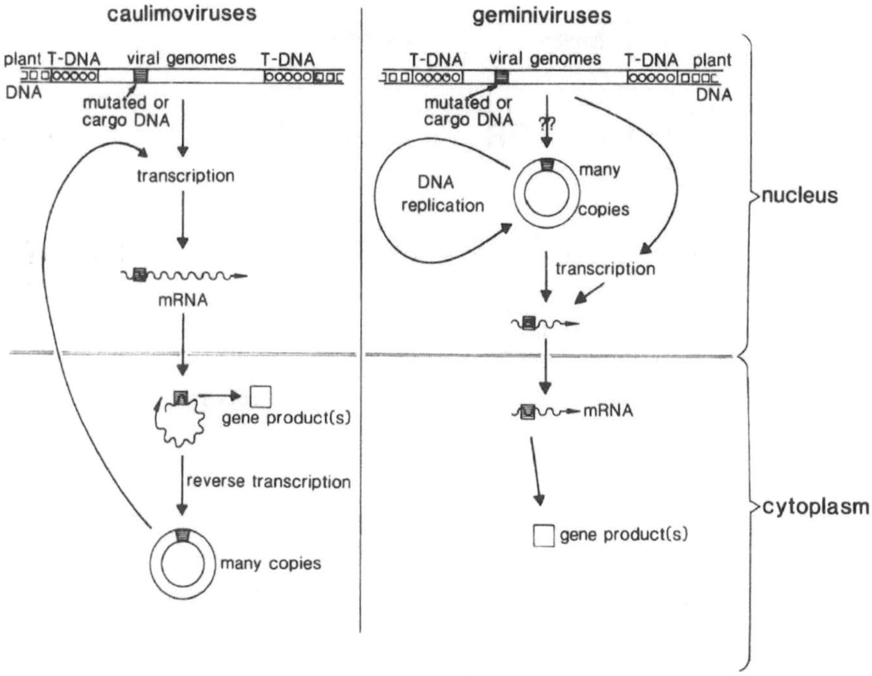

Fig. 3. Plants transgenic for oligomers of viral genomes: analysis of viral mutant strains and expression of foreign genes. The viral functions necessary for RNA/DNA or DNA/DNA replication are incorporated into the nuclear DNA of every plant cell and may allow the production of a molecule which can replicate independently. Inclusion of appropriate control signals would allow the copy number of these molecules to be regulated, in a manner analogous to the induction of a provirus in a bacterial cell. Two systems are compared; caulimoviruses (left) are known to escape from transforming DNA by transcription/reverse transcription (Grimsley *et al.*, 1986b), whereas the mechanism by which geminiviruses (right) escape from DNA integrated in the plant genome (Rogers *et al.*, 1986) is not yet understood. This approach might be used for analysis of *in vitro* produced viral mutant strains which are lethal when inoculated as naked nucleic acid, or for the development of plant expression vectors

meiotic transmissibility of the cargo DNA, the many free copies of the cargo DNA in the viral replicon should allow for high level expression of the cargo DNA gene product.

4. *Super-Infection of Plants Transgenic for Viral Sequences*

Plants transgenic for a fragment of a viral genome can be "challenged" with a different fragment of a viral genome in any of the following ways (see Fig. 4); (1) by inoculation of a particular part of a whole plant, for example with viral DNA or by aphids carrying viral particles; (2) by regeneration of plants from agroinfected cells, producing plants transgenic for both fragments; or (3) by crossing to other agroinfected plants. In those

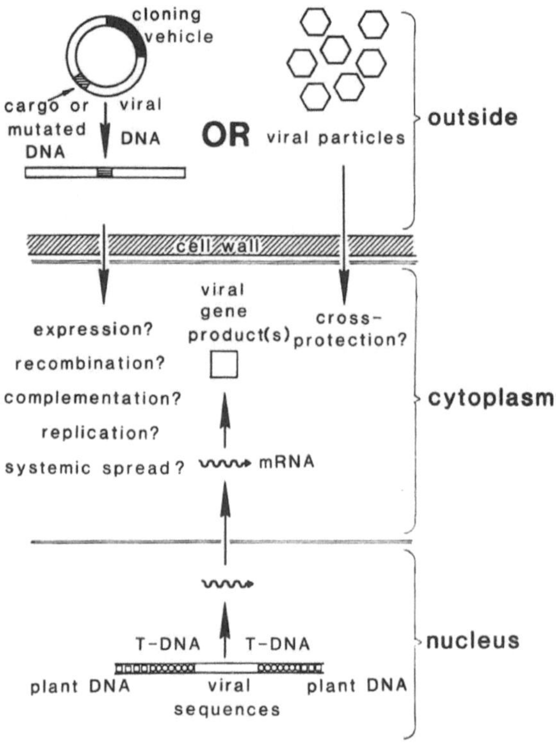

Fig. 4. Super-infection of plants transgenic for viral sequences. A variety of interesting questions can be approached by studying the interaction between such transformed plants and a "challenge" inoculation

cases where two or more defective parts of a viral genome are combined, systemic spread of the virus is an assay for recombination. Experiments described in section B 3 also relate to this approach.

i) Cross-Protection

By agroinfection with sequences important in the control of viral replication, for example, or by engineering sequences to produce anti-sense mRNA (Ecker and Davis, 1986) it may be possible to regenerate plants which are resistant to super-infection by related viruses. Plants transformed with viral components may also show resistance to viral infection (please refer to section B 3).

Tobacco plants transgenic for a CaMV 35S RNA promoter — TMV coat protein — nopaline synthase 3′ end hybrid gene construction have recently been reported to show significantly increased resistance to infection by TMV (Abel *et al.*, 1986; Beachy *et al.*, 1986). The biological basis of the superinfection immunity has yet to be elucidated, but it resembles the phenomenon described as "cross-protection" in some respects; infection of plants with viral RNA rather than with virions overcomes the immunity, lowered resistance is seen when the inoculum size of virions is increased, and in transgenic plants the number of chlorotic lesions is reduced on the inoculated leaf, probably due to fewer infection centres, leading to delayed symptom production.

ii) Complementation of Defective Viral Genes with Integrated Wild-Type Genes; Development of Complementation Vectors

Regenerated plants transgenic for particular wild-type viral genes can be tested as host plants for viruses carrying *in vitro* produced mutations which would normally be lethal. Such complementation studies would lead the way to the elaboration of viral vectors in which particular genes might be replaced with cargo DNA.

III. Perspectives

The term agroinfection has been proposed for the technique of introducing infectious viral genomes, genome components, or selected viral genes into plant cells using *Agrobacterium* Ti plasmid vectors. Two general methods have been outlined. In one, suspensions of bacteria which carry Ti plasmids containing inserted viral genomes are applied to the leaves or stems of intact plants in much the same way as mechanical inoculation of plants with virus preparations or nucleic acid is performed. The method is relatively simple and highly efficient and could prove to be valuable in a variety of applications. We expect this system will be particularly interesting for those systems where mechanical inoculation of cloned DNA has not yet been successful in producing viral infection.

The second agroinfection method involves the inoculation of cultured plant cells (such as protoplasts or leaf discs) with bacteria. This technique may be employed to obtain plants that contain the viral genome, genome components, or selected virus genes, in every cell if regeneration of plants is possible. The technique thus has the potential to permit analysis of viral gene function and in addition should allow the analysis of important aspects of the virus-host interaction.

Examples of the use of both agroinfection methods have been given in this review, and several other potential applications have been noted. Agroinfection provides a powerful new experimental system for the study of plant viruses and their pathology, and is a useful tool for plant molecular biology in a broader sense. This technique has, for example, already been exploited as a means for investigating the mechanism of T-DNA transfer to plants, and has been shown to be of use in producing plants which are resistant to virus infection. Other applications of interest to basic research or of agricultural importance (or both) will no doubt become apparent as the technique passes into general use.

Acknowledgements

We thank Roger Beachy, Richard Gardner, Jacques Hille, and G. Sastry for information provided prior to publication, and Barbara Hohn and Steve Rogers for critical reading and useful comments.

IV. References

Abel, P. P., Nelson, R. S., De., B., Hoffmann, N., Rogers, S. G., Fraley, R. T., Beachy, R. N., 1986: Delay of disease development in transgenic plants that express the tobacco mosaic virus coat protein gene. Science **232,** 738—743.

Baulcombe, D. C., Saunders, G. R., Bevan, M. W. Mayo, M. A., Harrison, B. D., 1986: Expression of biologically active viral satellite RNA from the nuclear genome of transformed plants. Nature **321,** 446—449.

Beachy, R. N., Abel, P. P., Nelson, R. S., Rogers, S. G., Fraley, R. T., 1986: Transgenic plants that express the coat protein gene of TMV are resistant to infection by TMV. In: Molecular Strategies for Crop Protection (UCLA Symposium). In press.

Bevan, M. W., Mason, S. E., Goelet, P., 1985: Expression of tobacco mosaic virus coat protein by a cauliflower mosaic virus promoter in plants transformed by *Agrobacterium.* EMBO J. **4,** 1921—1926.

Bisaro, D. M. Hamilton, W. D. O., Coutts, R. H. A., Buck, K. W., 1982: Molecular cloning and characterization of the two DNA components of tomato golden mosaic virus. Nucl. Acids Res. **10,** 4913—4922.

Bock, K. R., 1982: Geminivirus diseases in tropical crops. Plant Disease **66,** 266—270.

Branch A. D., Robertson, H. D., 1984: A replication cycle for viroids and other small infectious RNAs. Science **223,** 450—455.

Coutts, R. H. A., Buck, K. W., 1985: DNA and RNA polymerase activities of nuclei and hypotonic extracts of nuclei isolated from tomato golden mosaic virus infected tobacco leaves. Nucl. Acids Res. **13,** 7881—7897.

Covey, S. N., Hull, R., 1985: Advances in Cauliflower Mosaic Virus Research. Oxford Surveys of Plant Molecular and Cell Biology, **2,** 339—346.

Covey, S. N., Lomonossoff. G. P., Hull, R., 1981: Characterisation of cauliflower mosaic virus DNA sequences which encode major polyadenylated transcripts. Nucl. Acids Res. **9,** 6735—6747.

Cress, D. E., Kiefer, M. C., Owens, R. A., 1983: Construction of infectious potato spindle tuber viroid cDNA clones. Nucl. Acids Res. **11,** 6821—6835.

Daubert, S., Shepherd, R. J., Gardner, R. C., 1983. Insertional mutagenesis of cauliflower mosaic virus genome. Gene **25,** 201—208.

DeBlock, M., Herrera-Estrella, L., Van Montagu, M., Schell, J., Zambryski, P., 1984: Expression of foreign genes in regenerated plants and their progeny. EMBO J. **3,** 1681—1689.

Dixon, L. K., Koenig, I., Hohn, T., 1983: Mutagenesis of cauliflower mosaic virus. Gene **25,** 189—199.

Ecker, J. R., Davis, R. W., 1986: Inhibition of gene expression in plant cells by expression of antisense RNA. Proc. Natl. Acad. Sci. U.S.A. **83,** 5372—5376.

Fraley, R. T., Horsch, R. B., Matzke, A., Chilton, M.-D., Chilton, W. S., Sanders, P. R., 1984: In vitro transformation of petunia cells by an improved method of co-cultivation with *A. tumefaciens* strains. Plant Mol. Biol. **3,** 371—378.

Fraley, R. T., Rogers, S. G., Horsch, R. B., Eichholtz, D., Flick, J. S., Fink, C. L., Hoffmann, N. L., Sanders, P. R., 1985: The SEV system: A new disarmed Ti plasmid vector for plant transformation. Bio/technology **3**, 629—635.

Francki, R. I. B., Milne, R. G., Hatta, T., 1985: Geminivirus group. In: "Atlas of Plant Viruses", Vol. 1. CRC Press, Boca Raton, Florida, pp. 33—46.

Fromm, M. E., Taylor, L. P., Walbot, V., 1986: Stable transformation of maize after gene transfer by electroporation. Nature **319**, 791—793.

Gardner, R. C., Knauf, V. C., 1986a: Transfer of *Agrobacterium* DNA to plants requires a T-DNA border but not the *vir*E locus. Science **231**, 725—727.

Gardner, R. C., Chanoles, K. R., Owens, R. A., 1986b: Potato spindle tuber viroid infections mediated by the Ti plasmid of *Agrobacterium tumefaciens*. Plant Mol. Biol. **6**, 221—228.

Gheysen, G., Dhaese, P., Van Montagu, M., 1985: In "Genetic Flux in Plants", Hohn, B., Dennis, E. S., (eds.). Wien - New York: Springer.

Goodman, R. M., 1981: Geminiviruses. J. Gen. Virol. **54**, 9—21.

Grimsley, N., Hohn, B., Hohn, T., Walden, R. M., 1986a: Agroinfection, an alternative route for plant virus infection by using Ti plasmid. Proc. Nat. Acad. Sci. U.S.A. **83**, 3282—3286.

Grimsley, N., Hohn, T., Hohn, B., 1986b: Recombination in a plant virus: template-switching in cauliflower mosaic virus. EMBO J. **5**, 641—646.

Grimsley, N., Hohn, T., Davies, J. W., Hohn, B., 1987: *Agrobacterium*-mediated delivery of infectious maize streak virus into maize plants. Nature **324**, 177—179.

Guilley, H., Dudley, R. K., Jonard, G., Balazs, E., Richards, K. E., 1982: Transcription of cauliflower mosaic virus DNA: detection of promoter sequences, and characterisation of transcripts. Cell **30**, 763—773.

Hamilton, W. D. O., Sanders, R. C., Coutts, R. H. A., Buck, K. W., 1981: Characterization of tomato golden mosaic virus as a geminivirus. FEMS Microbiology Letters **11**, 263—267.

Hamilton, W. D. O., Bisaro, D. M., Buck, K. W., 1982: Identification of novel DNA forms in tomato golden mosaic virus infected tissue: Evidence for a two component viral genome. Nucl. Acids Res. **10**, 4901—4912.

Hamilton, W. D. O., Bisaro, D. M., Coutts, R. H. A., Buck, K. W., 1983: Demonstration of the bipartite nature of the genome of a single-stranded plant virus by infection with the cloned DNA components. Nucl. Acids Res. **11**, 7387—7396.

Hamilton, W. D. O., Stein, V. E., Coutts, R. H. A., Buck, K. W., 1984: Complete nucleotide sequence of the infectious cloned DNA components to tomato golden mosaic virus: Potential coding regions and regulatory sequences. EMBO J. **3**, 2197—2205.

Hernalsteens, J. P., Thia-Toong, L., Schell, J., Van Montagu, M., 1984: An *Agrobacterium*-transformed cell culture from the monocot *Asparagus officinalis*. EMBO J. **3**, 3039—3041.

Hille, J., Dekker, M., Luttighuis, H., van Kammen, A., Zabel, P., 1986: Detection of T-DNA transfer to plant cells by *A. tumefaciens* virulence mutants using agroinfection. Mol. gen. Genet. **205**, 411—416.

Hohn, T., Hohn, B., Pfeiffer, P. 1985: Reverse transcription in cauliflower mosaic virus. Trends Biochem. Sci. **5**, 205—209.

Hooykaas-Van Slogteren, G. M. S., Hooykas, P. J. J., Schilperoort, R. A., 1984: Expression of Ti plasmid genes in monocotyledonous plants infected with *Agrobacterium tumefaciens*. Nature **311**, 763—764.

Horsch, R. B., Fraley, R. T., Rogers, S. G., Sanders, P. R., Lloyd, A. R., Hoffmann,

N. L., 1984: Inheritance of functional foreign genes in plants. Science **223,** 496—498.

Horsch, R. B., Fry, J. E., Hoffmann, N. L., Wallroth, M., Eicholtz, D., Rogers, S. G., Fraley, R. T., 1985: A simple and general method for transferring genes into plants. Science **227,** 1229—1231.

Howell, S. H., Walker, L. L., Walden, R. M., 1982: Rescue of *in vitro* generated mutants of cloned CaMV genome on infected plants. Nature (London) **293,** 485—486.

Howell, S. H., 1985: The molecular biology of plant DNA viruses. CRC Critical Reviews in Plant Sciences **2,** 287—316.

Hutchins, C. J., Keese, P., Visavader, J. E., Rathjen, P. D., McInnes, J. L., Symons, R. H., 1985: Comparison of multimeric plus and minus forms of viroids and virusoids. Plant Mol. Biol. **4,** 293—304.

Ishikawa, M., Meshi, T., Ohno, T., Okada, Y., Sano, T., Ueda, I., Shikata, E., 1984: A revised replication cycle for viroids: the role of longer than unit length RNA in viroid replication.

Kaper, J. M., Tousignant, M. E., Lot, H., 1976: A low molecular weight replicating RNA associated with a divided genome plant virus: defective or satellite RNA? Biochem. Biophys. Res. Commun. **72,** 1237—1243.

Koukolíová-Nicola, Z., Shillito, R. D., Hohn, B., Wang, K., Van Montagu, M., Zambryski, P., 1985: Involvement of circular intermediates in the transfer of T-DNA from *Agrobacterium tumefaciens* to plant cells. Nature (London) **313,** 191—196.

Lebeurier, G., Hirth, L., Hohn, B., Hohn, T., 1982: *In vivo* recombination of cauliflower mosaic virus DNA. Proc. Nat. Acad. Sci. U.S.A. **79,** 2932—2936.

Matyis, J. G., Silva, D. M., Oliveira, A. R., Costa, A. S., 1975: Purificaco e morfologia do virus do mosaico dourado de tomateiro. Summa Phytopathol. **1,** 267—274.

Mossop, D. W., Francki, R. I. B., 1979: Comparative studies of two satellite RNAs of cucumber mosaic virus. Virology **95,** 395—404.

Mullineaux, P. M., Donson, J., Morris-Krsinich, B. A. M., Boulton, M. I., Davies, J. W., 1984: The nucleotide sequence of maize streak virus DNA. EMBO J. **3,** 3062—3068.

Odell, J. T., Dudley, R. K., Howell, S. H., 1981: Structure of the 19 S RNA transcript encoded by the cauliflower mosaic virus genome. Virology **111,** 377—385.

Odell, J. T., Nagy, F., Chua, N.-H., 1985: Identification of DNA sequences required for activity of the cauliflower mosaic virus 35 S promotor. Nature **313,** 810—812.

Ow, D. W., Wood, K. V., DeLuca, M., de Wet, J. R., Helsinki, D. R., Howell, S. H., 1986: Transient and stable expression of the firefly luciferase gene in plant cells and transgenic plants. Science **234,** 856—859.

Paszkowski, J., Shillito, R. D., Saul, M., Mandak, V., Hohn, T., Hohn, B., Potrykus, I., 1984: Direct gene transfer to plants. EMBO J. **3,** 2717—2722.

Peralta, E. G., Helmiss, R., Ream, W., 1986: *Overdrive,* a T-DNA transmission enhancer on the *A. tumefaciens* tumour-inducing plasmid. EMBO J. **5,** 1137—1142.

Polson, A., Von Wechmar, M. B., 1980: A Novel Way to Transmit Plant Viruses. J. Gen. Virol. **51,** 179—181.

Potrykus, I., Saul, M. W., Petruska, J., Paszkowski, J., Shillito, R. D., 1985: Direct gene transfer to cells of a graminaceous monocot. Mol. Gen. Genet. **199,** 183—188.

Rogers, S. G., Bisaro, D. M., Horsch, R. B., Fraley, R. T., Hoffman, N. L., Brand, L., Elmer, J. S., Lloyd, A. M., 1986: Tomato golden mosaic virus A component DNA replicates autonomously in transgenic plants. Cell **45**, 593—600.

Sastry, G. R. K., Miles, C. A., Miller, I. S., Borland, P. A., Saeed, N., May, C. A., 1986: Tryptophan auxotrophs for increasing safety of *Agrobacterium* Ti-based recombinant plasmid work. Plant Mol. Biol. Rep. **4**, 93—97.

Shaw, W., 1975: Chloramphenicol acetyltransferase from resistant bacteria. Methods Enzymol. **53**, 737—754.

Shewmaker, C. K., Caton, J. R., Houck, C. M., Gardner, R. C., 1985: Transcription of cauliflower mosaic virus integrated into plant genomes. Virology **140**, 281—288.

Stachel, S. E., Timmerman, B., Zambryski, P., 1986: Generation of single-stranded T-DNA molecules during the initial stages of T-DNA transfer from *Agrobacterium tumefaciens* to plant cells. Nature **322**, 706—712.

Stanley, D. M., 1985: The molecular biology of geminiviruses. Adv. Virus Res. **30**, 139—177.

Tabler, M., Sanger, H. L., 1984: Cloned single- and double-stranded DNA copies of potato spindle tuber viroid (PSTV) RNA and co-inoculated subgenomic DNA fragments are infectious. EMBO J. **3**, 3055—3062.

Walden, R. M., Howell, S. H., 1983: Uncut recombinant plasmids bearing nested cauliflower mosaic virus genomes infect plants by intragenomic recombination. Plant Mol. Biol. **2**, 927—937.

Waterworth, H. E., Kaper, J. M., Tousignant, M. E., 1979: CARNA 5, the small cucumber mosaic virus-dependent replicating RNA, regulates disease expression. Science **204**, 845—847.

Chapter 5

The Mechanism of T-DNA Transfer from *Agrobacterium tumefaciens* to the Plant Cell

Zdena Koukolíková-Nicola, Lisa Albright* and Barbara Hohn

Friedrich Miescher-Institut, P. O. Box 2543, CH-4002 Basel, Switzerland

* Department of Microbiology and Immunology, University of Washington, Seattle, WA 98195, U.S.A.

Present address: Department of Molecular Biology, Massachusetts General Hospital, Boston, MA 02114, U.S.A.

With 8 Figures

Contents

I. General Introduction

A. Scope of the Review

This review concentrates on the newest developments in our understanding of the first steps of transfer of the T-DNA from *Agrobacterium* to its host plant cell. We will present data about organisation, regulation and function of the virulence region. Furthermore, we will summarise current information about intermediates in the T-DNA transfer and discuss it in light of proposed models. A very short introduction to the biology of *Agrobacterium tumefaciens* will be presented. For further background the reader is referred to recent reviews (Hille *et al.,* 1984; Hooykaas and Schilperoort, 1984; Nester *et al.,* 1984; Gheysen *et al.,* 1985; Stachel and Zambryski, 1986b). *Agrobacterium rhizogenes* is discussed in this volume (White and Sinkar).

B. Crown Gall Disease

Grown gall tumours are neoplastic proliferations induced by the gram-negative bacterium *Agrobacterium tumefaciens* on wounded dicotyledonous (and some monocotyledonous) plants. In nature, the infection often occurs at or near the junction of the root and the stem, the crown of the plant. Explants from crown galls can be cultured *in vitro* in absence of the inciting bacteria: whereas normal plant cells need plant growth regulators such as cytokinin and auxin to grow *in vitro,* the transformed cells can proliferate on a hormone free medium (White and Braun, 1942; Braun and White, 1943; Braun and Mandle, 1948). Crown gall cells also produce nitrogenous compounds called opines which are absent in normal plant cells and which can be used by *Agrobacterium* as carbon and/or nitrogen source (Morel, 1956; Lioret, 1956; Tempé and Petit, 1982). There is a strict correlation between the type of opine synthesized by the tumour and the type of opine which can be catabolised by the inducing bacteria (Goldmann *et al.,* 1968; Petit *et al.,* 1970). It has been proposed that the inciting bacteria have a competitive advantage over other soil organisms in the rhizosphere of infected plants due to their ability to catabolize the opines released by the crown gall cells. This type of pathogenic interaction has been called genetic colonisation (Schell *et al.,* 1979) or opine concept (Petit *et al.,* 1978; for more details see Tempé *et al.,* 1984).

Hosts of *Agrobacterium* include most dicotyledonous and a few monocotyledonous plants (De Cleene and De Ley, 1976; Hooykaas-van Slogteren *et al.,* 1984; Hernalsteens *et al.,* 1984; De Cleene, 1985), but commercially important crop plants such as *gramineae* have not been thought to be susceptible (De Cleene, 1985). Graves and Goldman (1986) have reported detecting opines in tissue of maize seedlings inoculated with *Agrobacterium*; however, opine production may not always be sufficient evidence for transformation by *Agrobacterium* (Christou *et al.,* 1986). Recent demonstration by Grimsley *et al.* (1987) that cloned maize streak virus (MSV) DNA is infectious to maize plants, if *Agrobacterium*-mediated transformation is used, suggests that *Agrobacterium* might also be useful for genetic engineering of graminaceous plants.

C. Molecular Basis of the Neoplastic Transformation

a) The Ti Plasmid and Its Organisation

All virulent strains of *Agrobacterium* contain a very large (150—250 kb) plasmid, called the tumour-inducing (Ti) plasmid (Zaenen *et al.*, 1974; Van Larebeke *et al.*, 1975). There are three major categories of Ti plasmids: nopaline, octopine and agropine, classified by the type of opine whose synthesis they induce. Recently a new type of succinamopine strain has been defined (Chilton *et al.*, 1984; Blundy *et al.*, 1986). Other types of opines are also synthesized by the crown gall cells: for example, tumour cells induced by a nopaline strain synthesize not only nopaline but also agrocinopines A and B (Ellis and Murphy, 1981); cells transformed by octopine strains often additionally synthesize agropine and mannopine (Salomon *et al.*, 1984).

Two parts of the Ti plasmid are necessary for tumour formation:

1. A piece of the Ti plasmid, the T-DNA (transferred DNA) which is stably integrated into the plant nuclear genome (Chilton *et al.*, 1977, 1980; Thomashow *et al.*, 1980a; Willmitzer *et al.*, 1980). The expression within the plant cells of some of the T-DNA genes, also called *onc* genes, causes the tumourous growth. Possible integration into the chloroplast genome was also reported (De Block *et al.*, 1985).

Although the T-DNA is the only part of the Ti plasmid stably maintained in the plant genome, none of the T-DNA genes is responsible for T-DNA transfer and integration (Leemans *et al.*, 1982; Zambryski *et al.*, 1983).

2. A second essential region of the Ti plasmid, the virulence *(vir)* region (of about 35 kb), acts in *trans* during tumour induction (for example: Garfinkel and Nester, 1980; Holsters *et al.*, 1980; Ooms *et al.*, 1980; De Greve *et al.*, 1981). It is thought that *vir* functions are involved in the early steps of infection, such as transfer of T-DNA.

Other functions encoded by the Ti plasmid include:

1) Nopaline catabolism *(noc)* or octopine catabolism *(occ)* functions (Bomhoff *et al.*, 1976; Montoya *et al.*, 1977); 2) Functions for conjugal transfer of the Ti plasmid between bacteria (*tra* functions), which are inducible by opines (Genetello *et al.*, 1977; Kerr *et al.*, 1977; Ellis *et al.*, 1982); 3) Replication functions including the origin of replication (Ori) and the incompatibility functions (Koekman *et al.*, 1980, 1982; Gallie *et al.*, 1985). Circular maps of a nopaline and an octopine Ti plasmid are shown in Figure 1.

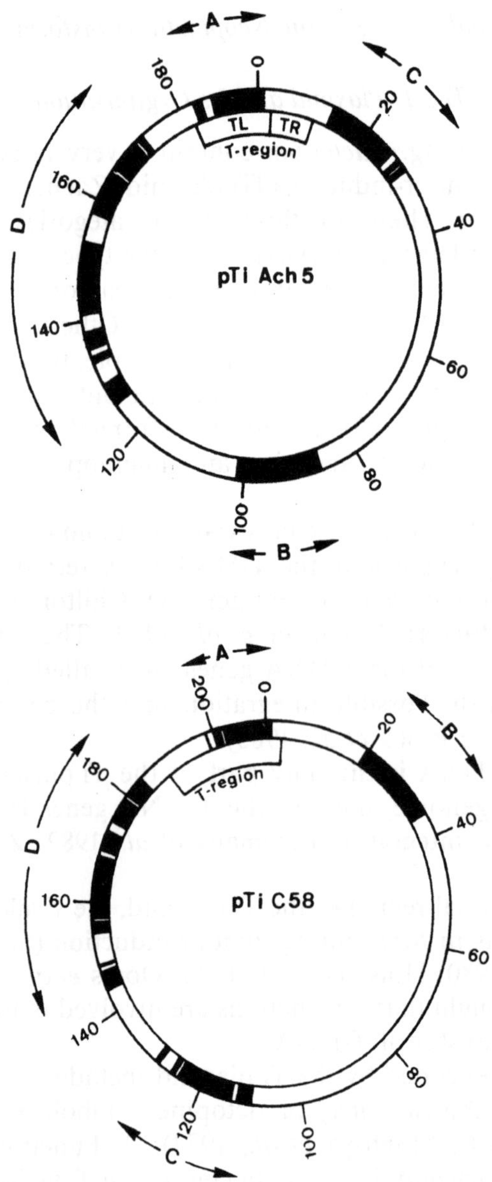

Fig. 1. Circular maps of an octopine Ti plasmid (pTiAch5) and a nopaline Ti plasmid (pTiC58). The homology regions (A, B, C and D) are indicated as black bars. Region A is part of the T-DNA, and is shown in more detail in Fig. 2. Region D is the *vir*-region. Map coordinates are in kb, starting from the common *Sma*I site within the T-DNA. With courtesy of G. Gheysen; from Gheysen *et al.*, 1985

b) Functional Organisation of the T-DNA

The functional organisation of the T-DNA of a nopaline and an octopine Ti plasmid is shown in Figure 2. The T-DNA of nopaline Ti plasmids is 24 kb in size (Lemmers *et al.*, 1980). Two noncontiguous segments of the

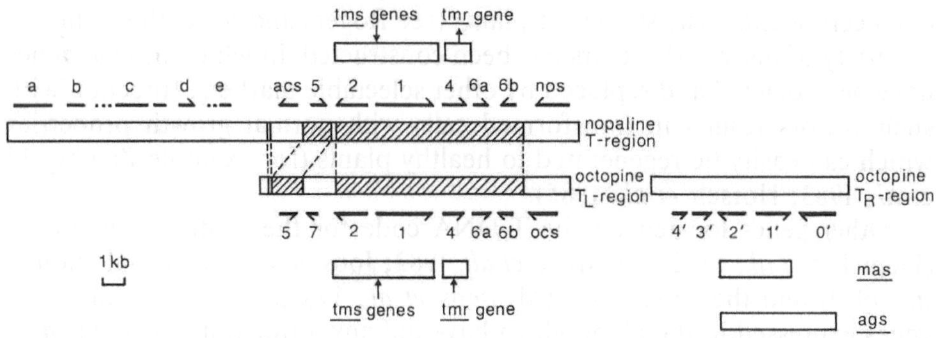

Fig. 2. Functional organisation of the T-DNA of nopaline (upper part) and octopine (lower part) Ti plasmids. The arrows indicate the different genes and their direction of transcription in the transformed plant cell. The gene nomenclature follows the original designations according to transcript size (Willmitzer *et al.*, 1982, 1983; Velten *et al.*, 1983). *Nos,* nopaline synthesis; *acs,* agrocinopine synthesis, *ocs,* octopine synthesis, *mas,* mannopine synthesis; *ags,* agropine synthesis; *tms, t*umour *m*orphology *s*hoots; *tmr, t*umour *m*orphology *r*oots. The dashed areas represent regions of homology between the two types of plasmids. With courtesy of G. Gheysen; modified after Gheysen *et al.*, 1985

Ti-plasmid are present in some transformed lines induced by the octopine strains (Thomashow *et al.*, 1980b; De Beuckeleer *et al.*, 1981). One segment, called T_L (14 kb), is always present and is necessary for tumour formation, whereas the other segment, called T_R (7 kb), is not found in all tumour lines. The fragment between the two T_L and T_R is sometimes called T_C-region.

The *onc* genes of the T-DNA (the T_L-DNA in the case of the octopine Ti plasmid) code for enzymes which are involved in the production of the phytohormones indoleacetic acid (IAA) and cytokinin (for recent review see Morris, 1986). Thus, plant cells transformed by wild-type *Agrobacteria* grow *in vitro* without addition of any phytohormones. Two loci, *tms* 1 and *tms* 2 (*t*umour *m*orphology *s*hoots, also called gene 1 and gene 2), code for the synthesis of IAA (Inzé *et al.*, 1984; Schröder *et al.*, 1984; Thomashow *et al.*, 1984, 1986). One locus, *tmr* (*t*umour *m*orphology *r*oots, also called gene 4), codes for an enzyme catalysing the first step of the cytokinin biosynthesis (Akiyoshi *et al.*, 1983; Barry *et al.*, 1984; Buchmann *et al.*, 1985). Based on high homology between *Pseudomonas savastanoi* genes involved in auxin (Yamada *et al.*, 1985; Follin *et al.*, 1985) and cytokinin (Powell and Morris, 1986) biosynthesis and the *onc* genes of *Agrobacterium* it has been suggested that these genes might have a common origin or that the T-DNA genes might have originated from those in *P. savastanoi* (Schell,

1986). Recently, it was proposed that these genes might have been transferred from *P. savastanoi* to *Agrobacterium* as a part of a transposon, whose remnants have been found within the T_C-region of the octopine Ti plasmid (Yamada *et al.*, 1986).

The property of transferring any foreign DNA inserted into the T-DNA is the basis for the great potential of the *Agrobacterium* system in genetic engineering and basic studies of plants (see Rogers and Klee, this volume). A variety of disarmed vectors has been constructed, in which the *onc* genes have been deleted and replaced by other selectable markers. Infection with such vectors results in transformed cells with normal growth properties which can easily be regenerated to healthy plants (for example Zambryski *et al.*, 1983; Horsch *et al.*, 1985).

Other genes located on the T-DNA code for the synthesis of opines (Depicker *et al.*, 1982; De Greve *et al.*, 1982; Joos *et al.*, 1983 a; Salomon *et al.*, 1984) and their excretion (Messens *et al.*, 1985). There are still other genes expressed in the plant which have unknown functions, as their inactivation results in apparently normal tumours. These genes may play a minor role in tumour formation or may be involved in the synthesis of yet unidentified opines. Interestingly, the promoter of one such gene of unknown function (gene 5 of T_L-DNA) was reported to be active in a tissue-specific fashion (Koncz and Schell, 1986). All T-DNA genes which have been sequenced contain eukaryotic expression signals such as promoters and polyadenylation sites (Barker *et al.*, 1983; Gielen *et al.*, 1984); they are equipped to be expressed in the plant cell. Thus, the T-DNA is a prokaryotic piece of DNA which is transferred to and expressed in a eukaryotic cell.

II. Early Events of Transformation

A. Virulence Functions

Mutations outside the T-DNA which affect virulence are located either on the bacterial chromosome (Garfinkel and Nester, 1980) or within the virulence region of the Ti plasmid (Garfinkel and Nester, 1980; Holsters *et al.*, 1980; Ooms *et al.*, 1980). The corresponding genes are responsible for the early events of transformation.

a) Chromosomal Virulence Region

An early step in transformation is the attachment of bacteria to plant cells (for recent reviews, see Matthysse, 1984; Pueppke and Hawes, 1985). Attachment is thought to be a prerequisite for T-DNA transfer to plant cells: first, tumour formation by virulent bacteria can be inhibited by preinoculation of plant tissue with an avirulent strain which still has the ability to attach (Lippincott and Lippincott, 1969; Glogowski and Galsky, 1978). Second, it has been shown that six avirulent chromosomal mutants are defective in their ability to attach to plant cells (Douglas *et al.*, 1982). The chromosomal virulence region of the octopine strain A6 contains two distinct but linked loci, 1.5 and 5 kb long, called *chv*A and *chv*B, respectively (Douglas *et al.*, 1985). Transposon insertions into these loci lead to avirulence and defectivity in attachment. These loci are expressed constitutively. Mutants in each locus show pleiotropic effects likely to be related to cell envelope changes. The *chv*B mutants lack the 2-linked-β-D-glucan, normally found in culture filtrates and phenol-water-extracts of wild-type *A. tumefaciens* (Puvanesarajah *et al.*, 1985; Zorreguieta and Ugalde, 1986). A clone containing the *chv*A and B loci, isolated from the A6 chromosome, has been shown to complement a chromosomal virulence mutant *ctu*1 of the C58 strain (Miller *et al.*, 1986). Recently, homologous genes called *ndv*A and B have been identified in *Rhizobium meliloti.* These loci have been shown to be functionally interchangeable with those of *A. tumefaciens* and to be essential for symbiosis (Dylan *et al.*, 1986; van Veen *et al.*, 1987). Thus, one of the first steps in the pathogenic and symbiotic interactions of *Agrobacterium* and *Rhizobium* with plant cells seems to be closely related.

It is highly probable that products of other chromosomally located genes, such as house-keeping genes, are also involved in early events of transformation, such as T-DNA processing and transfer. If, however, they normally perform basic functions within the cell, they cannot be identified by transposon mutagenesis.

b) Organisation of the Virulence Region

Mutational analyses have indicated that in addition to the T-DNA a continuous region of the Ti plasmid of about 35 kb is essential for virulence (Garfinkel and Nester, 1980; Holsters *et al.*, 1980; Ooms *et al.*, 1980). Despite extensive studies (for example Klee *et al.*, 1983; Hille *et al.*, 1984; Knauf *et al.*, 1984; Lundquist *et al.*, 1984; Hooykaas *et al.*, 1984) the region was not fully saturated by mutations and the assignment of the existing *vir* mutations to distinct *vir* complementation groups was not well established. Moreover, different research groups used different nomenclature for the *vir* loci. More recently, Stachel and Nester (1986) characterised in detail the genetic and transcriptional organisation of the *vir* loci of the octopine-type Ti plasmid pTiA6. This *vir* region contains six distinct *vir* complementation groups: *vir*A, *vir*B, *vir*C, *vir*D, *vir*E and *vir*G. Mutations in these loci eliminate (*vir*A, *vir*B, *vir*D and *vir*G) or significantly attenuate (*vir*C and *vir*E) the virulence. Since transposon mutagenesis, resulting in polar insertions, has been used to characterise the *vir* region, each of the *vir* complementation groups corresponds to a single *vir* transcription unit (Fig. 3). Another *vir* locus, *vir*F, was identified in the octopine Ti plasmid pTiAch5 about 7 kb rightward of *vir*E (Hooykaas *et al.*, 1984); such a locus was found neither in the octopine plasmid pTiA6 (Stachel and Nester, 1986) nor in the nopaline plasmid pTiC58 (Hirooka and Kado, 1986).

The high degree of homology of the octopine and nopaline Ti plasmids within the *vir*-region (Engler *et al.*, 1981) together with their respective complementability (Hooykaas *et al.*, 1984) and interchangeability (de Framond *et al.*, 1983; Hoekema *et al.*, 1983 b) indicated that the two *vir*

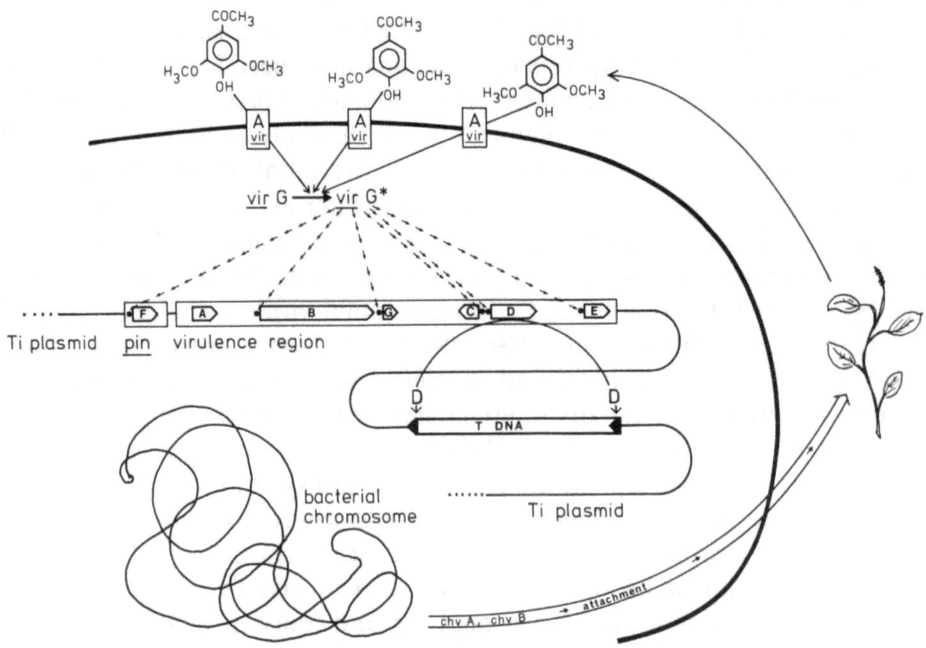

Fig. 3. Scheme representing the present view of the early events of transformation. Bacterial attachment to plant cells involves products of *chv*A and B loci; the inducer acetosyringone (chemical formula) interacts with the *vir*A product; the *vir*G product is activated and in turn activates transcription of the rest of the *vir* loci; the *vir*D products act at the T-DNA borders. The *vir* region (pTiA6) is schematically drawn according to Stachel and Nester (1986). Defined genetic boundaries and transcriptional orientation of the *vir* region loci *vir*A, *vir*B, *vir*G, *vir*C, *vir*D and *vir*E are indicated by arrows; the plant-inducible locus *pin*F is also indicated

regions are very similar. The nopaline *vir* region of pTiC58 shows the same overall organisation as the octopine *vir* region (Rogowsky *et al.*, submitted). All nopaline *vir* loci have their counterparts in the octopine *vir* region and the direction of transcription of all *vir* loci is conserved. Comparative sequence analysis of some of the nopaline and octopine *vir* loci (*vir*C, Yanofsky and Nester, 1986; Close *et al.*, submitted; *vir*E, Winans *et al.*, 1987; Hirooka *et al.*, in press) and genetic complementation studies (*vir*E, Hirooka *et al.*, in press) further substantiate this notion. It was agreed to adopt the nomenclature of Stachel and Nester (1986) also for the nopaline *vir* region.

c) Regulation of vir Region Expression

The Ti plasmid *vir* loci are transcriptionally activated in response to signal molecules produced by plant cells.

Use of a specially constructed Tn3-*lac*Z transposon (Stachel *et al.*, 1985 a) as mutagen provided a useful system to monitor the expression of the virulence region: the production of β-galactosidase, the *lac*Z gene product, is placed under control of the gene into which the transposon has inserted. The *vir* gene expression in *Agrobacterium* is specifically induced by co-cultivation with dicotyledonous plant cells: *vir*B, *vir*C, *vir*D and *vir*E are expressed only upon activation by plant cells. The pTiA6 *vir* region also contains another plant-inducible locus, *pin*F, non-essential for virulence.

The pTiA6 *vir* loci are organised as a single regulon whose induction by plant cells is controlled by the positive regulator elements encoded by *vir*A and *vir*G (Stachel and Zambryski, 1986 a; Winans *et al.*, 1986; Melchers *et al.*, 1986). Mutations in *vir*A attenuate (Stachel and Zambryski, 1986 a) or abolish (Winans *et al.*, 1986) the induction of other *vir* loci. This locus is monocistronic, codes for a protein of 92,000 daltons (Leroux *et al.*, 1987) and is constitutively transcribed at low level. Mutations in *vir*G eliminate *vir* induction. This locus contains one open reading frame coding for a protein of about 30,000 daltons, it is vegetatively transcribed at low level, but plant-inducible to high level and is self-regulated in a complex fashion. S1 nuclease protection analysis of transcripts showed that two distinct messages, one vegetative and one inducible are produced from the *vir*G gene (Stachel and Zambryski, 1986 a).

Sequence analysis of the *vir*G and *vir*A loci revealed that the products of these two genes are closely related to other bacterial two-component positive regulatory systems: for example *omp*R/*env*Z and *ntr*C/*ntr*B (Winans *et al.*, 1986; Melchers *et al.*, 1986; Nixon *et al.*, 1986; Leroux *et al.*, 1987). As a response to altered osmolarity *omp*R and *env*Z gene products regulate the expression of *E. coli* outer membrane porins coded by *omp*C and *omp*F. Similarly, nitrogen-regulated genes are positively controlled by the *ntr*C/*ntr*B two-component regulatory system. Thus it seems that prokaryotes modified the same two-component regulatory system to positively respond to different environmental stimuli. By analogy to the other better studied regulatory systems, the following model has been proposed: The *vir*A product is a membrane-bound chemo-receptor protein, transmitting the extracellular signal. After interacting with the signal compound, the *vir*A protein transforms the *vir*G product to an active form which, in turn, activates transcription of the rest of the *vir* loci (see Fig. 3; Winans *et al.*, 1986; Melchers *et al.*, 1986; Rogowsky *et al.*, submitted). Two conserved hexameric sequences have been found in the promoter regions (or transcribed regions) of several *vir* transcription units, which might play a role in the positive regulation by *vir*G product (Das *et al.*, 1986).

A subset of the *vir* loci is regulated by yet another regulatory system: a spontaneous chromosomal mutant has been isolated, the so-called *ros* (*ro*ugh *s*urface) mutant, which expresses the *vir*C and *vir*D genes at a high vegetative level (Close *et al.*, 1985). The *ros* mutant is highly pleiotropic. It has an unusual colony morphology: the colonies have rough (non-mucoid) surfaces due to lack of the major extracellular polysaccharide. Since the proximal *vir*D genes are involved in T-DNA processing (see III. B. a) Close *et al.* (submitted) speculate that the *ros* phenotype may represent a physiological state of the bacterium which normally occurs during *Agrobacterium*-plant cell interaction and which signals favorable conditions for T-DNA transfer. This would occur at a later stage than the plant-induced *vir* gene expression, for example as a consequence of bacterial surface changes upon close contact of the bacterium with the plant cell. Two independent promoters can direct transcription of *vir*D. One promoter requires the presence of the plant inducer and *vir*A and *vir*G products, whereas the other promoter functions under the conditions caused by the *ros* mutation (Tait *et al.*, unpublished).

d) The Nature of the Inducer

Vir gene expression is specifically activated by plant cell metabolites present in exudates of a variety of plant cell cultures. The induction does not require plant-bacterial contact and is mediated by small soluble compounds (Stachel *et al.*, 1985b and 1986a). Two major inducing factors present in conditioned medium of *N. tabacum* root cultures have been identified: they are phenolic compounds called acetosyringone (AS; 4-acetyl-2,6-dimethoxyphenol; for chemical formula, see Fig. 3) and α-hydroxyacetosyringone (OH-AS; Stachel *et al.*, 1985b). Synthetic AS or OH-AS have been shown to be sufficient for *vir* induction. Related compounds such as sinapinic acid have also been shown to induce *vir* gene espression (Stachel *et al.*, 1985b).

A variety of other commercially available chemicals, most of which are known plant metabolites, proved to be slight inducers individually, but in combination acted as potent inducers (Bolton *et al.*, 1986). This indicates that in nature the *vir* gene expression might be induced by a variety of phenolic compounds which could explain the wide host range of *Agrobacterium.*

Production of AS and OH-AS is stimulated by wounding. Active plant cell metabolism is required for their production. AS and OH-AS might be part of the wound response of the plant cell or of the lignin synthesis by damaged cells, since these compounds are likely products of the shikimic acid biosynthetic pathway (Stachel *et al.*, 1985b, 1986a).

Agrobacterium is attracted towards a range of phenolic plant compounds. Interestingly, chemotaxis towards AS is determined by the Ti plasmid and occurs at an about 100-fold lower concentration than that required for efficient *vir* induction (Ashby *et al.*, in press). This suggests that AS acts at two steps of the parasitic attack: at low concentration it directs the bacterium towards the wounded plant cells. In the vicinity of the plant cells the higher concentration of AS causes *vir* gene expression and ultimately T-DNA transfer.

There are two indications that the limited host range of the Ti plasmid pTiAg162 might partly be due to inability of its *vir*A product to recognise the common inducers such as AS. First, in addition to other factors the wide-host-range *vir*A gene is necessary to increase the host range of the limited host range Ti plasmid pTiAg162 (Yanofsky *et al.*, 1985). Second, the limited host-range *vir*A product gives only a limited response to AS or co-cultivation with tobacco plant cells (Stachel and Zambryski, 1986a; Leroux *et al.*, 1987).

The expression of *Rhizobium* nodulation genes, which are required for early symbiotic interactions, is induced by another family of plant compounds present in root exudates, the flavonoids (Peters *et al.*, 1986; Redmond *et al.*, 1986; Firmin *et al.*, 1986; Downie and Johnston, 1986). In contrast to AS, whose synthesis is increased by wounding, the flavonoids are normal plant metabolites. This difference probably reflects different requirements for symbiotic and pathogenic invasions.

B. T-DNA Transfer

We will use the term "T-DNA transfer" not only for the transfer of the T-DNA into the plant cell but also for its integration into the plant genome. The words "T-DNA processing" will be used for the steps preparing the T-DNA for transfer across the bacterial membrane.

Much of our knowledge on T-DNA transfer has come from identification of *cis* acting elements within and in the vicinity of the T-DNA in the Ti plasmid and from analysis of T-DNA/plant DNA junctions (see below).

a) The 25-bp Terminal Sequence

Comparison of T-DNA/plant DNA junctions with respective sequences of the nopaline Ti plasmids pTiT37 and pTiC58 (Yadav *et al.*, 1982; Zambryski *et al.*, 1982) revealed an imperfect direct repeat of 25 bp at both ends of the T-region: 21 out of 25 bp are conserved between the left and the right terminal repeat. Similar repeats were found at the ends of the T_L (called A and B) and the T_R-regions (called C and D) of octopine Ti plasmids (Simpson *et al.*, 1982; Holsters *et al.*, 1983; Barker *et al.*, 1983; Gielen *et al.*, 1984) and more recently in the Ri plasmids of *A. rhizogenes* (Slightom *et al.*, 1985, 1986). A comparison of all identified terminal repeats shows that a core of total 14 bp grouped in two blocks (of 10 and 4 bp) is fully conserved (Fig. 4).

Genetic analysis has shown the importance of the nopaline right terminal sequence: deletion mutants of the left border are still oncogenic, while deletion mutants of the right border are unable to induce tumours (Holsters *et al.*, 1980; Joos *et al.*, 1983a; Shaw *et al.*, 1984; Wang *et al.*, 1984). Oncogenicity of a right-border deletion mutant of a nopaline Ti plasmid could be restored by a synthetic 25-bp right terminal repeat, in an orientation-dependent manner (Wang *et al.*, 1984). Thus the right 25-bp terminal sequence is sufficient to promote transfer and furthermore this sequence is acting in a polar fashion, from the right to the left border. The polar activity of the 25-bp sequence might in part explain why the right border is essential, since it is the right border which due to its location and orientation transfers the *onc* genes first. Any requirement for the left border might be obscured by presence of pseudoborders: within the T-region of the octopine Ti plasmids several sequences have been found which show partial homology to the 25-bp repeat (Barker *et al.*, 1983; Gielen *et al.*, 1984). Hepburn and White (1985) hypothesized that a pseudoborder can also be used as a right border. Use of pseudoborders might be the explanation for the aberrant T-DNAs that are sometimes observed (Thomashow *et al.*, 1980b; De Beuckeleer

A	SEQUENCE	SOURCE			REFERENCE
1.	TGGCAGGATATATTGTGGTGTAAAC	Nop	pTiT37	LB	1,2
2.	TGACAGGATATATTGGCGGGTAAAC	Nop	pTiT37	RB	1,2
3.	TGACAGGATATATTGGCGGGTAAAC	Nop	pTiC58	RB	1
4.	CGGCAGGATATATTCAATTGTAAAT	Oct	pTiA6	T_L LB(A)	3
5.	CGGCAGGATATATTCAATTGTAAAT	Oct	pTiAch5	T_L LB(A)	4,5
6.	TGGCAGGATATATACCGTTGTAATT	Oct	pTiAch5	T_L RB(B)	4,5
7.	CGGCAGGATATATTCAATTGTAAAC	Oct	pTi15955	T_L LB(A)	6
8.	TGGCAGGATATATACCGTTGTAATT	Oct	pTi15955	T_L RB(B)	6
9.	TGGCAGGATATATCGAGGTGTAAAA	Oct	pTi15955	T_R LB(C)	6
10.	TGGCAGGATATATGCGGTTGTAATT	Oct	pTi15955	T_R RB(D)	6

```
                  C GT      T
B   TGGCAGGATATATT X  TGTAAA      Consensus sequence   25-bp repeat
                  G AG      C

                  6 56      5
    7 8          6    8    7
    10      10    4 34  10  4
```

Fig. 4. A. Comparison of published 25-bp repeat sequences (modified after Slightom *et al.,* 1985). The sequences of only the upper strand are shown. The different Ti plasmids in which the given repeats are located are indicated. Nop and Oct designate nopaline and octopine; LB and RB, the left and right border, respectively. A, B, C and D designate the four octopine repeats according to Barker *et al.* (1983). References: 1. Zambryski *et al.,* 1982; 2. Yadav *et al.,* 1982; 3. Simpson *et al.,* 1982; 4. Holsters *et al.,* 1983; 5. Gielen *et al.,* 1984; 6. Barker *et al.,* 1983.

B. Consensus 25-bp sequence (modified after Slightom *et al.,* 1985). Numbers indicate how many times out of 10 cases a particular nucleotide is found

et al., 1981; De Greeve *et al.,* 1982; Hepburn *et al.,* 1983; Van Lijsebettens *et al.,* 1986; Simpson *et al.,* 1986; Chyi *et al.,* 1986).

The functional importance of the border sequence was further tested using binary vector systems consisting of: 1) a small plasmid containing a plant selectable marker such as an *onc* gene or a prokaryotic antibiotic resistance gene fused to a plant promoter, in *cis* with the border(s) to be tested and 2) a helper plasmid carrying the octopine *vir* region in *trans.* The following conclusions can be made:

1) Only DNA linked to at least one border is transferred. Depending on the assay system and the border used, one border is usually sufficient for transfer (An *et al.,* 1985; Rubin, 1986; Gardner and Knauf, 1986; Jen and Chilton, 1986 a, b; Horsch and Klee, 1986).

2) In general, the right border is more efficient for transfer than the left border (Rubin, 1986; Gardner and Knauf, 1986; Jen and Chilton, 1986 a, b).

This generalisation is contradicted only by the report by Horsch and Klee (1986; for possible explanation, see II. B. b).

3) For small binary vectors, the orientation of the border with respect to the selectable marker does not have any influence on efficiency (Rubin, 1986; Gardner and Knauf, 1986; Jen and Chilton, 1986a, b; Horsch and Klee, 1986).

Jen and Chilton (1986a, b) have mapped the T-DNA transferred to the plant cell by one-border-vectors which are about 50 kb in size: sequences to the left of the border up to and including the selectable genes *(onc)* form a part of the T-DNA, whereas sequences beyond the *onc* genes are not always found.

4) In binary vectors containing two borders not only the DNA located inbetween the borders but also outside the borders is transferred (inside and outside refer to the T-DNA in the Ti plasmid configuration; Rubin, 1986; Gardner and Knauf, 1986). Transfer of DNA located outside the borders occurs at much lower frequency for the Ti plasmid (Joos *et al.*, 1983b).

In principle, these observations can be explained by hypothesizing that transfer is a polar process and that small vectors are transferred more efficiently than large vectors. Thus, one border can efficiently direct transfer of a small vector, and the orientation of the border on a small plasmid does not make any difference. The entire Ti plasmid, due to its large size, is transferred only at a very low frequency. Therefore, on a Ti plasmid, the right border must be in the correct orientation such that the *onc* genes are transferred first.

b) Overdrive

Fine structure analysis of the octopine T_L right border and its neighbouring sequences revealed an element of 24 bp that enhances T-DNA transfer efficiency (Peralta and Ream, 1985; Peralta *et al.*, 1986). This element, called *overdrive,* is located about 17 bp to the right of both right border repeats (T_LB and T_RD) in the octopine-type plasmid pTiA6NC (Barker *et al.*, 1983) and consists of the following consensus sequence:

5′TAAPuTPyXCTGTPuTXTGTTTGTTTG3′
(Pu = purine; Py = pyrimidine; X = any base)

A similar sequence is situated next to the right border of the pRiA4 T_L-DNA and a less conserved sequence can be found to the right of the right border repeat of the nopaline pTiT37 plasmid (Peralta *et al.*, 1986). A synthetic *overdrive* is functional in either orientation, even if slightly moved from its normal position. It was hypothesized that *overdrive* greatly enhances interaction between the right border and the appropriate *vir* proteins, most probably at the T-DNA processing step: the *overdrive* is located outside the T-DNA and must thus act prior to excision or copy of the T-DNA. Experiments in *Agrobacterium* and *E. coli* indicate that *overdrive* is actually necessary for efficient T-DNA processing (border nicking assay, see III. A. b) in the wild type situation. If, however, the octopine *vir* genes are overproduced, the *overdrive* is not needed (E. Nester, pers. comm.).

The need for *overdrive* to constitute a fully functional right border contradicts the results of Wang *et al.* (1984). To reconcile these findings the authors (Peralta *et al.*, 1986) speculate that for efficient T-DNA transfer octopine *vir* proteins might need the enhancing function of the *overdrive* whereas the corresponding nopaline proteins might not. This might, for example, reflect different levels of expression of the respective octopine and nopaline *vir* genes. Different efficiencies in T-DNA processing by nopaline versus octopine *vir* regions have also been suggested by the experiments of Machida *et al.* (1986). On the other hand, if the *overdrive* acts at long distance, the context of the actual construction might be very important. In addition, differences in the oncogenic systems of the nopaline and octopine Ti plasmids, for example in the number of initially transformed plant cells required to develop a normal tumour, might also in part explain the discrepancy.

The presence of an *overdrive* sequence in proximity of the right 25-bp border repeat is at least in part responsible for the higher transfer efficiency of right border versus left border fragments when tested with octopine *vir*-region in *trans* (described in II. B. a). The right border fragments used in these studies all included the *overdrive*. The only exception (Horsch and Klee, 1986) where the left octopine border behaved like a right one might be explained by the following recent observations (H. Klee and W. Ream, personal communication): there are two sequences highly homologous to the *overdrive* in the vicinity of the left border A of the octopine Ti plasmid (Barker *et al.*, 1983). Contrary to all other published constructions used to assay for border efficiency the actual construction by Horsch and Klee (1986) contains both these sequences.

The T-DNA border repeat-*overdrive* system resembles site-specific inversion systems (Plasterk and van de Putte, 1985; Kahmann *et al.*, 1985; Johnson and Simon, 1985; Craig, 1985; Huber *et al.*, 1985) and promoter-enhancer systems (Serfling *et al.*, 1985). For example, specific flanking sequences are required in *cis* for efficient site-specific inversion. These elements are functionally independent of their position and orientation in respect to the inverted repeats.

c) Analysis of the T-DNA/Plant DNA Junctions

Several fragments containing T-DNA/plant DNA junction were cloned and sequenced.

End points of the T-DNA within the sequenced junction fragments are shown in Figure 5. The right T-DNA border seems to be recognised very precisely: it is located within or less than 10 bp apart from the right 25-bp repeat. The left border junction, however, seems to be more variable. In only three of the studied cases the T-DNA ends within the 25-bp repeat, whereas in the four other cases the T-DNA ends as far as 100 bp inside of the left terminal repeat. In one case, a considerable scrambling of the target — and the left T-DNA sequences was observed at the left border (Simpson *et al.*, 1982). One has to bear in mind that some of the studied tumour lines were kept in culture for many years, so that secondary rearrangements might have hap-

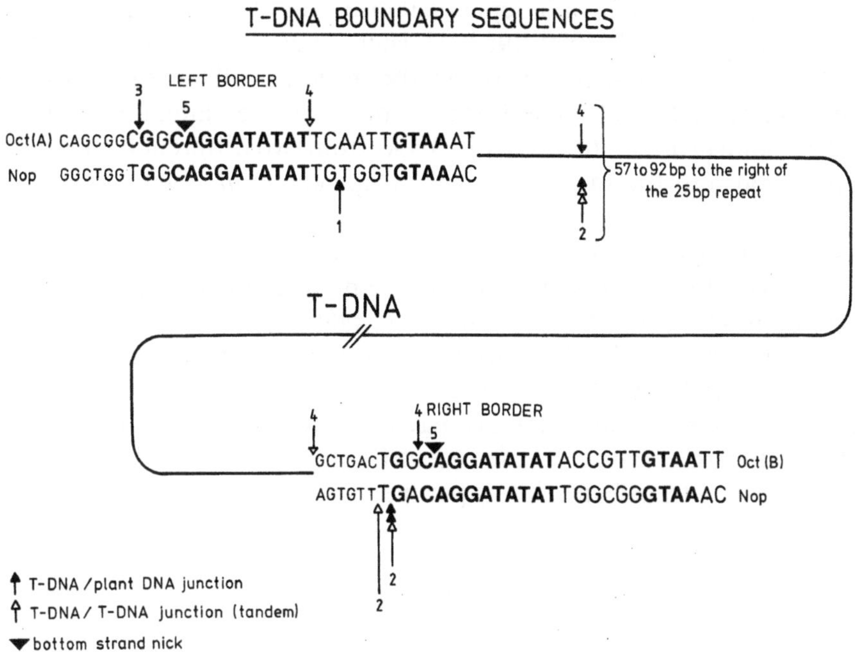

Fig. 5. Plant DNA/T-DNA boundary sequences. The T-DNA is shown from left to right; the sequence of the upper strand only is shown. A and B in parentheses designate the actual octopine border repeats according to Barker *et al.* (1983). Big letters depict the 25-bp border sequences, small letters the surrounding sequences, and bold letters the fully conserved sequences. Different types of arrows indicate the positions of T-DNA/plant DNA or T-DNA/T-DNA junctions of the octopine or nopaline T-DNA, respectively. The position of the bottom strand nick in the octopine Ti plasmid is also shown. References: 1. Yadav *et al.*, 1982; 2. Zambryski *et al.*, 1982; 3. Simpson *et al.*, 1982; 4. Holsters *et al.*, 1983; 5. Albright *et al.*, 1987

pened during this time. In some of the studied cases, the T-DNA was found to be arranged in tandem arrays. The findings that the junction sequence separating the tandem T-DNA copies can contain plant DNA sequences suggest that the amplification must have happened after an initial interaction of the T-DNA with plant DNA (Holsters *et al.*, 1983; Van Lijsebettens *et al.*, 1986; Peerbolte *et al.*, 1986 a, b). Thus, some of the analysed junctions might not represent primary integration events.

d) T-DNA Localisation and Structure Within the Plant Genome

Genetic analysis of regenerated plants from transformed tissue revealed random integration of T-DNA at the chromosome level (Wallroth *et al.*, 1986; Chyi *et al.*, 1986; Spielmann and Simpson, 1986). In five out of six cases the T-DNA was found to be located at more than one genetic locus. The number of genetic loci is less or equal to the number of physical loci reflecting silent T-DNA copies (due, for example, to methylation) or genetic linkage of several copies (Spielmann and Simpson, 1986). Multiple T-DNA copies can either be found at independent insertion sites or linked as tandem copies (Zambryski *et al.*, 1980, 1982; Holsters *et al.*, 1983; Van Lijsebettens *et al.*, 1986; Wallroth *et al.*, 1986; Peerbolte *et al.*, 1986 a, b) or inverted repeats centered either at the left or right border (Spielmann and Simpson, 1986; Peerbolte *et al.*, 1986 b). Two T-DNAs from two different bacteria may be transferred into the same plant cell (Depicker *et al.*, 1985; Petit *et al.*, 1986).

As more transformed lines are studied in detail, it is becoming clear that aberrant T-DNA structures are found in the plant genome at quite high frequency (up to 40%; Van Lijsebettens *et al.*, 1986; Spielmann and Simpson, 1986; Simpson *et al.*, 1986; Peerbolte *et al.*, 1986 a, b; for a summary consult Simpson *et al.*, 1986). Such aberrant T-DNAs might be due to use of pseudo-borders, defective transfer or rearrangements inside the plant cell either before or after integration. These recent findings should be kept in mind when performing transformation experiments.

III. T-DNA Processing

A. *The Search for Processing Intermediates*

Since *Agrobacterium* itself is thought not to enter the plant cell, there are in principle two possibilities for the T-DNA transfer:

1) The entire plasmid could be transported across the bacterial and plant cell membranes. Only the T-DNA would be stably integrated into the plant genome and the rest of the Ti plasmid would get lost.

2) In the bacterial cell, the T-DNA could be excised or copied from the Ti plasmid and this part only would be transferred to the plant cell.

In early experiments by de Framond *et al.* (1983) and Hoekema *et al.*, (1983 a) the T-DNA and the virulence functions were placed onto two different replicons within the *Agrobacterium* cell. The fact that this so-called binary system was fully oncogenic showed that the T-DNA needs not be physically linked to the rest of the Ti plasmid in order to be transferred to the plant cell. It was speculated (de Framond, 1983) that such a binary system might in fact mimic the natural transfer process, i. e. that the T-DNA might be excised or copied from the Ti plasmid within the bacterial cell prior to transfer. More recent results by Hoekema *et al.* (1984) which show that the T-DNA inserted into the chromosome of *Agrobacterium* is efficiently transferred to the plant cell, if the virulence genes are provided in *trans*, also supported this hypothesis.

Although the mechanism of T-DNA transfer is still unknown, recent experiments have started to shed light on the first steps of the transfer process, the T-DNA processing. Two kinds of approaches have been used to study the processing steps of the transfer: 1) genetic assays strongly selective for the expected type of T-DNA intermediate molecules, and 2) physical analysis of the T-DNA after induction.

The first approach has the advantage of being very sensitive but the molecules isolated may not represent major species of the processed T-DNA. The advantage of the second approach is that the entire population of molecules is studied. Both approaches attempt to identify T-DNA structures resulting from induction of *vir* genes.

a) Genetic Assays

Early attempts to isolate and select supercoiled T-DNA circles from vegetative *Agrobacterium* failed, possibly because of lack of *vir* gene induction (de Framond, 1983). Koukolíková-Nicola *et al.* (1985) designed two independent methods to specifically search for T-DNA structures induced by co-cultivation with plant cells. Both are based on the assumption that T-DNA intermediates are double-stranded DNA molecules. While the first method (cosmid rescue) detects both circular and tandem structures, the second method (plasmid rescue) allows detection of circular molecules only.

The cosmid approach is based on the very sensitive technique of λ *in vitro* packaging. The T-DNA region or the nopaline Ti plasmid C 58 was enlarged to 47 kb in size by inserting the cosmid p3030 and the transposon Tn 7. The resulting T-region, being the size of λ DNA, should be packageable if tandem arrays of T-DNA or circular (covalently or non-covalently closed) T-DNAs are formed upon co-cultivation. Bacteria containing this specially constructed Ti plasmid were cocultivated with plant protoplasts; total DNA was prepared and used for λ *in vitro* packaging. Packaged particles were transduced into *E. coli* and detected by selection for the resistance carried by the cosmid. Formation of packageable T-DNA stuctures is specifically induced by co-cultivation of bacteria with regenerating tobacco protoplasts. The rescued structures are comprised only of the T-DNA. Each molecule contains a single 25-bp terminal repeat, a hybrid of the left and right repeat sequence, which lies at the junction of the two ends of the T-DNA (see Fig. 6). The same T-DNA structures were also isolated by the second approach which is based on the Ti plasmid pGV3850 (Zambryski *et al.*, 1983). It has a short T-DNA region containing the left and right borders flanking the cloning vector pBR322. Total "induced" DNA was used for direct DNA transformation of *E. coli* selecting for the resistance carried by the T-DNA.

More recent experiments have shown that these T-DNA stuctures are formed within the bacteria upon induction by either a filtrate of plant cells or acetosyringone (Stachel *et al.*, 1985 b, 1986 a; Koukolíková-Nicola *et al.*, in preparation). Circular T-DNA molecules could be selected *in vivo* in *Agrobacterium* using a very strong selection (Machida *et al.*, 1986). These structures contain the hybrid 25-bp repeat, described above. Moreover, Alt-Moerbe *et al.* (1986) have shown that T-DNA circles are formed *in vivo* in *E. coli* strains carrying subclones of the *vir*-region (see III. B. a).

The fact that T-DNA structures could be recovered both by DNA packaging and DNA transformation was initially interpreted as an indication that they were circular molecules and that they were formed by site-specific recombination within the left and right 25-bp sequences (Koukolíková-

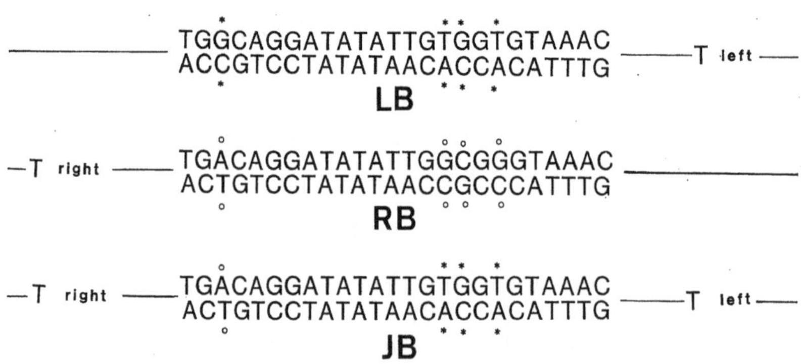

Fig. 6. Diagram of the left and right nopaline border sequences as well as the junction 25-bp sequences of the rescued nopaline cosmid clones (Koukolíková-Nicola *et al.*, 1985). Nucleotides specific for the left and right border sequences are designated by asteriscs or circles, respectively. In the junction sequence such nucleotides are also indicated. LB, RB and JB designate left, right and joined border sequences, respectively. Tleft and Tright designate the sequences at either the left or the right T-DNA end. Unlabelled lines represent Ti plasmid DNA outside the T-DNA

Nicola *et al.*, 1985). However, the fact that the junctions of T-DNA and plant DNA lie within or very near to the terminal repeats (see II. B. c) would rather suggest a linear form for the transferred molecules.

More recent experiments indicate that the T-DNA molecules identified by cosmid rescue are in fact linear (or non-covalently closed circles):

1) The biological relevance of the hybrid sequence subcloned from one of the rescued cosmid clones was tested using a binary vector system with the octopine *vir* region as helper. If the T-DNA transfer pathway would involve a covalently closed double-stranded circular molecule, the binary vector containing the hybrid border should be efficient in transfer. The hybrid sequence, however, was found to be inactive for the transfer or integration (Koukolíková-Nicola *et al.*, in preparation). Also the hybrid 25-bp sequence differs only by one base pair from the left 25-bp sequence (although the sequence to the left of the 25-bp sequence is different in the hybrid and the left border fragment; see Fig. 6) and no *overdrive* is known in the vicinity. Therefore, it might be expected to be as inefficient as the left border (Jen and Chilton, 1986 b) if no processing steps are bypassed.

2) The cosmid rescue approach (Koukolíková-Nicola *et al.*, 1985) has revealed that heating the total DNA prior to the packaging reaction drastically decreases the number of rescued cosmid clones. Reannealing subsequent to the heating step, however, restores the original number. By analogy to bacteriophage λ DNA, the packageable T-DNA structures were proposed to be linear molecules (or non-covalently closed circles) *in vivo*, assuming a circular structure *in vitro* (i. e. after DNA extraction) by annealing at cohesive ends. The melting temperature of the "sticky" ends has been experimentally estimated to be about 24 °C (Koukolíková-Nicola *et al.*, in preparation). The authors proposed that during processing staggered double-

stranded cuts are introduced within the core sequence in both right and left borders. This results in complementary single-stranded protruding ends, the cohesive ends.

Based on the different identified hybrid border sequences (see Fig. 6, for one example) the junction of the T-DNA ends must have occurred within the 12 (or 13) base pairs absolutely conserved between the left and the right border sequences (Koukolíková-Nicola *et al.*, 1985; Alt-Moerbe *et al.*, 1986; Machida *et al.*, 1986). As pointed out above (II. B. a), comparison of all sequenced terminal repeats revealed a core of 10 absolutely conserved base pairs (see Fig. 4). Thus, the length of the single-stranded ends might be as long as 10 bases. The experimentally determined melting temperature is consistent with this hypothesis.

The fact that the double-stranded T-DNA molecules are actually linear could explain the very low efficiency of recovery of T-DNA circles *in vivo* and by plasmid rescue (Machida *et al.*, 1986; Koukolíková-Nicola *et al.*, 1985). In addition, some mechanism must be postulated to prevent the T-DNA ends from ligating *in vivo* (see below).

b) Physical Assays

Double-stranded cleavage at each border was shown by Southern analysis: upon induction a double-stranded cut occurs within each border sequence, producing in addition to the original restriction fragments containing the borders new restriction fragments of predicted sizes. Up to 10 % of the borders appear to be cut upon induction of wild-type or *ros* strains carrying only the nopaline Ti plasmid (Stachel *et al.*, 1986 b; Koukolíková-Nicola *et al.*, in preparation; T. Close and C. Kado, personal communication). Other groups, studying the octopine Ti plasmid, reported that strains merodiploid for the octopine *vir* genes G, C, D and E (Albright *et al.*, 1987; Veluthambi *et al.*, 1987) or merodiploid for the *vir* G gene alone (Yanofsky *et al.*, 1986) were needed to reproducibly see about the same amount of cutting at the octopine as compared to the nopaline borders. This could be due to differences in efficiency of induction, in procedures of total DNA extraction and possibly in efficiency of nopaline and octopine *vir* functions.

The cleavage site at the left nopaline border was mapped to be within about 50 bp to the left or to the right of the 25-bp sequence at the most (Koukolíková-Nicola *et al.*, in preparation). Both "inside" and "outside" T-DNA fragments are present in about the same amount.

Unless the DNA preparation is protease digested, some of the fragments indicating border cleavage do not enter agarose gels. This suggests that a protein could be attached either covalently or very tightly (not removed by phenol extraction) to the right T-DNA end but not to the end of the fragment "outside" of the T-DNA. The opposite was shown for the left border. The result is a free T-DNA molecule with a protein(s) attached at its right end and the rest of the Ti plasmid carrying a protein(s) bound at the left border (Z. Koukolíková-Nicola *et al.*, unpublished).

In addition to double-stranded cleavage at each border, border nicking and production of single-stranded T-DNA have been detected. Albright *et al.* (1987) used a nuclease protection assay to look for single-stranded nicks within the border sequences. Strand-specific RNA probes were hybridised in solution to *Agrobacterium* DNA, and the hybrids were digested with RNase. *Agrobacterium* DNA protected the RNA according to the length of the hybrids: intact DNA protected the full length of the probe, while DNA containing breaks protected smaller regions. Specific nicks were detected within borders A and B of the octopine pTiA6 plasmid. Using a strain merodiploid for the octopine *vir* genes G, C, D and E about 50 % of the border fragments were nicked within the bottom strand (in relation to the conventional T-DNA map, see Figs. 2 and 5), whereas several-fold less contained a nick within the top strand (Albright *et al.*, 1987). These single-stranded nicks may correspond to the S 1-sensitive T-DNA border sites detected in the nopaline pGV3850 plasmid by Southern analysis (Stachel *et al.*, 1986b): digestion with S 1 nuclease of AS-induced DNA results in double-stranded cleavages at the borders, most probably due to S 1 cutting opposite to the single-stranded nicks. The sites of the bottom and top strand-specific nicks within the octopine borders were mapped (Albright *et al.*, 1987).

The bottom strand nick within each border is located at a unique site. The sites within the left and right (A and B) octopine border repeats are identical (see Fig. 5). The location of the top strand nick within 5 bp to the bottom strand one (Albright *et al.*, 1987) is somewhat inconsistent with the prediction from melting profiles of the cohesive ends (Koukolíková-Nicola *et al.*, in prep.). More precise mapping of the putative staggered nicks is essential to reconcile these findings.

Single-stranded linear T-DNA molecules of bottom strand polarity (T-strands) were found upon induction (Stachel *et al.*, 1986b). Southern hybridisation techniques enabled the authors to characterise large molecules encompassing the entire T-DNA of the nopaline pGV3850 plasmid. Undigested non-denatured *Agrobacterium* DNA was transferred to nitrocellulose. These conditions select for single-stranded or protein-bound DNA. Since the DNA preparations were digested with pronase, the DNA detected by this method was most probably single-stranded. Sensitivity of this DNA species to single-strand specific exonuclease VII is consistent with linearity and single-strandedness.

Single-stranded molecules comprised approximately 10% of the T-homologous DNA. Only T-DNA sequences were found as single-stranded molecules. In the case of octopine Ti plasmid similar techniques were used to demonstrate single-stranded DNA hybridising to the T_L-DNA (Veluthambi *et al.*, 1987).

Albright *et al.* (1987) provided evidence for single-stranded bottom strand DNA using the RNase protection protocol, by hybridising DNA without prior denaturation to the RNA probe. This method had a much higher resolution (to within a few nucleotides) than did the Southern hybridisations, but was limited by the probe length, which allowed analysis of DNA within only a few hundred base pairs of each border. Virtually all of the single-stranded DNA molecules ended at a location identical to that of the bottom strand nicks. Moreover, single-stranded DNA at each side of both borders A and B was detected. Thus it seems that not only the right border B but also the left border A generates single-stranded DNA, for example by serving as origin of replication (see Fig. 8). Furthermore, the authors suggested that the two additional borders (C and D) to the right of border B could have generated single-stranded DNA outside of this border. Stachel *et al.* (1986b) would not have detected single-stranded DNA outside the borders by their technique unless both ends were determinate.

c) Summary and Discussion

Several points should be considered in any discussion about intermediates of the T-DNA transfer process. First, several intermediate forms could be involved. Second, the quantity of a given T-DNA structure cannot be used as the only argument for its biological relevance. On the contrary, intermediates of a pathway are often present in low concentrations. Thus, accumulation of high amounts of a given structure might be an indication for a side-product. If the T-DNA transfer is blocked, as it is probably the case in free, non plant attached bacterial cells, side-products are likely to accumulate. The fact that these structures arise only upon induction does not qualify them as intermediates.

Genetic and physical data indicate that a variety of T-DNA structures are induced by *vir* gene expression: duplex Ti plasmid DNA containing nicks and double-stranded cuts at the borders, double-stranded linear T-DNA, single-stranded linear T-DNA, and T-DNA circles. It is difficult to estimate the relative frequencies of each of these structures because of the different strains, conditions for induction, and assays used. It is probably fair to say that the *in vivo* formation of covalently closed T-DNA circles is a rare event in *Agrobacterium* (Machida *et al.*, 1986), detected only if a strong genetic selection is used. The frequencies of the other types of cleavages and formation of T-strands are all within the same order of magnitude, and represent a significant fraction of T-DNA molecules after induction. At this point, all of the structures are potentially biologically relevant.

All analysed structures are interrelated because the *vir*D locus is involved in their formation (see III. B. b) and because each of them is derived from strand breaks at the 25-bp repeat sequence. Thus two criteria for biological relevance, i. e. involvement of *vir* products and border sequences in their formation, holds true for all the identified structures. *Vir*-mutants producing or accumulating only some of the physically or genetically detectable molecules might help to distinguish between intermediates and side-products.

At this point, however, it is difficult to reconcile all phenomena into a coherent picture of T-DNA processing. Double-stranded cleavages, nicks within the bottom as well as the top strand and linear double-stranded T-DNA molecules containing cohesive ends suggest that the T-DNA might be excised from the Ti plasmid by staggered nicks, generating a linear double-stranded T-DNA (Koukolíková-Nicola *et al.*, in prep.; see Fig. 7). On the other hand, border nicks within the bottom strand and T-strands

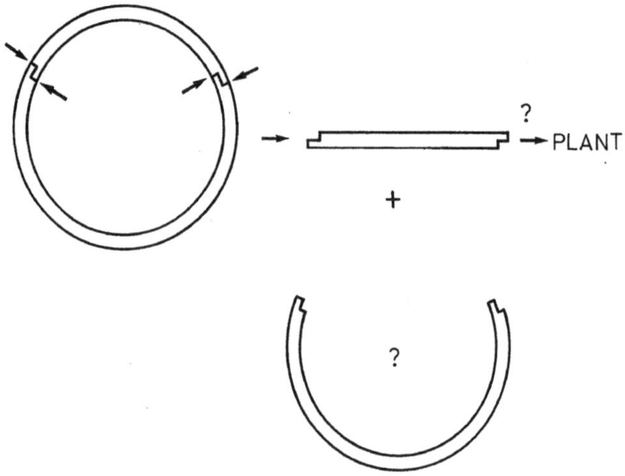

Fig. 7. *Excision model.* This model has been proposed by Koukolíková-Nicola *et al.* (in preparation) and parts of it have been suggested by Jen and Chilton (1986a, b): Upon induction *vir*D proteins introduce staggered doublestranded cuts at specific sites within both the left and the right 25-bp sequences and thus excise the T-DNA from the Ti plasmid. The double-stranded linear T-DNA might be transported as such into the plant cell and the rest of the Ti plasmid might eventually get degraded

suggest that border nicking might be followed by single-stranded DNA production (Stachel *et al.*, 1986b; Albright *et al.*, 1987; see Fig. 8).

Excision of the T-DNA from the Ti plasmid would have a precedent in conservative transposition or site-specific recombination (or inversion) in bacteria. Recently, Bender and Kleckner (1986) have provided strong genetic evidence for the non-replicative "cut-and-paste" model proposed for transposition of the transposon Tn10, in which the transposon is excised by double-stranded cuts at its ends and the excised duplex segment is integrated into the target site (for example Grindley and Reed, 1985). It has been demonstrated that site-specific recombinases introduce staggered double-stranded cuts (for example Reed and Grindley, 1981; Craig and Nash, 1983; Iida and Hiestand-Nauer, 1986). Moreover, one case is known in prokaryotes, where upon a developmental stimulus, a piece of DNA is excised by site-specific recombination (Haselkorn *et al.*, 1986).

An excisive mechanism must provide a means of retaining T-DNA in the population of bacteria. One possibility is that T-DNA excision occurs only when multiple copies of T-DNA are present. A precedent for this type of mechanism exists in Tn10/IS10 (Roberts *et al.*, 1985). IS10 transposes preferentially when it is hemimethylated, as it would be when newly replicated. The authors suggested that activation of only one hemimethylated element might be particularly advantageous for elements like IS10 that move by non-replicative excision without direct religation of the broken ends of the donor molecule. This would provide a way to keep an intact copy of the donor

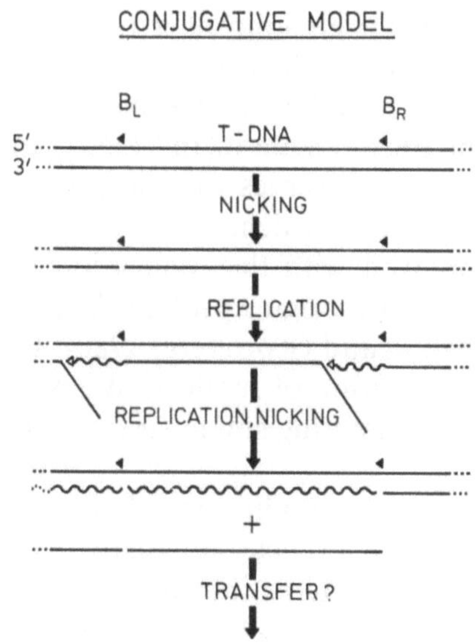

Fig. 8. *Conjugative model.* This model was independently proposed by Stachel and Zambryski (1986) and Albright *et al.* (1987) based on their respective findings: upon induction single-stranded endonucleolytic cuts within the bottom strand of the left and right T-DNA border sequences are produced. These cleavages provide free 3′OH groups from which DNA synthesis can be primed. Using the top strand as template, DNA synthesis initiates at one right border and proceeds to the next left one displacing the bottom strand 5′ to 3′. This generates a linear, single-stranded molecule which is transferred out of the bacterial cell. Alternatively, strand-displacement by a 5′→3′ single-stranded helicase or degradation of the top strand would also free the single-stranded bottom strand

molecule. Another possibility is that conjugation of the Ti plasmid, induced by opines synthesized by transformed plant cells, provides a mechanism of replenishing a bacterial population with the Ti plasmid. Still another possibility is that the frequency of double-stranded cuts is maintained at less than 100% efficiency to prevent loss of the T-DNA, and potentially the Ti plasmid, from the bacterial population.

There is no obvious role for T-strands in a strictly excisive mechanism. One could imagine that T-strand formation might be a side-product initiated by the site-specific action of the *vir*D proteins (see III. B. a). Following site-specific nicking the single-strand might, for example, be displaced by for example the action of helicase and of single-strand binding (SSB) proteins which are normally present in the bacterial cell or which might be overexpressed from the transfer-constitutive pGV3850 plasmid (Golub and Low, 1986).

Single-stranded DNA replication primed by border nicks has precedents in processing of DNA for conjugal transfer between bacteria (Willets and Wilkins, 1984) as well as in replication of single-stranded bacteriophages (Kornberg, 1980). Albright *et al.* (1987) specifically suggested that each border nick might serve as an origin of replication, which implies that the newly synthesized T-strand originating at a right border would proceed through the left border, which itself would have primed another round of synthesis (see Fig. 8). To release the newly synthesized strand, some mechanism must maintain the nicked state of the DNA. In the single-stranded DNA phage ΦX174, protein A serves this purpose. After nicking at the origin of replication it remains covalently attached to the 5′ end of the displaced strand and complexed with the replication machinery. When the replication fork returns to the origin, protein A introduces another nick and releases the displaced strand (Kornberg, 1980). In the T-DNA system, however, no direct demonstration of replication has been made until now, and other mechanisms of generating single-stranded DNA cannot be ruled out.

There is no obvious role for double-stranded linear T-DNA molecules excised from the Ti plasmid in a strictly replicative mechanism. One might, however, imagine that *vir*D proteins introduce a specific top strand nick only as a side-product. Alternatively, the borders might preferentially get cleaved at the top strand during DNA preparation due to mechanical or endonucleolytic degradation opposite or in the vicinity of the bottom strand nick.

One way to combine these two viewpoints into a more unified model is to speculate that excision of the T-DNA from the Ti plasmid is followed by single-stranded DNA synthesis from a T-DNA rolling circle. However, to account for all data, one would have to hypothesize that the rest of the Ti plasmid (after T-DNA excision) can also generate single-stranded DNA.

The polar action of the 25-bp repeat (Wang *et al.*, 1984; Peralta and Ream, 1985) is consistent with either an excisive or replicative mechanism. The asymmetric sequence of the 25-bp repeat implies that any protein interacting with that sequence does so in an oriented fashion. In the excisive model, depending on the orientation of the right border to the T-DNA, this interaction could determine whether the "outside" or the "inside" of the T-DNA is transferred. For example, protein(s) attached at the right end of the excised T-DNA (see III. A. b) might act as "pilot" protein(s) to promote transfer. In the replicative model, unidirectional replication of single-stranded DNA would provide polarity.

The ability of a single-border to direct transfer of small binary vectors is consistent with both the replicative and excisive mechanism. In the replicative model, a single border would serve as an origin of replication for the entire replicon. In the excision model, staggered nicks at the single border would generate a double-stranded T-DNA identical to that derived from the Ti plasmid.

The T-DNA circle formation *in vivo* in both *E. coli* (Alt-Moerbe *et al.*, 1986; J. Alt-Moerbe and J. Schröder, pers. comm.) and *Agrobacterium* (Machida *et al.*, 1986) might result from homologous recombination induced by the *vir*D promoted site-specific nicks or might result from religation of the ends of the excised double-stranded T-DNA molecules or from second-strand synthesis from the T-strands.

All data are consistent in their implication that linear T-DNA is transferred to the plant cell, whether single- or double-stranded. This is in agreement with the location of the ends of the T-DNA in tumour lines the DNA of which has been sequenced (see Fig. 5). However, only a few T-DNA/plant DNA junctions have been sequenced so far and, as was discussed earlier (II. B. c), they might not be representative of primary integration events. Analysis of junctions in recently established tumour lines should yield more information.

The recent demonstration that single-stranded DNA can be used for DNA transformation in animal cells makes transfer of single-stranded T-DNA to the plant cell a reasonable possibility (Rauth *et al.*, 1986). On the other hand, it was shown in the case of direct DNA-mediated plant transformation that linear DNA transforms at higher efficiency than circular DNA (Shillito *et al.*, 1986). Thus linear double-stranded T-DNA might constitute a good substrate for integration.

Identification of the actual molecule which is transferred to the plant cell may await isolation of such molecules directly from plant cells. Experiments attempting to do so are technically extremely difficult and so far have yielded no conclusive results (Virts and Gelvin, 1985).

Agroinfection using maize streak virus (MSV) DNA also failed to give an indication about the T-DNA molecule actually transferred to the plant cell (Grimsley *et al.*, 1987). MSV has a single-stranded circular genome. Depending on the orientation of the cloned tandem genome of MSV with respect to T-DNA borders, either the + or the − strand of the MSV DNA would initially enter the plant cell if T-strands as such would be transferred. However, both orientations are equally infectious indicating either that the transferred molecule is not single-stranded or that it does not matter for the escape of the virus which of the two strands is initially entering the plant cell. In fact, if the virus escapes via homologous recombination, second strand synthesis might be necessary (see also Grimsley and Bisaro, this volume).

B. Proteins Involved in T-DNA Processing

The general organisation and regulation or the *vir*-region was discussed in chapters II. A. b) and c). Here we discuss the *vir* loci whose products were shown or suggested to play a role in the T-DNA transfer.

a) Vir D Locus

Several authors, using different assay systems, have shown that this locus is involved in T-DNA processing. Analysis of a number of *vir* mutant strains of *Agrobacterium* demonstrated that the 5′ region of the *vir*D locus is essential for *in vivo* T-DNA circle formation (Yamamoto *et al.*, 1987), site-specific single-stranded nicks within the 25-bp terminal sequences (Yanofsky *et al.*, 1986) and double-stranded cuts at the octopine and nopaline T-DNA borders, respectively (Veluthambi *et al.*, 1987; T. Close and C. Kado, personal communication).

Alt-Moerbe *et al.* (1986) have shown that a 3.6 kb fragment of the *vir*C/D locus of the nopaline plasmid pTiC58 is sufficient for *in vivo* T-DNA circle formation in *E. coli.* Sequence analysis of one junction fragment of the T-DNA circles revealed exactly the same hybrid sequence as found in the rescued T-DNA molecules from *Agrobacterium* (Koukolíková-Nicola *et al.,* 1985; Machida *et al.,* 1986). The smallest fully active *vir* subclone codes for three polypeptides of 15, 56 and 43 kd; the last being probably a degradation product of the 56 kd protein (J. Alt-Moerbe and J. Schröder, personal communication). The two proteins of 15 and 56 kd correspond to the 16 and 56 kd proteins encoded by the first two open reading frames of the octopine *vir*D locus, called *vir*D1 and D2 (Yanofsky *et al.,* 1986; Jayaswal *et al.,* in preparation). These two octopine proteins have been shown in *E. coli* to be sufficient to direct site-specific nicking at the borders at exactly the same site as was found in *Agrobacterium* (Yanofsky *et al.,* 1986) In *E. coli,* the same two proteins were also responsible for double-stranded cleavage at the borders (Yanofsky *et al.,* 1986; Jayaswal *et al.,* in preparation) and formation of single-stranded DNA hybridising to the octopine T_L-DNA (Jayaswal *et al.,* in preparation). Both proteins are necessary, although truncated products of the *vir*D2 protein seem still to be active. The octopine 16 kd protein shares a region of homology with a previously characterised DNA-binding domain (Drummond *et al.,* 1986) suggesting that it may also bind DNA (Yanofsky *et al.,* 1986).

In summary, the protein products of the 5′ end of the *vir*D locus have been shown by various groups to be essential for the following reactions: T-DNA circle formation in *E. coli* and *Agrobacterium,* and production of single-stranded and double-stranded cuts within the 25-bp border sequences both in *Agrobacterium* and *E. coli.* At present it is not possible to decide whether all structures promoted by the *vir*D proteins are biologically relevant for the transfer process. They all involve the T-DNA border sequences and are specifically induced by the action of the *vir*D products. They might constitute different intermediate structures or side-products of the T-DNA processing pathway or might even be the result of *in vivo* or *in vitro* artefacts induced by the site-specific action of the *vir*D proteins (for discussion, see III. A. c). Assessment of the real function of *vir*D proteins might have to await *in vitro* experiments.

b) Vir C Locus

Several lines of indirect evidence have accumulated that this locus might be involved in the efficiency of T-DNA transfer.

Mutations in the *vir*C locus result in an alteration of host range: on some plants *vir*C mutants show an attenuated or avirulent phenotype, but are virulent on other plant hosts (Hille *et al.,* 1984; Hooykaas *et al.,* 1984; Yanofsky *et al.,* 1985; Close *et al.,* submitted). In addition, the *vir*C locus is involved in eliciting a hypersensitive response on certain *Vitis* cultivars: whereas the wild-type strain induces a hypersensitive response and is avirulent, *vir*C mutants no longer induce hypersensitivity and become virulent

(Yanofsky *et al.*, 1985). This indicates that the *vir*C locus may be necessary for efficient T-DNA transfer but is not essential for tumourigenesis.

Molecular and genetic analyses of the octopine and nopaline *vir*C loci have been performed by Yanofsky and Nester (1986) and Close *et al.* (submitted), respectively. In both cases, nucleotide sequence analysis revealed two open reading frames, designated *vir*C1 and *vir*C2, encoding proteins of about 25.7 kd (25.5 for nopaline), and 22.7 kd (22.1 for nopaline). *Vir*C1 proteins of pTiC58 and pTiA6NC are identical at 87.4 % of amino acid residues. In *vir*C2, there are two blocks of higher homology (84.6 % and 78.5 %) separated by a 41 residue segment of much lower homology. In both cases, only two nucleotides separate the stop codon of *vir*C1 from the start codon of *vir*C2, indicating that these genes may be translationally coupled (Yanofsky and Nester, 1986). Mutations within either of the two open reading frames create the characteristic *vir*C phenotype. Although this demonstrates a role for *vir*C2 in virulence, the role of *vir*C1 is more difficult to establish since mutations in this gene may be polar.

The results of Horsch *et al.* (1986) can be interpreted in a similar way as above. They compared the transfer efficiency of different markers located within the T-DNA of a binary vector and of a Ti plasmid using a set of *vir* mutants. The behaviour of *vir*C mutants was striking: the binary vector markers (*nos* and plant kanamycin-resistance) were transferred at much higher frequency than the Ti plasmid markers (*ocs* and *onc* genes). Due to smaller size and/or higher copy number of the binary vector versus the Ti plasmid, the markers located on the binary vector might be transferred more efficiently and thus be less affected by a defect in the *vir*C product(s). Thus, *vir*C may function in increasing the overall efficiency of the transfer process.

Yanofsky *et al.* (1986) and Yamamoto *et al.* (1987) have shown that *vir*C mutants were no less efficient in site-specific nicking and T-DNA circle formation than wild-type, thus indicating that *vir*C products act at a later stage of the transfer process.

Clearly, more data is needed before a precise function for *vir*C can be assigned.

c) Vir E Locus

There is a lot of controversy about the function of this locus. This is partly due to the fact that different experimentators use different host plants. It is well documented that *vir*E mutants are so-called host range mutants; i. e. they are avirulent on some species, whereas they induce attenuated tumours on others (Stachel and Nester, 1986, Horsch *et al.*, 1986; Hirooka and Kado, 1986; K. McBride, personal communication). Therefore, depending on the host plant used these mutants might be scored as virulent or avirulent. Two different assays have been used to investigate the function of this locus in more detail: opine synthesis and kanamycin-resistance in infected leaf discs (Horsch *et al.*, 1986) and agroinfection using potato tuber spindle viroid (PSTV) as infectious agent (Gardner and Knauf, 1986; see also Grimsley and Bisaro, this volume).

The first approach has clearly demonstrated that the *vir*E mutants are able to transfer and stably integrate T-DNA into the genome of *Petunia* although at an extremely low frequency (Horsch *et al.,* 1986). Gardner and Knauf (1986) have shown that although the *vir*E mutants were completely avirulent on tomato, they gave rise to systemic viroid infection. Similar results were obtained by Hille *et al.* (1986) using cauliflower mosaic virus (CaMV) for agroinfection. Since in theory there is no need for prior integration of the T-DNA carrying the multimeric complementary DNA copies in order to get infection, Gardner and Knauf (1986) interpreted their results as indication that the *vir*E gene product(s) are involved in the integration of the T-DNA into the plant host chromosome. However, the fact that *vir* mutants do induce attenuated tumours on other host plants together with some considerations about the agroinfection approach, invite to caution. Agroinfection is a very sensitive assay. In principle, only one infectious molecule has to be transferred to the plant to cause systemic infection. Although the *vir*E mutants are able to give viroid symptoms, the efficiency of T-DNA transfer might not be sufficient to form a tumour. Thus there seems to be no need at the moment to postulate that the *vir*E locus encodes an integrase to explain the results.

In any case, a tumour produced by an integrase-mutant would be predicted to contain only rearranged T-DNA. To our knowledge, for *vir*E mutants there is no data available. On the other hand, the T-DNA borders might function only in the bacteria to define and generate the ends of the T-DNA. This means that a *vir*-encoded "integrase" function might not be necessary at all, since plant cells have been shown to assimilate foreign DNA by means other than *Agrobacterium* (Krens *et al.,* 1982; Paszkowski *et al.,* 1984; Paszkowski and Potrykus, this volume; Shillito *et al.,* 1985; Fromm *et al.,* 1986). Thus, the T-DNA might get integrated into the genome using the recombination and repair enzymes of the plant cells. In fact, the T-DNA structures induced by *Agrobacterium*-mediated transformation mapped by Peerbolte *et al.* (1986 a, b) are similar to the integration patterns which are found in the case of direct gene-mediated transformation (Czernilofsky *et al.,* 1986), although the T-DNA might still be better protected against degradation during transfer to the nucleus.

The nopaline *vir*E locus is 2.0 kb long and encodes at least one protein of 69 kd (Hirooka and Kado, 1986); the octopine *vir*E locus was shown to encode two polypeptides of 7 and 66 kd (Winans *et al.,* 1987). Comparative DNA sequence analysis showed that *vir*E of pTiC58 has strong homology to *vir*E of pTiA6NC. Interestingly, mutations truncating the nopaline *vir*E protein narrow the host range of the mutated *A. tumefaciens* strains and lower the frequency of infection, depending on the size of the protein (Hirooka and Kado, 1986). This protein therefore might have several domains that function in increasing the efficiency of the infection process.

Any hypothesis for the function of the *vir*E locus must also accommodate the intriguing results by Otten *et al.* (1984) who showed that *vir*E mutants can successfully transfer their T-DNA to plants if they are coinoculated with a wild-type strain. This unusual form of interbacterial complementation prompted the authors to suggest that the *vir*E gene product is secreted.

d) VirF Locus

A 5 kb region of pTiB6S3 containing the *vir*F locus has been suggested to be implicated in efficiency of T-DNA transfer (Otten *et al.,* 1985). This interpretation is based on the assumption that the level of opine synthesis reflects the number of the initially transformed cells. This function was not found in the nopaline Ti plasmid pTiC58.

IV. Conclusions

In the past few years our knowledge about the early events of transformation has considerably increased: the virulence region has been characterised in more details, plant substances inducing the *vir* gene expression have been isolated and we are in the process of unravelling the T-DNA processing.

However, many questions are still unanswered. What is the function of all the proteins encoded by the virulence region? The details of the T-DNA processing have to be worked out. We now have the tools in our hands to address these questions directly. Still nothing is known about the actual transfer of the T-DNA: bacterial/plant cell contacts leading to transformation have not been adequately described, the way by which the T-DNA moves into the plant cell and is targeted into the nucleus is still a mystery. Thus, this unique system of DNA transfer from a prokaryotic to an eukaryotic organism will continue to occupy and fascinate us.

Acknowledgements

The authors thank all colleagues who provided information prior to publication, Vera Knesel for tireless preparation of this manuscript and T. Hohn for his patience.

V. References

Akiyoshi, D. E., Morris, R. O., Hinz, R., Mischke, B. S., Kosuge, T., Garfinkel, D., Gordon, M. P., Nester, E. W., 1983: Cytokinin auxin balance in crown gall tumors is regulated by specific loci in the T-DNA. Proc. Natl. Acad. Sci., U.S.A. **80,** 407—411.

Albright, L. M., Yanofsky, M. F., Leroux, B., Ma, D., Nester, E. W., 1987: Processing of the T-DNA of *Agrobacterium tumefaciens* generates border nicks and linear, single-stranded DNA. J. Bacteriol. **169,** 1046—1055.

Alt-Moerbe, J., Rak, B., Schröder, J., 1986: A 3.6 kbp segment from the *vir* region of Ti plasmids contains genes responsible for border sequence-directed production of T-region circles in *E. coli.* EMBO J. **5,** 1129—1135.

An, G., Watson, B. D., Stachel, S., Gordon, M. P., Nester, E. W., 1985: New cloning vehicles for transformation of higher plants. EMBO J. **4,** 277—284.

Ashby, A. M., Watson, M. D., Shaw, C. H., FEMS Micro. Letts., in press.

Barker, R. F., Idler, K. B., Thompson, D. V., Kemp, J. D., 1983: Nucleotide sequence of the T-DNA region from the *Agrobacterium tumefaciens* octopine Ti plasmid pTi15955. Plant Mol. Biol. **2,** 335—350.

Barry, G., Rogers, S. G., Fraley, R. T., Brand, L., 1984: Identification of a cloned cytokinin biosynthetic gene. Proc. Natl. Acad. Sci., U.S.A. **81,** 4776—4780.

Bender, J., Kleckner, N., 1986: Genetic evidence that Tn10 transposes by a nonreplicative mechanism. Cell **45,** 801—815.

Blundy, K. S., White, J., Firmin, J. L., Hepburn, A. G., 1986: Characterisation of the T-region of the SAP-type Ti plasmid pTiAT181: identification of a gene involved in SAP synthesis. Mol. Gen. Genet. **202,** 62—67.

Bolton, G. W., Nester, E. W., Gordon, M. P., 1986: Plant phenolic compounds induce expression of the *Agrobacterium tumefaciens* loci needed for virulence. Science **232,** 983—985.

Bomhoff, G., Klapwijk, P. M., Kester, H. C. M., Schilperoort, R. A., Hernalsteens, J. P., Schell, J., 1976: Octopine and nopaline synthesis and breakdown is genetically controlled by a plasmid of *Agrobacterium tumefaciens.* Mol. Gen. Genet. **145,** 177—181.

Braun, A. C., Mandle, R. J., 1948: Studies on the inactivation of the tumor-inducing principle in crown gall. Growth **12,** 255—269.

Braun, A. C., White, P. R., 1943: Bacteriological sterility of tissues derived from secondary crown gall tumors. Phytopathol. **33,** 85—100.

Buchmann, I., Marner, F. J., Schröder, G., Waffenschmidt, S., Schröder, J., 1985: Tumor genes in plants: T-DNA encoded cytokinin biosynthesis. EMBO J. **4,** 853—859.

Chilton, M.-D., Drummond, M. H., Merlo, D. J., Sciaky, D., Montoya, A. L., Gordon, M. P., Nester, E. W., 1977: Stable incorporation of plasmid DNA into higher plant cells: the molecular basis of crown gall tumorigenesis. Cell **11,** 263—271.

Chilton, M.-D., Saiki, R. K., Yadav, N., Gordon, M. P., Quetier, F., 1980: T-DNA from *Agrobacterium* Ti plasmid is in the nuclear DNA fraction of crown gall tumor cells. Proc. Natl. Acad. Sci., U.S.A. **77,** 4060—4064.

Chilton, W. S., Tempé, J., Matzke, M., Chilton, M.-D., 1984: Succinamopine: a new crown gall opine. J. Bacteriol. **157,** 357—362.

Christou, P., Platt, S. G., Ackerman, M. C., 1986: Opine synthesis in wild-type plant tissue. Plant Physiol. **82,** 218—221.

Chyi, J.-S., Jorgensen, R. A., Goldstein, D., Tanksley, S. D., Loaiza-Figueroa, F., 1986: Localization and stability of *Agrobacterium*-mediated T-DNA insertions in the *Lycopersicon* genome. Mol. Gen. Genet. **204,** 64—69.

Close, T. J., Tait, R. C., Kado, C. I., 1985: Regulation of Ti plasmid virulence genes by a chromosomal locus of *Agrobacterium tumefaciens.* J. Bacteriol. **164,** 774—781.

Close, T. J., Tait, R. C., Hirooka, T., Kim, L., Kado, C. I.; Molecular characterisation of the *vir*C genes of the Ti plasmid. Submitted.

Craig, N. L., Nash, H. A., 1983: The mechanism of phage λ site-specific recombination: site-specific breakage of DNA by Int topoisomerase. Cell **35,** 795—803.

Craig, N. L., 1985: Site-specific inversion: Enhancers, recombination proteins, and mechanism. Cell **41,** 649—650.

Czernilofsky, A. P., Hain, R., Herrera-Estrella, L., Lörz, H., Goyvaerts, E., Baker, B. J., Schell, J., 1986: Fate of selectable marker DNA integrated into the genome of *Nicotiana tabacum.* DNA **5,** 101—113.

Das, A., Stachel, S., Ebert, P., Allenza, P., Montoya, A., Nester, E., 1986: Promoters of *Agrobacterium tumefaciens* Ti-plasmid virulence genes. Nucl. Acids Res. **14,** 1355—1364.

De Beuckeleer, M., Lemmers, M., De Vos, G., Willmitzer, L., Van Montagu, M., Schell, J., 1981: Further insight on the transferred-DNA of octopine crown gall. Mol. Gen. Genet. **183,** 283—288.

De Block, M., Schell, J., Van Montagu, M., 1985: Chloroplast transformation by *Agrobacterium tumefaciens.* EMBO J. **4,** 1367—1372.

De Cleene, M., De Ley, J., 1976: The host range of crown gall. Bot. Rev. **42,** 389—466.

De Cleene, M., 1985: The susceptibility of monocotyledons to *Agrobacterium tumefaciens.* Phytopath. Z. **113,** 81—89.

de Framond, A. J., Barton, K. A., Chilton, M.-D., 1983: Mini-Ti: a new vector strategy for plant genetic engineering. Bio/technology **1,** 262—269.

de Framond, A. J., 1983: Construction de plasmides dérivés du plasmide Ti à nopaline pTiT37 d'*Agrobactérium tumefaciens* en vue de leur utilisation comme vecteurs de génie génétique chez les plantes. Ph. D. thesis, Inst. Nat. Agr., Paris - Grignon.

De Greve, H., Decraemer, H., Seurinck, J., Van Montagu, M., Schell, J., 1981: The functional organization of the octopine *Agrobacterium tumefaciens* plasmid pTiB6S3. Plasmid **6,** 235—248.

De Greve, H., Dhaese, P., Seurinck, J., Lemmers, M., Van Montagu, M., Schell, J., 1982: Nucleotide sequence and transcript map of the *Agrobacterium tumefaciens* Ti plasmid-encoded octopine synthase gene. J. Mol. Appl. Genet. **1,** 499—512.

Depicker, A., Stachel, S., Dhaese, P., Zambryski, P., Goodman, H. M., 1982: Nopaline synthase: transcript mapping and DNA sequence. J. Mol. Genet. **1,** 561—574.

Depicker, A., Herman, L., Jacobs, A., Schell, J., Van Montagu, M., 1985: Frequencies of simultaneous transformation with different T-DNAs and their relevance to the *Agrobacterium*/plant cell interaction. Mol. Gen. Genet. **201,** 477—484.

Douglas, C. J., Halperin, W., Nester, E. W., 1982: *Agrobacterium tumefaciens* mutants affected in attachment to plant cells. J. Bacteriol. **152,** 1265—1275.

Douglas, C. J., Staneloni, R. J., Rubin, R. A., Nester, E. W., 1985: Identification and genetic analysis of an *Agrobacterium tumefaciens* chromosomal virulence region. J. Bacteriol. **161,** 850—860.

Downie, J. A., Johnston, A. W. B., 1986: Nodulation of legumes by *Rhizobium:* The recognized root? Cell **47,** 153—154.

Drummond, M., Whitty, P., Wootton, J., 1986: Sequence and domain relationships of *ntr*C and *nif*A from *Klebsiella pneumoniae:* Homologies to other regulatory proteins. EMBO J. **5,** 441—447.

Dylan, T., Ielpi, L., Stanfield, S., Kashyap, L., Douglas, C., Yanofsky, M., Nester, E., Helinski, D. R., Ditta, G., 1986: *Rhizobium meliloti* genes required for nodule development are related to chromosomal virulence genes in *Agrobacterium tumefaciens.* Proc. Natl. Acad. Sci., U.S.A. **83,** 4403—4407.

Ellis, J. G., Murphy, P. J., 1981: Four new opines from crown gall tumours — their detection and properties. Mol. Gen. Genet. **181,** 36—43.

Ellis, J. G., Kerr, A., Petit, A., Tempé, J., 1982: Conjugal transfer of nopaline and agropine Ti-plasmids. The role of agrocinopines. Mol. Gen. Genet. **186,** 269—274.

Engler, G., Depicker, A., Maenhaut, R., Villarroel-Mandiola, R., Van Montagu, M.,

Schell, J., 1981: Physical mapping of DNA base sequence homologies between an octopine and nopaline Ti-plasmid of *Agrobacterium tumefaciens*. J. Mol. Biol. **152,** 183—208.

Firmin, J. L., Wilson, K. E., Rossen, L., Johnston, A. W. B., 1986: Flavonoid activation of nodulation genes in *Rhizobium* reversed by other compounds present in plants. Nature **324,** 90—92.

Follin, A., Inzé, D., Budar, F., Genetello, C., Van Montagu, M., Schell, J., 1985: Genetic evidence that the tryptophan 2-mono-oxygenase gene of *Pseudomonas savastanoi* is functionally equivalent to one of the T-DNA genes involved in plant tumour formation by *Agrobacterium tumefaciens*. Mol. Gen. Genet. **201,** 178—185.

Fromm, M. E., Taylor, L. P., Walbot, V., 1986: Stable transformation of maize after gene transfer by electroporation. Nature **319,** 791—793.

Gallie, D. R., Hagiya, M., Kado, C. I., 1985: Analysis of *Agrobacterium tumefaciens* plasmid pTiC58 replication region with a novel high-copy-number derivative. J. Bacteriol. **161,** 1034—1041.

Gardner, R. C., Knauf, V. C., 1986: Transfer of *Agrobacterium* DNA to plants requires a T-DNA border but not the *vir*E locus. Science **231,** 725—727.

Garfinkel, D. J., Nester, E. W., 1980: *Agrobacterium tumefaciens* mutants affected in crown gall tumorigenesis and octopine catabolism. J. Bacteriol. **144,** 732—743.

Genetello, C., Van Larebeke, N., Holsters, M., De Picker, A., Van Montagu, M., Schell, J., 1977: The Ti-plasmids of *Agrobacterium* as conjugative plasmids. Nature **265,** 561—563.

Gheysen, G., Dhaese, P., Van Montagu, M., Schell, J., 1985: DNA flux across genetic barriers: the crown gall phenomenon. In: Hohn, B., Dennis, E. S. (eds.), Genetic Flux in Plants (Plant Gene Research, Vol. 2), pp.11—47. Wien - New York: Springer.

Gielen, J., De Beuckeleer, M., Seurinck, J., Deboeck, F., De Greve, H., Lemmers, M., Van Montagu, M., Schell, J., 1984: The complete nucleotide sequence of the T_L-DNA of the *Agrobacterium tumefaciens* plasmid pTiAch5. EMBO J. **3,** 835—846,

Glogowski, W., Galsky, A. G., 1978: *Agrobacterium tumefaciens* site attachment as a necessary prerequisite for crown gall tumor formation on potato discs. Plant Physiol. **61,** 1031—1033.

Goldmann, A., Tempé, J. Morel, G., 1968: Quelques particularités de diverses souches d'*Agrobacterium tumefaciens*. C. R. Soc. Biol. (Paris) **162,** 623—631.

Golub, E. I., Low, K. B., 1986: Derepression of single-stranded DNA-binding protein genes on plasmids derepressed for conjugation, and complementation of an *E. coli ssb*$^-$ mutation by these genes. Mol. Gen. Genet. **204,** 410—416.

Graves, A. C. F., Goldman, S. L., 1986: The transformation of *Zea mays* seedlings with *Agrobacterium tumefaciens*. Detection of T-DNA specific activities. Plant Mol. Biol. **7,** 43—50.

Grimsley, N., Hohn, T., Davies, J. W., Hohn, B., 1987: *Agrobacterium*-mediated delivery of infectious maize streak virus into maize plants. Nature **325,** 177—179.

Grindley, N. D. F., Reed, R. R., 1985: Transpositional recombination in prokaryotes. Ann. Rev. Biochem. **54,** 863—896.

Haselkorn, R., Golden, J. W., Lammers, P. J., Mulligan, M. E., 1986: Developmental rearrangement of cyanobacterial nitrogen-fixation genes. Trends in Genetics **2,** 255—259.

Hepburn, A. G., Clarke, L. E., Blundy, K. S., White, J., 1983: Nopaline Ti plasmid, pTiT37, T-DNA insertions into a flax genome. J. Mol. Appl. Genet. **2,** 211—224.

Hepburn, A. G., White, J., 1985: The effect of right terminal repeat deletion on the oncogenicity of the T-region of pTiT37. Plant Mol. Biol. **5,** 3—11.

Hernalsteens, J.-P., Thia-Toong, L., Schell, J., Van Montagu, M., 1984: An *Agrobacterium*-transformed cell culture from the monocot *Asparagus officinalis.* EMBO J. **3,** 3039—3041.

Hille, J., Hoekema, A., Hooykaas, P., Schilperoort, R. A., 1984: Gene organisation of the Ti-plasmid. In: Verma, D. P. S., Hohn, Th. (eds.), Genes Involved in Microbe-Plant Interactions (Plant Gene Research, Vol. 1), pp. 287—309. Wien - New York: Springer.

Hille, J., Dekker, M., Luttighuis, H. O., van Kammen, A., Zabel, P., 1986: Detection of T-DNA transfer to plant cells by *A. tumefaciens* virulence mutants using agroinfection. Mol. Gen. Genet. **205,** 411—416.

Hirooka, T., Kado, C. I., 1986: Location of the right boundary of the virulence region on *Agrobacterium tumefaciens* plasmid pTiC58 and a host specifying gene next to the boundary. J. Bacteriol. **168,** 237—243.

Hirooka, T., Rogowsky, P., Kado, C. I., 1987: Characterization of the *vir*E locus of *Agrobacterium tumefaciens* plasmid pTiC58. J. Bacteriol., in press.

Hoekema, A., Hirsch, P. R., Hooykaas, P. J. J., Schilperoort, R. A., 1983a: A binary plant vector strategy based on separation of *vir*- and T-region of the *Agrobacterium tumefaciens* Ti plasmid. Nature (London) **303,** 179—181.

Hoekema, A., van Haaren, M. J. J., Hille, J., Hoge, J. H. C., Hooykaas, P. J. J., Krens, F. A., Wullems, G. J., Schilperoort, R. A., 1983b: *Agrobacterium tumefaciens* and its Ti plasmid as tools in transformation of plant cells. In: Goldberg, R. B. (ed.), Plant Molecular Biology. New York: Alan R. Liss.

Hoekema, A., Roelvink, P. W., Hooykaas, P. J. J., Schilperoort, R. A., 1984: Delivery of T-DNA from the *Agrobacterium tumefaciens* chromosome into plant cells. EMBO J. **3,** 2485—2490.

Holsters, M., Silva, B., Van Vliet, F., Genetello, C., De Block, M., Dhaese, P., Depicker, A., Inzé, D., Engler, G., Villarroel, R., Van Montagu, M., Schell, J., 1980: The functional organization of the nopaline *A. tumefaciens* plasmid pTiC58. Plasmid **3,** 212—230.

Holsters, M., Villarroel, R., Gielen, J., Seurinck, J., De Greve, H., Van Montagu, M., Schell J., 1983: An analysis of the boundaries of the octopine T_L DNA in tumors induced by *Agrobacterium tumefaciens.* Mol. Gen. Genet. **190,** 35—41.

Hooykaas, P. J. J., Hofker, M., den Dulk-Ras, H., Schilperoort, R. A., 1984: A comparison of virulence determinants in an octopine Ti plasmid, a nopaline Ti plasmid and an Ri plasmid by complementation analysis of *Agrobacterium tumefaciens* mutants. Plasmid **11,** 195—205.

Hooykaas, P. J., Schilperoort, R. A., 1984: The molecular genetics of crown gall tumorigenesis. In: Scandalios, J. G., (ed.), Molecular Genetics of Plants (Advances in Genetics, Vol. 22), pp. 209—283. New York: Academic Press.

Hooykaas-van Slogteren, G. M. S., Hooykaas, P. J., Schilperoort, R. A., 1984: Expression of Ti plasmid genes in monocotyledonous plants infected with *Agrobacterium tumefaciens.* Nature **311,** 763—764.

Horsch, R. B., Fry, J. E., Hoffmann, N. L., Eichholtz, D., Rogers, S. G., Fraley, R. T., 1985: A simple and general method for transferring genes into plants. Science **227,** 1229—1231.

Horsch, R. B. Klee, H. J., Stachel, S., Winans, S. C., Nester, E. W., Rogers, S. G.,

Fraley, R. T., 1986: Analysis of *Agrobacterium tumefaciens* virulence mutants in leaf discs. Proc. Natl. Acad. Sci. U.S.A. **83,** 2571—2575.

Horsch, R. B., Klee, H. J., 1986: Rapid assay of foreign gene expression in leaf discs transformed by *Agrobacterium tumefaciens*: Role of T-DNA borders in the transfer process. Proc. Natl. Acad. Sci. U.S.A. **83,** 4428—4432.

Huber, H. E., Iida, S., Arber, W., Bickle, T. A., 1985: Site-specific DNA inversion is enhanced by a DNA sequence element in *cis*. Proc. Natl. Acad. Sci. U.S.A. **82,** 3776—3780.

Iida, S., Hiestand-Nauer, R., 1986: Localized conversion at the crossover sequences in the site-specific DNA inversion system of bacteriophage P1. Cell **45,** 71—79.

Inzé, D., Follin, A., Van Lijsebettens, M. Simoens, C., Genetello, C., Van Montagu, M., Schell, J., 1984: Genetic analysis of the individual T-DNA genes of *Agrobacterium tumefaciens;* further evidence that two genes are involved in indole-3-acetic acid synthesis. Mol. Gen. Genet. **194,** 265—274.

Jen, G. C., Chilton, M.-D., 1986a: The right border region of pTiT37 T-DNA is intrinsically more active than the left border region in promoting T-DNA transformation. Proc. Natl. Acad. Sci. U.S.A. **83,** 3895—3899.

Jen, G. C., Chilton, M.-D., 1986b: Activity of T-DNA borders in plant cell transformation by Mini-T plasmids. J. Bacteriol. **166,** 491—499.

Johnson, R. C., Simon, M. I., 1985: *Hin*-mediated site-specific recombination requires two 26 bp recombination sites and a 60 bp recombinational enhancer. Cell **41,** 781—791.

Joos, H., Inzé, D., Caplan, A., Sormann, M., Van Montagu, M., Schell, J., 1983a: Genetic analysis of T-DNA transcripts in nopaline crown galls. Cell **32,** 1057—1067.

Joos, H., Timmermann, B., Van Montagu, M., Schell, J., 1983 b: Genetic analysis of transfer and stabilization of *Agrobacterium* DNA in plant cells. EMBO J. **2,** 2151—2160.

Kahmann, R., Rudt, F., Koch, C., Mertens, G., 1985: G inversion in bacteriophage Mu DNA is stimulated by a site within the invertase gene and a host factor. Cell **41,** 771—780.

Kerr, A., Manigault, P., Tempé, J., 1977: Transfer of virulence *in vivo* and *in vitro* in *Agrobacterium.* Nature (London) **265,** 560—561.

Klee, H. J., White, F. F., Iyer, V. N., Gordon, M. P., Nester, E. W., 1983: Mutational analysis of the virulence region of an *Agrobacterium tumefaciens* Ti plasmid. J. Bacteriol. **153,** 878—883.

Knauf, V. C., Yanofsky, M. F., Montoya, A., Nester, E. W., 1984: Physical and functional map of an *Agrobacterium tumefaciens* tumor-inducing plasmid that confers a narrow host range. Bacteriol. **160,** 564—568.

Koekman, B. P., Hooykaas, P. J. J., Schilperoort, R. A., 1980: Localization of the replication control region on the physical map of the octopine Ti plasmid. Plasmid **4,** 184—195.

Koekman, B. P., Hooykaas, P. J. J., Schilperoort, R. A., 1982: A functional map of the replicator region of the octopine Ti plasmid. Plasmid **7,** 119—132.

Koncz, C., Schell, J., 1986: The promoter of T_L-DNA gene 5 controls the tissue-specific expression of chimaeric genes carried by a novel type of *Agrobacterium* binary vector. Mol. Gen. Genet. **204,** 383—396.

Kornberg, A., 1980: DNA Replication, Freeman, San Francisco.

Koukolíková-Nicola, Z., Shillito, R. D., Hohn, B., Wang, K., Van Montagu, M., Zambryski, P., 1985: Involvement of circular intermediates in the transfer of T-DNA from *Agrobacterium tumefaciens* to plant cells. Nature **313,** 191—196.

Krens, F. A., Molendijk, L., Wullems, G. J., Schilperoort, R. A., 1982: *In vitro* transformation of plant protoplasts with Ti-plasmid DNA. Nature **296**, 72—74.

Leemans, J., Deblaere, R., Willmitzer, L., De Greve, H., Hernalsteens, J. P., Van Montagu, M., Schell, J., 1982: Genetic identification of functions of TL-DNA transcripts in octopine crown galls. EMBO J. **1**, 147—152.

Lemmers, M., De Beuckeleer, M., Holsters, M., Zambryski, P., Depicker, A., Hernalsteens, J. P., Van Montagu, M., Schell J., 1980: Internal organization, boundaries and integration of Ti plasmid DNA in nopaline crown gall tumours. J. Mol. Biol. **144**, 353—376.

Leroux, B., Yanofsky, M. F., Winans, S. C., Ward, J. E., Ziegler, S. E., Nester, E. W., 1987: Characterization of the *vir* A locus of *Agrobacterium tumefaciens:* a transcriptional regulator and host range determinant. EMBO J. **6**, 849—856.

Lioret, C., 1956: Sur la mise en évidence d'un acide aminé non identifié particulier aux tissus de crown gall. Bull. Soc. Fr. Physiol. vég. **2**, 76.

Lippincott, B. B. and Lippincott, J. A., 1969: Bacterial attachment to a specific wound site as an essential stage in tumor initiation by *Agrobacterium tumefaciens.* J. Bacteriol. **97**, 620—628.

Lundquist, R. C., Close, T. J., Kado, C. I., 1984: Genetic complementation of *Agrobacterium tumefaciens* Ti plasmid mutants in the virulence region. Mol. Gen. Genet. **193**, 1—7.

Machida, Y., Usami, S., Yamamoto, A., Niwa, Y., Takebe, I., 1986: Plant-inducible recombination between the 25-bp border sequences of T-DNA in *Agrobacterium tumefaciens.* Mol. Gen. Genet. **204**, 374—382.

Matthysse, A., 1984: Interaction of *Agrobacterium tumefaciens* with the plant cell surface. In: Genes Involved in Microbe-Plant Interactions, Verma, D. P. S., Hohn, Th. (eds.), Plant Gene Research, Vol. 1, pp. 33—54. Wien - New York: Springer.

Melchers, L. S., Thompson, D. V., Idler, K. B., Schilperoort, R. A., Hooykaas, P. J. J., 1986: Nucleotide sequence of the virulence gene *vir*G of the *Agrobacterium tumefaciens* octopine Ti plasmid: significant homology between *vir* G and the regulatory genes *omp*R, *pho*B and *dye* of *E. coli.* Nucl. Acids Res. **14**, 9933—9942.

Messens, E., Lenaerts, A., Van Montagu, M., Hedges, R. W., 1985: Genetic basis for opine secretion from crown gall tumour cells. Mol. Gen. Genet. **199**, 344—348.

Miller, I. S., Fox, D., Saeed, N., Borland, P. A., Miles, C. A., Sastry, G. R. K., 1986: Enlarged map of *Agrobacterium tumefaciens* C58 and the location of chromosomal regions which affect tumorigenicity. Mol. Gen. Genet. **205**, 153—159.

Montoya, A., Chilton, M.-D., Gordon, M. P., Sciaky, D., Nester, E. W., 1977: Octopine and nopaline metabolism in *Agrobacterium tumefaciens* and crown-gall tumor cells: Role of plasmid genes. J. Bacteriol. **129**, 101—107.

Morel, G., 1956: Métabolisme de l'arginine par les tissus de crown gall de topinambour. Bull. Soc. Fr. Physiol. vég. **2**, 75.

Morris, R. O., 1986: Genes specifying auxin and cytokinin biosynthesis in phytopathogens. Ann. Rev. Plant Physiol. **37**, 509—538.

Nester, E. W., Gordon, M. P., Amasino, R. M., Yanofsky, M. F., 1984: Crown gall: a molecular and physiological analysis. Ann. Rev. Plant Physiol. **35**, 387—413.

Nixon, B. T., Ronson, C. W., Ausubel, F. M., 1986: Two-component regulatory systems responsive to environmental stimuli share strongly conserved domains with the nitrogen assimilation regulatory genes *ntr* B and *ntr* C. Proc. Natl. Acad. Sci. U.S.A. **83**, 7850—7854.

Ooms, G., Klapwijk, P. M., Poulis, J. A., Schilperoort, R. A., 1980: Characterization of Tn904 insertions in octopine Ti plasmid mutants of *Agrobacterium tumefaciens.* J. Bacteriol. **144,** 82—91.

Otten, L., De Greve, H., Leemans, J., Hain, R., Hooykaas, P. J. J., 1984: Restoration of virulence of *vir* region mutants of *Agrobacterium tumefaciens* strain B6S3 by coinfection with normal and mutant *Agrobacterium* strains. Mol. Gen. Genet. **195,** 159—163.

Otten, L., Piotrowiak, G., Hooykaas, P., Dubois, M., Szegedi, E., Schell, J., 1985: Identification of an *Agrobacterium tumefaciens* pTiB6S3 *vir* region fragment that enhances the virulence of pTiC58. Mol. Gen. Genet. **199,** 189—193.

Paszkowski, J., Shillito, R. D., Saul, M., Mandák, V., Hohn, T., Hohn, B., Potrykus, I., 1984: Direct gene transfer to plants. EMBO J. **3,** 2717—2722.

Peerbolte, R., Leenhouts, K., Hooykaas-van Slogteren, G. M. S., Hoge, J. H. C., Wullems, G. J., Schilperoort, R. A., 1986a: Clones from a shooty tobacco crown gall tumor I: deletions, rearrangements and amplifications resulting in irregular T-DNA structures and organizations. Plant Mol. Biol. **7,** 265—284.

Peerbolte, R., Leenhouts, K., Hooykaas-van Slogteren, G. M. S., Wullems, G. J., Schilperoort, R. A., 1986b: Clones from a shooty tobacco crown gall tumor II: irregular T-DNA structures and organization, T-DNA methylation and conditional expression of opine genes. Plant Mol. Biol. **7,** 285—299.

Peralta, E. G., Ream, L. W., 1985: T-DNA border sequences required for crown gall tumorigenesis. Proc. Natl. Acad. Sci., U.S.A. **82,** 5112—5116.

Peralta, E. G., Hellmiss, R., Ream, W., 1986: *Overdrive,* a T-DNA transmission enhancer on the *A. tumefaciens* tumour-inducing plasmid. EMBO J. **5,** 1137—1142.

Peters, N. K., Frost, J. W., Long, S. R., 1986: A plant flavone, luteolin, induces expression of *Rhizobium meliloti* nodulation genes. Science **233,** 977—980.

Petit, A., Delhaye, S., Tempé, J., Morel, G., 1970: Recherches sur les guanidines des tissus de crown gall. Mise en évidence d'une relation biochimique spécifique entre les souches d'*Agrobacterium* et les tumeurs qu'elles induisent. Physiol. vég. **8,** 205—213.

Petit, A., Dessaux, Y., Tempé, J., 1978: The biological significance of opines. I. A study of opine catabolism by *Agrobacterium tumefaciens.* Proc. 4th Inter. Conf. Plant Pathol. Bact., M. Ridé, ed., INRA-Angers, 143—152.

Petit, A., Berkaloff, A., Tempé, J., 1986: Multiple transformation of plant cells by *Agrobacterium* may be responsible for the complex organization of T-DNA in crown gall and hairy root. Mol. Gen. Genet. **202,** 388—393.

Plasterk, R. H. A., van de Putte, P., 1985: The invertible P-DNA segment in the chromosome of *Escherichia coli.* EMBO J. **4,** 237—242.

Powell, G., K., Morris, R. O., 1986: Nucleotide sequence and expression of a *Pseudomonas savastanoi* cytokinin biosynthetic gene: homology with *Agrobacterium tumefaciens tmr* and *tzs* loci. Nucl. Acids Res. **14,** 2555—2565.

Pueppke, S. G., Hawes, M. C. 1985: Understanding the binding of bacteria to plant surfaces. Trends in Biotechnol. **3,** 310—313.

Puvanesarajah, V., Schell, F. M., Stacey, G., Douglas, C. J., Nester, E. W., 1985: Role of 2-linked-β-D-glucan in the virulence of *Agrobacterium tumefaciens.* J. Bacteriol. **164,** 102—106.

Rauth, S., Song, K.-J., Ayares, D., Wallace, L., Moore, P. D., Kucherlapati, R., 1986: Transfection and homologous recombination involving single-stranded DNA substrates in mammalian cells and nuclear extracts. Proc. Natl. Acad. Sci., U.S.A. **83,** 5587—5591.

Redmond, J. W., Batley, M., Djordjevic, M. A., Innes, R. W., Kuempel, P. L., Rolfe,

B.-G., 1986: Flavones induce expression of nodulation genes in *Rhizobium.* Nature **323,** 632—635.

Reed, R. R., Grindley, N. D. F., 1981: Transposon-mediated site-specific recombination in vitro: DNA cleavage and protein-DNA linkage at the recombination site. Cell **25,** 721—728.

Roberts, D., Hoopes, B. C., McClure, W. R., Kleckner, N., 1985: IS 10 transposition is regulated by DNA adenine methylation. Cell **43,** 117—130.

Rogowsky, P., Chimera, J. A., Close, T. J., Shaw, J. J., Kado, C. I.; Regulation of the *vir* genes of the *Agrobacterium tumefaciens* plasmid pTiC58. J. Bacteriol., submitted.

Rubin, R. A., 1986: Genetic studies on the role of octopine T-DNA border regions in crown gall tumor formation. Mol. Gen. Genet. **202,** 312—320.

Salomon, F., Deblaere, R., Leemans, J., Hernalsteens, J.-P., Van Montagu, M., Schell, J., 1984: Genetic identification of functions of T_R-DNA transcripts in octopine crown galls. EMBO J. **3,** 141—146.

Schell, J., Van Montagu, M., De Beuckeleer, M., De Block, M., Depicker, A., De Wilde, M., Engler, G., Genetello, C., Hernalsteens, J.-P., Holsters, M., Seurinck, J., Silva, B., Van Vliet, F., Villarroel, R., 1979: Interactions and DNA transfer between *Agrobacterium tumefaciens,* the Ti plasmid and the plant host. Proc. R. Soc. Lond. B **204,** 251—266.

Schell, J., 1986: T-DNA genes from *Agrobacterium* plasmids. In: Gustafson, J. P., Stebbins, L., Ayala, F. J. (eds.), Proceedings of the 17th Stadler Genetic Symposium: Genetics, Development and Evolution, Vol. 17, pp. 343—353. Plenum Publishing Corporation.

Schröder, G., Waffenschmidt, S., Weiler, E. W., Schröder, J., 1984: The T-region of Ti plasmids codes for an enzyme synthesizing indole-3-acetic acid. Eur. J. Biochem. **138,** 387—391.

Serfling, E., Jasin, M., Schaffner, W., 1985: Enhancers and eukaryotic gene transcription. Trends in Genetics **1,** 224—230.

Shaw, C. H., Watson, M. D., Carter, G. H., Shaw, C. H., 1984: The right hand copy of the nopaline Ti-plasmid 25-bp repeat is required for tumour formation. Nucl. Acids Res. **12,** 6031—6041.

Shillito, R. D., Saul, M. W., Paszkowski, J., Müller, M., Potrykus, I., 1985: High efficiency direct gene transfer to plants. Bio/technology **3,** 1099—1103.

Simpson, R. B., O'Hara, P. J., Kwok, W., Montoya, A. L., Lichtenstein, C., Gordon, M. P., Nester, E. W., 1982: DNA from the A6S/2 crown gall tumors contains scrambled Ti-plasmid sequences near its junctions with the plant DNA. Cell **29,** 1005—1014.

Simpson, R. B., Spielmann, A., Margossian, L., McKnight, T. D., 1986: A disarmed binary vector from *Agrobacterium tumefaciens* functions in *Agrobacterium rhizogenes.* Plant Mol. Biol. **6,** 403—415.

Slightom, J. L., Jouanin, L., Leach, F., Drong, R. F., Tepfer, D., 1985: Isolation and identification of T_L-DNA/plant junctions in *Convolvulus arvenis* transformed by *Agrobacterium rhizogenes* strain A4. EMBO J. **4,** 3069—3077.

Slightom, J. L., Durand-Tardif, M., Jouanin, L., Tepfer, D., 1986: Nucleotide sequence analysis of T_L-DNA of *Agrobacterium rhizogenes* agropine type plasmid. J. Biol. Chem. **261,** 108—121.

Spielmann, A., Simpson, R. B., 1986: T-DNA structure in transgenic tobacco plants with multiple independent integration sites. Mol. Gen. Genet. **205,** 34—41.

Stachel, S. E., An, G., Flores, C., Nester, E. W., 1985a: A Tn3 lacZ transposon for the random generation of β-galactosidase gene fusions: application to the analysis of gene expression in *Agrobacterium.* EMBO J. **4,** 891—898.

Stachel, S. E., Messens, E., Van Montagu, M., Zambryski, P., 1985 b: Identification of the signal molecules produced by wounded plant cells that activate T-DNA transfer in *Agrobacterium tumefaciens*. Nature **318,** 624—629.

Stachel, S. E., Nester, E. W., 1986: The genetic and transcriptional organization of the *vir* region of the A6 Ti plasmid of *Agrobacterium tumefaciens*. EMBO J. **5,** 1445—1454.

Stachel, S. E., Zambryski, P. C., 1986 a: Vir A and vir G control the plant-induced activation of the T-DNA transfer of A. tumefaciens. Cell **46,** 325—333.

Stachel, S. E., Nester, E. W., Zambryski, P. C., 1986 a: A plant cell factor induces *Agrobacterium tumefaciens vir* gene expression. Proc. Natl. Acad. Sci., U.S.A. **83,** 379—383.

Stachel, S. E., Timmerman, B., Zambryski, P., 1986 b: Generation of single-stranded T-DNA molecules during the initial stages of T-DNA transfer from *Agrobacterium tumefaciens* to plant cells. Nature **322,** 706—712.

Stachel, S. E., Zambryski, P. C., 1986 b: Agrobacterium tumefaciens and the susceptible plant cell: A novel adaptation of extracellular recognition and DNA conjugation. Cell **47,** 155—157.

Tempé, J., Petit, A., 1982: Opine utilization by *Agrobacterium*. In: Kahl, G., Schell, J. (eds.), Molecular Biology of Plant Tumors, pp. 451—459. New York: Academic Press.

Tempé, J., Petit, A., Farrand, S. K., 1984: Induction of cell proliferation by *Agrobacterium tumefaciens* and *Agrobacterium rhizogenes*: a parasite's point of view. In: Verma, D. P. S., Hohn, Th. (eds.), Genes Involved in Microbe-Plant Interactions (Plant Gene Research, Vol. 1), pp. 271—286. Wien - New York: Springer.

Thomashow, M. F., Nutter, R., Postle, K., Chilton, M.-D., Blattner, F. R., Powell, A., Gordon, M. P., Nester, E. W., 1980 a: Recombination between higher plant DNA and the Ti plasmid of *Agrobacterium tumefaciens*. Proc. Natl. Acad. Sci., U.S.A. **77,** 6448—6452.

Thomashow, M. F., Nutter, R., Montoya, A. L., Gordon, M. P., Nester, E. W., 1980b: Integration and organisation of Ti plasmid sequences in crown gall tumors. Cell **19,** 729—739.

Thomashow, L. S., Reeves, A., Thomashow, M. F., 1984: Crown gall oncogenesis: Evidence that a T-DNA gene from the *Agrobacterium* Ti plasmid pTiA6 encodes an enzyme that catalyzes synthesis of indole-acetic acid. Proc. Natl. Acad. Sci., U.S.A. **81,** 5071—5075.

Thomashow, M. F., Hugly, W. G., Buchholz, L. S., Thomashow, L. S., 1986: Molecular basis for the auxin independent phenotype of crown gall tumor tissues. Science **231,** 616—618.

Van Larebeke, N., Genetello, C., Schell, J., Schilperoort, R. A., Hermans, A. K., Hernalsteens, J. P., Van Montagu, M., 1975: Acquisition of tumour-inducing ability by non-oncogenic agrobacteria as a result of plasmid transfer. Nature (London) **255,** 742—743.

Van Lijsebettens, M., Inzé, D., Schell, J., Van Montagu, M., 1986: Transformed cell clones as a tool to study T-DNA integration mediated by *Agrobacterium tumefaciens*. J. Mol. Biol. **188,** 129—145.

van Veen, R. J. M., den Dulk-Ras, H., Schilperoort, R. A., Hooykaas, P. J. J., 1987: Chromosomal nodulation genes: Sym-plasmid containing *Agrobacterium* strains need chromosomal virulence genes (*chv*A and *chv*B) for nodulation. Plant Mol. Biol. **8,** 105—108.

Velten, J., Willmitzer, L., Leemans, J., Ellis, J., Deblaere, R., Van Montagu, M., Schell, J., 1983: T_R-genes involved in agropine production. In: Pühler, A. (ed.),

Molecular genetics of the bacteria-plant interaction, pp. 303—312. Berlin - Heidelberg - New York: Springer.

Veluthambi, K., Jayaswal, R. K., Gelvin, S. B., 1987: The virulence genes A, G and D mediate the double-stranded border cleavage of the T-DNA from the *Agrobacterium tumefaciens* Ti plasmid. Proc. Natl. Acad. Sci., U.S.A. **84,** 1881—1885.

Virts, E. L., Gelvin, S. B., 1985: Analysis of transfer of tumor-inducing plasmids from *Agrobacterium tumefaciens* to *Petunia* protoplasts. J. Bacteriol. **162,** 1030—1038.

Wallroth, M., Gerats, A. G. M., Rogers, S. G., Fraley, R. T., Horsch, R. B., 1986: Chromosomal localization of foreign genes in *Petunia hybrida.* Mol. Gen. Genet. **202,** 6—15.

Wang, K., Herrera-Estrella, L., Van Montagu, M., Zambryski, P., 1984: Right 25bp terminus sequence of the nopaline T-DNA is essential for and determines direction of DNA transfer from Agrobacterium to the plant genome. Cell **38,** 455—462.

White, P. R., Braun, A. C., 1942: A cancerous neoplasm of plants. Autonomous, bacteria-free crown-gall tissue. Cancer Res. **2,** 597—617.

Willetts, N., Wilkins, B., 1984: Processing of plasmid DNA during bacterial conjugation. Microbiol. Rev. **48,** 24—41.

Willmitzer, L., De Beuckeleer, M., Lemmers, M., Van Montagu, M., Schell, J., 1980: DNA from Ti plasmid is present in the nucleus and absent from plastids of plant crown-gall cells. Nature (London) **287,** 359—361.

Willmitzer, L., Simons, G., Schell, J., 1982: The T_L-DNA in octopine crown gall tumours codes for seven well-defined polyadenylated transcripts. EMBO J. **1,** 139—146.

Willmitzer, L., Dhaese, P., Schreier, P. H., Schmalenbach, W., Van Montagu, M., Schell, J., 1983: Size, location, and polarity of T-DNA-encoded transcripts in nopaline crown gall tumors; evidence for common transcripts present in both octopine and nopaline tumors. Cell **32,** 1045—1056.

Winans, S. C., Ebert, P. R., Stachel, S. E., Gordon, M. P., Nester, E. W., 1986: A gene essential for *Agrobacterium* virulence is homologous to a family of positive regulatory loci. Proc. Natl. Acad. Sci., U.S.A. **83,** 8278—8282.

Winans, S. C., Allenza, P., Stachel, S. E., McBride, K. E., Nester, E. W., 1987: Characterization of the *vir*E operon of the *Agrobacterium* Ti plasmid pTiA6. Nucl. Acids Res. **15,** 825—837.

Yadav, N. S., Vanderleyden, J., Bennet, D. R., Barnes, W. M., Chilton, M.-D., 1982: Short direct repeats flank the T-DNA on a nopaline Ti plasmid. Proc. Natl. Acad. Sci., U.S.A. **79,** 6322—6326.

Yamada, T., Palm, C. J., Brooks, B., Kosuge, T., 1985: Nucleotide sequences of the *Pseudomonas savastanoi* indoleacetic acid genes show homology with *Agrobacterium tumefaciens* T-DNA. Proc. Natl. Acad. Sci. U.S.A. **82,** 6522—6526.

Yamada, T., Lee, P.-D., Kosuge, T., 1986: Insertion sequence elements of *Pseudomonas savastanoi:* Nucleotide sequence and homology with *Agrobacterium tumefaciens* transfer DNA. Proc. Natl. Acad. Sci., U.S.A. **83,** 8263—8267.

Yamamoto, A. Iwahashi, M., Yanofsky, M. F., Nester, E. W., Takebe, I., Machida, Y., 1987: The promoter proximal region in the *vir*D locus of *Agrobacterium tumefaciens* is necessary for the plant-inducible circularization of T-DNA. Mol. Gen. Genet. **206,** 174—177.

Yanofsky, M., Lowe, B., Montoya, A., Rubin, R., Krul, W., Gordon, M., Nester, E., 1985: Molecular and genetic analysis of factors controlling host range in *Agrobacterium tumefaciens.* Mol. Gen. Genet. **201,** 237—246.

Yanofsky, M. F., Nester, E. W., 1986: Molecular characterization of a host-range-determining locus from *Agrobacterium tumefaciens*. J. Bacteriol. **168,** 244—250.

Yanofsky, M. F., Porter, S. G., Young, C., Albright, L. M., Gordon, M. P., Nester, E. W., 1986: The *vir*D operon of *Agrobacterium tumefaciens* encodes a site-specific endonuclease. Cell **47,** 471—477.

Zaenen, I., Van Larebeke, N., Teuchy, H., Van Montagu, M., Schell, J., 1974: Supercoiled circular DNA in crown gall inducing *Agrobacterium* strains. J. Mol. Biol. **86,** 109—127.

Zambryski, P., Holsters, M., Kruger, K., Depicker, A., Schell, J., Van Montagu, M., Goodman, H. M., 1980: Tumor DNA structure in plant cells transformed by *A. tumefaciens*. Science **209,** 1385—1391.

Zambryski, P., Depicker, A., Kruger, K., Goodman, H., 1982: Tumor induction by *Agrobacterium tumefaciens:* analysis of the boundaries of T-DNA. J. Mol. Appl. Genet. **1,** 361—370.

Zambryski, P., Joos, H., Genetello, C., Leemans, J., Van Montagu, M., Schell, J., 1983: Ti plasmid vector for the introduction of DNA into plant cells without alteration of their normal regeneration capacity. EMBO J. **2,** 2143—2150.

Zorreguieta, A., Ugalde, R. A., 1986: Formation in *Rhizobium* and *Agrobacterium* spp. of a 235-kilodalton protein intermediate in β-D(1-2)-glucan synthesis. J. Bacteriol. **167,** 947—951.

Chapter 6

Molecular Analysis of Root Induction by *Agrobacterium rhizogenes*

Frank F. White[1] and Vilas P. Sinkar[2]

[1] Department of Plant Pathology, Throckmorton Hall, Kansas State University, Manhattan, KS 66506, U.S.A.

[2] Department of Biochemistry, University of Washington, Seattle, WA 98195, U.S.A.

With 8 Figures

Contents

I. Introduction

Agrobacterium rhizogenes incites hairy root disease of plants. The disease is closely related to crown gall disease due to strains of *Agrobacterium tumefaciens*. The ability of both bacteria to cause plant tumors is encoded by the large virulence plasmids in the respective strains. The virulence plasmid of *A. rhizogenes* has become known as the root-inducing (Ri) plasmid, to distinguish the plasmid from the tumor-inducing (Ti) plasmid of *A. tumefaciens*. The underlying mechanisms for hairy root and crown gall disease are very similar. Therefore, this review will not discuss the extensive literature pertaining to the Ti plasmid except where necessary for comparison to the Ri plasmid. However, an understanding of the Ri plasmid as permitted by present knowledge is incomplete without consideration of the Ti plasmid, and the reader is referred to other chapters in this book and recent reviews (Zambryski *et al.*, 1983; Nester *et al.*, 1984).

The Ri plasmid can be thought as another class of Ti plasmid, and the basic structures of the Ri plasmids conform to many of the Ti plasmid characteristics. Both plasmids confer on the bacterium the ability to transfer distinct portions of the plasmids called T-DNA to plant genomes. Expression of genes contained on the T-DNA regions are responsible for many of the aberrant growth morphologies observed on plants infected with either Ri or Ti plasmid containing strains. The Ri plasmids also conform to the opine hypothesis which states that plasmid genes involved in the synthesis of compounds called opines are introduced into the plant where synthesis occurs. In turn, the inciting bacteria contain genes which enable them to catabolize the opines. Thus, the bacterium has diverted metabolites from the plant and engineered an ecological niche for itself in the plant tumor (for review see Tempe *et al.*, 1984). Ri plasmids are known to encode for the synthesis of compounds of at least two classes of opines. One class is represented by opines of the agropine group, and the other opines are members of the agrocinopine group (Tepfer and Tempe, 1981; Petit *et al.*, 1983).

The basis for the morphogenic effects of three Ti T-DNA genes is now known. The *tms*1 and *tms*2 loci (also referred to as shooty, *iaa, aux,* or genes 1 and 2) are involved in the synthesis of the plant auxin indole-3-acetic acid (IAA) (Inze *et al.*, 1984; Schröder *et al.*, 1985; Thomashow *et al.*, 1985, 1986). The *tms*1 locus encodes a tryptophan monoxygenase, converting tryptophan to indole acetamide (IAM), and *tms*2 encodes an indole acetamide hydrolase, converting IAM to IAA. The *tmr* locus (also referred in the literature as rooty, root inhibition, *cyt,* or gene 4) encodes a dimethylallylpyrophosphate: AMP (DMA) transferase which is involved in the synthesis of cytokinins (Akiyoshi *et al.*, 1984). In some respects strains with a Ri plasmid behave similarly to strains with a mutated Ti *tmr* gene. Since there is a superficial resemblance in plant growth responses from strains with Ti *tmr* mutations and strains with an Ri plasmid. Both strain types cause extensive root formation on the stems of certain plants such as tobacco *(N. tabacum)*. However, the responses of many plants clearly indicate that the Ri plasmid possesses unique features. The goal of this review will be to clarify the relationship of the Ri plasmid to the Ti plasmids and to highlight, as well as provide speculations, regarding the unique and interesting features of the Ri plasmids.

II. Taxonomy

Natural isolates of *Agrobacterium* give rise to crown galls on wounded plants which develop distinctive morphologies depending on the inciting strain and the site of inoculation. While varying degrees of differentiation may be seen with particular strains and host plants, three morphological types are generally noted. The first two tumor morphologies are associated with infections by strains of *A. tumefaciens.* The tumor morphologies are either shoot-forming teratomas or relatively undifferentiated callus. The plasmids of the strains which cause these types of tumor morphologies have been studied extensively, and are represented by the nopaline type and octopine type Ti plasmids, repectively (Fig. 1a). The third variation is the pronounced root induction which occurs around the inoculated wound (Fig. 1a and 1b). The root morphology was distinctive enough for early

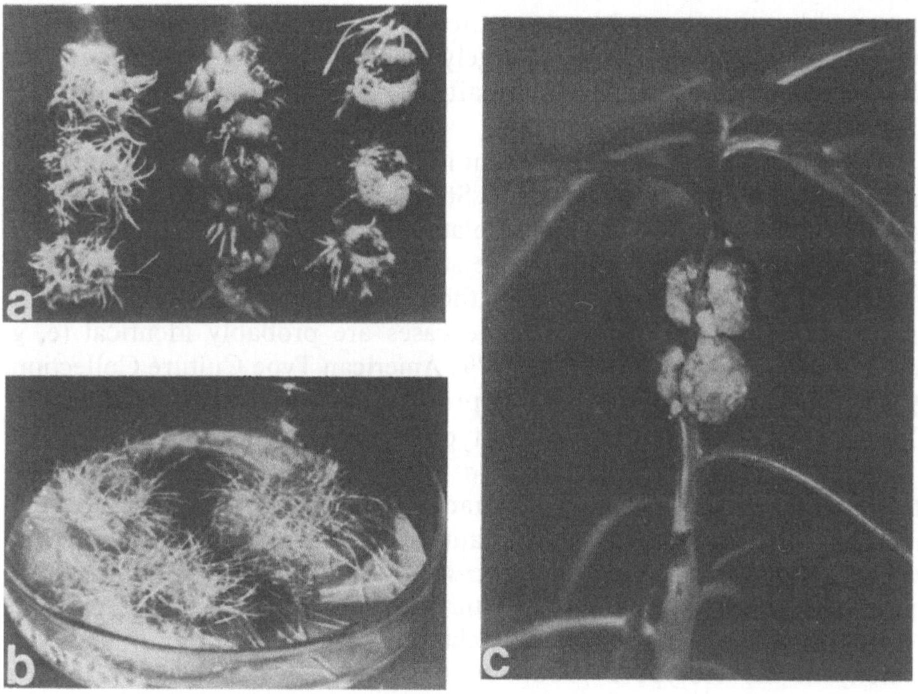

Fig. 1. a) Tumors on the stem of *Kalanchoë* after inoculation with (left to right) strains A4 (hairy root, agropine type Ri plasmid), T37 (teratoma, nopaline type Ti plasmid), and A6 (unorganized, octopine type Ti plasmid). b) Root proliferation on basal side of carrot disks after inoculation with strain containing pRiA4. c) Callus growth observed on *N. glauca* as an example of an alternative response to a strain containing an Ri plasmid. Plant was inoculated with strain A4

plant pathologists to define the symptoms as hairy root disease (Hedgecock, 1905). The hairy root tumor morphology and apparent differences between the hairy root bacterium and *A. tumefaciens,* led Riker *et al.* (1930) to classify the bacterium as a seperate species, *A. rhizogenes.* This nomenclature was accepted, although the close relationship between hairy root and crown gall was clearly recognized, and despite several reports that hairy root and crown gall diseases should be considered variant symptoms of *A. tumefaciens* infections (Siegler, 1928; Brown 1929). The nomenclature has been the source of much confusion since that time. In retrospect, it is clear that there were two types of hairy root bacteria. The original strains differed depending on whether the strain was isolated from apple or rose (Riker *et al.,* 1930; Munnecke *et al.,* 1963). The apple isolate was either described as weakly virulent on tobacco *(Nicotiana tabacum)* and tomato *(Lycopersicon esculentum),* or avirulent on these two plant species. The rose isolate resulted in profuse rooting on tobacco and tomato. The two groups most likely correspond to the two best studied Ri strains 8196 and A4. The morphology of the tumor incited by strains of either group also varies depending on the plant species (Anderson and Moore, 1979; De Cleene and De Ley, 1981). A4 type strains, for example, cause galls on *N. glauca* stems (Fig. 1c; White *et al.,* 1982). *Agrobacterium* taxonomy has since been improved but the names have remained the same. The maintenance of *A. rhizogenes* nomenclature is now largely due to convenience, and one now often wonders what to call a strain after Ri plasmid transfer into an avirulent strain of *A. tumefaciens.*

The actual number of independent isolates of *A. rhizogenes* in strain collections is undoubtedly very small. Strains 15834, A4, 1855, and TR105 (A4 type) all appear to have been isolated from hairy root disease on roses which occurred in California. These strains were possibly collected from the same plants by different plant pathologists (Munnecke *et al.,* 1963; Ark and Thompson, 1961), and in some cases are probably identical (e. g. ATCC 15834=A4; Moore *et al.,* 1979; American Type Culture Collection, 1982). The second group of strains probably share the same relationship. Thus, 8196, Tr7, and TR101 (8196 type) may be different names for one original isolate from apple (Riker *et al.,* 1930). The A4 and 8196 type strains differ in their virulence characteristics. The 8196 strains are relatively weakly virulent on many plants including tobacco, tomato, *Kalanchoë tubiflora,* and carrot *(Daucus carota)* while A4 strains induce extensive rooting on these hosts (Petit *et al.,* 1983).

All of the above strains are now classified by criteria other than pathogenicity in the biotype 2 group of *Agrobacterium* (Keane *et al.,* 1970; Kersters *et al.,* 1973; Holmes and Roberts, 1981). The biotype 2 group includes crown gall strains as well as hairy root strains. Virulence and tumor morphology are unquestionably associated with the presence of the Ri or Ti plasmid in all the known cases, and tumor symptoms appear similar when biotype 1 and 2 strains containing the same plasmid are inoculated on plants under the appropriate conditions. However, chromosomal genes also play a role in the infection process (Garfinkel and Nester, 1980), and the question of whether a biotype is more commonly associated with a plasmid type and/or a host plant is largely unexplored. Most experiments in *Agrobacterium* research do not measure the effect of genetic differences, which may have little or no effect on tumor initiation, on long term survival of the bacterium in the field.

A number of biotype 1 strains isolated in Great Britain reportedly give the hairy root response upon plant inoculation (Costantino *et al.,* 1981; Holmes and Roberts, 1981). Relatively little research has been performed on these strains, although DNA hybridization studies indicated a high degree of relatedness between the plasmids and the Ri plasmids from the biotype 2 strains (Costantino *et al.,* 1981). Technically, any strain of *Agrobacterium* which induces roots on infected plants could be classified as *A. rhizogenes.* For example, some limited host range strains (LHR) induce roots on *Nicotiana rustica* (Yanofsky *et al.,* 1985). This example also serves to illustrate that the hairy root phenotype could reflect a host range phenotype.

III. Ri Plasmid Structure

The association of virulence with the presence of a large plasmid in *A. rhizogenes* was largely assumed by analogy to the situation in *A. tumefaciens* (Currier and Nester, 1976; Schell *et al.,* 1976). However, the situation proved slightly more complex than for the Ti plasmid. *In planta* conjugation experiments (Kerr cross) clearly suggested that virulence was transferred via a plasmid from *A. rhizogenes* to an avirulent *Agrobacterium* spp. (Albinger and Biederbeck, 1977; Moore *et al.,* 1979; White and Nester, 1980a). The transconjugants were capable of inciting hairy root tumors which were indistinguishable from the tumors incited by the parent strain. After conjugation, the recipient bacteria appear to have received as many as three plasmids (White and Nester, 1980a). The amount of each plasmid in an individual preparation would vary due to the dissociation of the largest plasmid. Subsequently, both A4 and 8196 type strains have been shown to possess multiple plasmids, and agarose gel analyses suggested the smaller plasmids may be the dissociation products of a single large plasmid (White and Nester, 1980b; Chilton *et al.,* 1982; Petit *et al.,* 1983). Several strains of the A4 type were found to only have the medium sized plasmid (250 kb), and the logical assumption is that the smaller plasmid (160 kb) has been lost during culture (White and Nester, 1980b; Petit *et al.,* 1983).

The individual components of the large plasmid have been isolated in seperate strains, and virulence has always been associated with the medium sized plasmid, which is now commonly labeled pRi to conform with the Ti plasmid nomenclature (Chilton *et al.,* 1982). The multicomponent nature of the Ri plasmid in A4 type strains represents an interesting subject for further study. Little is known regarding the factors which control the Ri plasmid configuration. Recently, Jouanin *et al.* (1985) have cloned the individual origins of replication from the component plasmids and found that each plasmid possesses a unique origin of replication. Also, since the plasmids are compatible with each other and with other *Agrobacterium* plasmids, the cloned origins represent potential new shuttle vectors (Jouanin *et al.,* 1985).

The opines that are synthesized by hairy root tumors and utilized by the bacterium further suggest that the Ri plasmids can be grouped into two main classes. These groups correspond to the A4 and 8196 strain types. A4 type strain incite roots and callus tissue which may produce all of the members of the agropine class of opines (Petit *et al.,* 1983; Tepfer and Tempe, 1981; White *et al.,* 1982; Willmitzer *et al.,* 1982). The 8196 type strains induced growths which produced all of the agropine class opines with the exception of agropine itself (Petit *et al.*, 1983). Utilization of the opines corresponds to opine synthesis, and strains generally can utilize the same opines that are produced in the roots and tumors (Petit *et al.*, 1983). The differences in opine utilization and synthesis allow a convenient classification; A4 types are referred to as agropine type Ri plasmids, while 8196 types are referred to as mannopine type Ri plasmids. The possibility remains that more Ri plasmid specific opines might be found.

For the sake of simplicity, much of the work involving characterization of the Ri plasmid has employed strains harboring only the medium sized virulence plasmid. The fact should be considered, however, that the larger plasmid (pArA4c) probably represents the complete genetic complement necessary for the bacteria to compete in the natural environment. Virulent strains which are missing the pArA4a form or portion of the plasmid complement can be obtained by selecting resistance to agrocin 84, a bacteriocin produced by *Agrobacterium* strain K84 (Kerr and Roberts, 1976). Uptake of agrocin 84 is thought to occur via a permease intended to transport nutrients, specifically, the agrocinopines, present in the tumor (Ellis and Murphy, 1981). Thus, a strain without pArA4a would not be able to utilize the agrocinopines of the tumor. Further support for this idea comes from Petit *et al.* (1983) who have shown that pArA4a is also required for catabolism of agropine-related opines other than agropine. A similar arrangement of opine catabolism genes may exist on plasmids of the 8196 type since differences were noted in the utilization of agropinic acid by wild type (three plasmids) and transconjugant (one plasmid) strains (Petit *et al.*, 1983). Therefore, strains without the full complement of plasmids would not be able to fully utilize the opine nutrients of a tumor and, therefore, not compete well with strains that could.

Restriction fragment maps for both the agropine and mannopine type Ri plasmids have been constructed in several laboratories (Fig. 2). Three maps for the agropine type plasmid have been reported (Pomponi *et al.*, 1983; Huffman *et al.*, 1984; Jouanin, 1984), and all are very similar. The relatively minor differences in fragments may represent divergence of the plasmids. Alternatively, differences may have been generated by imprecise dissociation of the large plasmid (pArA4c) or by transposition of insertion elements. The map for pRiA4b is representative of the agropine type plasmids (Fig. 2; Huffman *et al.*, 1984; Jouanin, 1984). A map is currently being determined for pArA4a (Jouanin *et al.*, 1985). A restriction map for the mannopine type plasmid pRi8196 was determined by Koplow *et al.* (1985). The heterogeneity of fragment patterns from gel analyses (Costantino *et al.*, 1981) indicate that significant divergence of the agropine- and mannopine-type plasmids has occurred. Relatedness has not been assessed in detail, although gross hybridization studies indicate that they contain extensive regions of homology (Costantino *et al.*, 1981).

Several comparisons between agropine type Ri plasmids and Ti plasmids have been made using DNA hybridization techniques (White and Nester, 1980b; Riseulo *et al.*, 1982; Willmitzer *et al.*, 1982; Huffman *et

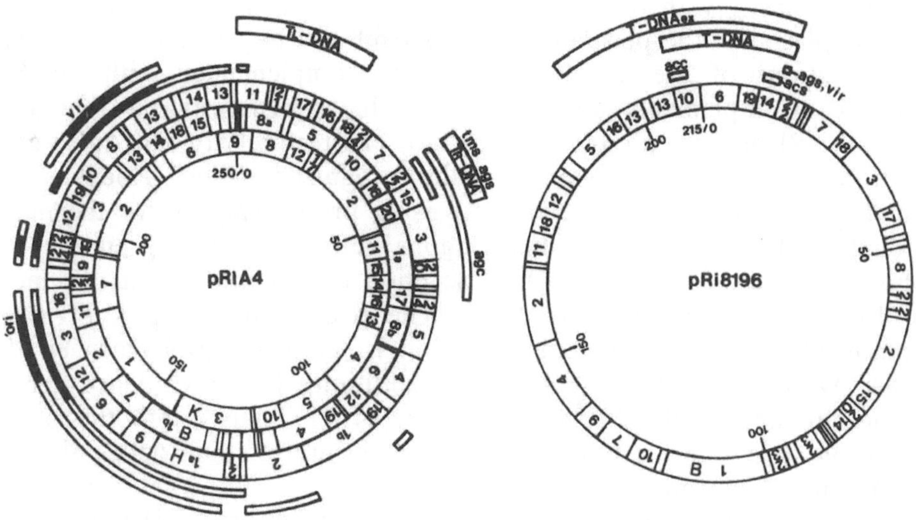

Fig. 2. Endonuclease restriction maps of an agropine type (pRiA4) and mannopine type (pRi8196) Ri plasmids. Data for pRiA4 is taken from Huffman *et al.* (1984) and Jouanin (1984). Data for pRi8196 is from Koplow *et al.* (1984). The outer most thin bars represent homology to an octopine type Ti plasmid. The inner bar represents homology to a nopaline type Ti plasmid. Dark regions represent a relatively strong degree of relatedness as indicated by hybridization signal intensity. Symbols are *vir,* virulence region; *tms,* auxin genes; *ags,* agropine synthesis; *agc,* agropine catabolism; *acs,* agrocinopine synthesis; *acc,* agrocinopine catabolism; *ori,* origin of replication; B, *Bam*HI; H, *Hin*dIII; K, *Kpn*I. Size markings are in kilobase pairs

al., 1984; Jouanin, 1984). These experiments have confirmed the close evolutionary relationship of the various plasmids. The region of the agropine Ri plasmid which appears to be most closely related to the Ti plasmid is the region of the virulence genes. While the positions of the *vir* genes of the Ri plasmid have not been precisely determined, the general position can be reasonably placed by hybridization to the fragments containing the Ti plasmid *vir* genes (Huffman *et al.,* 1984; Jouanin, 1984). The origin of replication has been cloned from the region of *Bam*HI-11 (Jouanin *et al.,* 1985) and can now be assigned a position with greater accuracy. The positions of the agropine synthesis genes (*ags;* Huffman *et al.,* 1984; Jouanin, 1984) and agropine catabolism genes (*agc;* Jouanin, 1984) were initially assigned on the basis of homology to the regions of the corresponding genes on the octopine type Ti plasmid (which also directs the synthesis of agropine type opines). Recent hybridization data using complementary DNA (cDNA) isolated from agropine-synthesizing carrot lines further confirms that several of the *ags* genes are located in the right T-DNA region (De Paolis *et al.,* 1985). The finding that the dissociated form of pRiA4 (=pArA4b) does not confer the ability to utilize other agropine type opines besides agropine (Petit *et al.,* 1983) suggests that not all of the catabolism genes are located in the region of homology to *agc* of the Ti plasmid.

The physical structure and hybridization data strongly suggest a functional similarity with the Ti plasmid. Due to the compatibility of the Ri and Ti plasmids, the plasmids can be introduced to the same strain and *vir* mutations of the Ti plasmid can be complemented with the Ri plasmid. Complementation experiments have demonstrated that the agropine Ri plasmid pRi1855 (=pRiA4b) can restore virulence to strains with mutations in most if not all of the recognized *vir* loci (Hooykaas *et al.*, 1984). Hybridization experiments using probes specific for the individual *vir* loci also suggest homologous genes are present on the Ri plasmid (Unger *et al.*, 1985). One exception to this finding was the lack of hybridization to the *vir*E probe. The reason for the lack of *vir*E homology despite *vir*E complementation is unknown (Hooykaas *et al.*, 1984; White, unpublished results). In addition, the Ri virulence genes alone can apparently direct the transfer of Ti T-DNA into the plant when the Ti T-DNA is introduced separately into a strain with only the Ri plasmid (Hoekema *et al.*, 1984).

While Ri are clearly structurally and functionally related to Ti plasmids, Ri plasmids have diverged significantly and should not be simply considered as mutant or recombinant Ti plasmids. Large segments of pRiA4 show little or no detectable degree of relatedness to Ti plasmids by DNA hybridization criteria (Huffmann *et al.*, 1984). Particularly notable is the near lack of homology in one of the T-DNA regions (Fig. 2; Huffman *et al.*, 1984). Less information is available for the mannopine-type plasmid, however, the situation is likely to be similar. Using probes which covered the pRi8196 single T-DNA region, Lahners *et al.* (1984) could only detect weak hybridization signals to octopine or nopaline type Ti plasmids. One curious region of homology was the octopine Ti plasmid *vir* region present in the pRi8196 right T-DNA border (Lahners *et al.*, 1984).

The accumulated data supports the hypothesis that the Ri plasmids evolved from a common ancestral plasmid of the Ti plasmids (White and Nester, 1980b). Both the mannopine and the agropine type Ri plasmids share features which are characteristic of both octopine and nopaline Ti plasmids. The Ri plasmids (in the cointegrate form) confer sensitivity to agrocin 84 and also direct the synthesis of agrocinopines, features shared by the nopaline Ti plasmids. Regions of homology between the loci of the nopaline Ti plasmid T-DNA involved in agrocinopine production and less well-defined regions of the Ri plasmid T-DNA have been detected (Huffman *et al.*, 1984; Lahners *et al.*, 1985; Willmitzer *et al.*, 1982). On the other hand the Ri plasmids direct the synthesis and catabolism of opines of the agropine family, a trait common to the octopine Ti plasmid, and share sequence homology to the genes involved in the synthesis and catabolism of agropine-like compounds (De Paolis *et al.*, 1984; Huffman *et al.*, 1984; Jouanin, 1984; Petit *et al.*, 1983). While the two Ri plasmid types have also diverged from each other, homology within the core of the left T-DNA region of the agropine plasmid and the single T-DNA region of the mannopine plasmid does exist (Spano *et al.*, 1982), suggesting a closer degree of relatedness between the plasmids than to other Ti plasmids. Homology to the Ri plasmid left T-DNA has only been detected in one other plasmid besides pRi8196. Hybridization to one of the large plasmids present in the limited host range strain AB2/73 was observed with a clone from pRiA4b (Unger *et al.*, 1985).

IV. Ri T-DNA Organization

Two T-DNA regions have been identified in agropine Ri plasmids (Fig. 2; Spano *et al.*, 1982; De Paolis *et al.*, 1985; White *et al.*, 1985) and are referred to as the left T-DNA (T_L-DNA) and right T-DNA (T_R-DNA). The regions are separated by approximately 15 kb of non-integrated DNA (White *et al.*, 1985). The removal of either T-DNA affects tumor morphology on some plants, particularly *Kalanchoë* (Fig. 3, White *et al.*, 1985). The T_R-DNA contains genes which are homologous with T-DNA from Ti

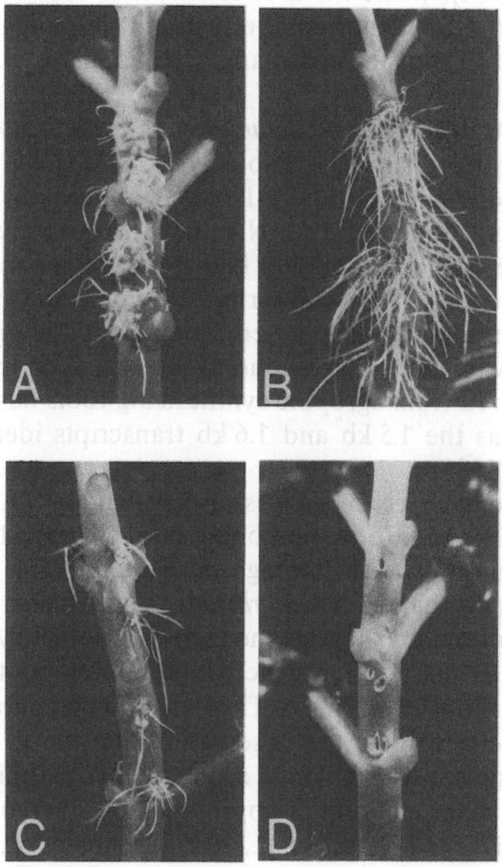

Fig. 3. Effects of T_L-DNA and T_R-DNA deletion on *Kalanchoë* stems. A. Wild type response with strain R1000. B. Deletion of T_L-DNA in strain R1200. C. Deletion of T_R-DNA in strain R1205. D. Deletion of both T_L-DNA and T_R-DNA in strain R1240. Data is from White *et al.* (1985)

plasmids. Most important for the growth and tumorigenic effects are the two genes involved in the synthesis of indole acetic acid. IAA production by the Ri *tms* genes has not been measured directly. However, the Ri and Ti regions are highly homologous by DNA hybridization studies (Huffman *et al.*, 1984; Willmitzer *et al.*, 1982). Hybridization with probes specific for *tms*1 and *tms*2 indicate that the organization of the genes is identical to the Ti plasmid genes (Huffman *et al.*, 1984). Insertions into the Ri *tms* genes severely reduce the virulence of a strain (Cardarelli *et al.*, 1985; White *et al.*, 1985). Insertion into either *Tms* 1 or *Tms* 2 severely attenuates virulence on *Nicotiana glauca* stems, *N. tabacum* stems, and *Kalanchoë* leaves (White *et al.*, 1985). Thus, the effect of *tms* mutations indicate that for many of the plants tested auxins play an important role in root and tumor induction. The *tms* mutations do not completely eliminate growth effects on all plants. Some residual root formation is observed on stems (Fig. 3; White *et al.*, 1985) and apical carrot disk surfaces (Cardarelli *et al.*, 1985).

Hybridization experiments suggested that genes involved in agropine synthesis are also located in the T_R-DNA region (Willmitzer *et al.*, 1982; Huffman *et al.*, 1984). While the individual genes involved in synthesis of agropine have yet to be identified, the region of homology to the Ti genes has been detected in transformed plants (De Paolis *et al.*, 1985; Taylor *et al.*, 1985a; White *et al.*, 1985).

Studies of Ri transformed *N. glauca* tumor tissue (Taylor *et al.*, 1985b) and regenerated carrot plants (De Paolis *et al.*, 1985) have revealed discrete mRNA species derived from the T_R-DNA. Tumorous tissue yielded transcripts from across the entire T_R-DNA. Two of the transcripts appear to correspond to the *tms* loci in the left half of the region, and are of similar size to the homologous Ti *tms* transcripts (Taylor *et al.*, 1985b; Willmitzer *et al.*, 1983). Additional transcripts were localized to the right portion of the T_R-DNA region which may correspond to the genes for agropine synthesis. The cDNA prepared from agropine synthesizing roots has been mapped to the same region as the 1.5 kb and 1.6 kb transcripts identified in tumors (De Paolis *et al.*, 1985).

The morphogenic affect of IAA is also dependent on the concomitant level of cytokinin — root formation being favored when the ratio of auxin to cytokinin is relatively high (Skoog and Miller, 1957). The pronounced tendency of root formation by *A. rhizogenes* on many plant species as opposed to gall formation suggests the lack of a potent cytokinin synthesis gene such as the Ti plasmid *tmr* gene which inhibits root formation (Garfinkel *et al.*, 1981; Joos *et al.*, 1983). The lack of a comparable *tmr* locus on the Ri plasmid helps explain the observation that the Ri plasmid behaves under some circumstances similar to a mutated Ti plasmid.

The T_L-DNA spans a region of approximately 20 kb of the agropine Ri plasmid (Fig. 4; Durand-Tardif *et al.*, 1985: Taylor *et al.*, 1985b; White *et al.*, 1982, 1983, 1985). Some 21,126 base pairs of the T_L-DNA have been sequenced (Slightom *et al.*, 1986). The T-DNA has been examined in detail in relatively few individual tumor, root or plant lines (Durand-Tardif *et al.*, 1985; White *et al.*, 1985). Therefore, generalizations regarding the boundaries of T_L-DNA region may be premature. However, the evidence from different laboratories is generally consistent, and the sequenced portion covers the region which is transferred in all known cases. The sequence indicates the potential for at least 18 open reading frames with greater than 255 nucleotides (Slightom *et al.*, 1986; Fig. 4). Both eukaryotic-like and procaryotic-like control sequences were associated with some of the open reading frames (Slightom *et al.*, 1986). However, there is no real consensus regarding what constitutes an effective promoter in *Agrobacterium* T-DNA since a wide variety of structures have been found (Barker *et al.*, 1983). Although the general features of the T_L-DNA appear similar to the sequence of the Ti plasmid T-DNA, the sequence essentially confirms the low degree of relatedness to the other T-DNA regions (Barker *et al.*, 1983; Slighton *et al.*, 1986).

The T_L-DNA varies in different transformed plant species and individual plants, and the variability can result from the amount of leftward or rightward sequences which are incorporated. In one line of *N. glauca* the T-DNA extended into *Hin*dIII-11; and in a second line, *Hin*dIII-11 was not detected (White *et al.*, 1985). The border in the second line was presumably just within *Hin*dIII-11 and was not detected by hybridization. Alternatively, the T-DNA only extended into the 0.8 kb fragment separating *Hin*dIII-11 and *Hin*dIII-21, since a *Hin*dIII-21 internal fragment was observed (White *et al.*, 1985). Left borders that occur in *Hin*dIII-11 have been reported for *D. carota* (carrot) and *Convolvulus arvensis* (morning glory) T_L-DNA (Durand-Tardif *et al.*, 1985). These borders occurred specifically within the small *Bam*HI fragment (*Bam*HI-32; Durand-Tardif *et al.*, 1985). The left borders in *N. glauca* may also occur in *Bam*HI-32, but this possibility has not been ascertained (White *et al.*, 1985). In two *N. tabacum* lines, the leftward extent of T-DNA occurred within *Eco*RI- 40 (Durand-Tardif *et al.*, 1985). *Eco*RI-40 lies within *Hin*dIII-21, indicating that a third left border is possible. Whether the *Eco*RI-40 border is a *N. tabacum* specific event is open to question since the *Hin*dIII-21 internal fragment was present in three of four *N. tabacum* regenerated plants from another study (Taylor *et al.*, 1985a). Thus, there may exist several sites where left T_L-DNA plant/plasmid junctions can occur.

The extent of the rightward region of the T_L-DNA also can vary, although in most transformed plant lines examined in detail the T_L-DNA extend into *Hin*dIII-16. More precisely, the borders were located between the *Kpn*I site at the right end of *Hin*dIII-16 since the entire *Kpn*I-12 fragment is consistently observed in transformed plant DNA (Durand-Tardif *et al.*, 1985; White *et al.*, 1982, 1985). Examination of the nucleotide sequence of the Ri T_L-DNA reveals a sequence within 60 bp of the *Kpn*I-12 site which is partially related to the consensus border sequences of Ti plasmid T-DNA regions (Slightom *et al.*, 1986). A similar sequence was located within 26 base pairs of a left T_L-DNA/plant junction (Slightom *et al.*, 1985). Possible alternative right borders for three *C. arvensis* transgenic plants have been reported (Slightom *et al.*, 1985), but the exact placement is not known. If the Ri plasmid employs the consensus border sequence the fact that 27 consensus sequences were located throughout the T_L-DNA (Slightom *et al.*, 1986) would suggest that a number of alternate borders are possible. The transfer of Ti T-DNA in the presence of the Ri virulence region strongly suggests that these sequences are involved in the transfer and integration of the Ri T-DNA (Hoekema *et al.*, 1984).

The agropine Ri T_L-DNA region is comparable in size to other T-DNA regions, but does not appear to be closely related to the well characterized Ti T-DNA regions with the possible exception of limited regions of homology to the T-DNA of nopaline type Ti plasmids (Huffman *et al.*, 1984). One or both regions of homology are presumed to be involved with the synthesis of agrocinopines on the basis of the proximity of the homologous sequences to the agrocinopine gene of the nopaline type T-DNA (Willmitzer *et al.*, 1982). The agropine Ri plasmids direct the synthesis of agrocinopine A in transformed roots (Petit *et al.*, 1983).

The T_L-DNA region is transcribed in both callus tissues and regenerated plants (Taylor *et al.*, 1985b; Durand-Tardif *et al.*, 1985; Ooms *et al.*, 1986b). Although individual transcripts are not well characterized, some of the transcripts correspond roughly to open reading frames of appropriate size and direction (Fig. 4). Discrepancies exist in the different sets of data which hopefully will be resolved as more transformed plant lines are examined and individual transcripts are characterized. The relative amounts of the different transcripts vary in a culture and tissue specific manner (Durand-Tardif *et al.*, 1985; Ooms *et al.*, 1986b); whether the differences reflect transcriptional or posttranscriptional control is not known. The capacity of the plant to regulate the expression of the acquired DNA would have intriguing evolutionary implications.

The T_L-DNA has also been subjected to extensive mutagenesis, and four loci which affect tumorigenesis have been tentatively identified (Fig. 4; White *et al.*, 1985). Each locus loosely corresponds to an open reading frame of Slightom *et al.* (1986): ORF 10, *rol*A; ORF 11, *rol*B; ORF 12, *rol*C; and ORF 15, *rol*D. Whether the transposon insertions have caused disruptions in ORF expression remains to be demonstrated. The effects of the *rol* mutations are best observed on leaves of *Kalanchoë* (White *et al.*, 1985; Fig. 5). Mutations in *rol*A cause a change in root morphology. Roots which emanate from the wound appear to grow straighter, giving the tumor a less compact appearance. The *rol*B mutations eliminate root formation completely on leaves. The *rol*C and *rol*D mutations are more subtle than the previous two mutations. In a normal hairy root response on *Kalanchoë* leaves, slight callus growth occurs, followed by extensive root proliferation. The *rol*C mutation appears to arrest or slow the process such that root initiation is retarded. The *rol*D mutations, on the other hand, appear to accentuate callus growth until the tumorous response resembles a crown gall tumor. The subtle effects on tumor morphology, particularly those of *rol*C and *rol*D, are reminiscent of the *tml* mutations of the Ti plasmid T-DNA (Garfinkel *et al.*, 1981). However, there is no evidence that the loci are related (Slightom *et al.*, 1986). Additional mutations in the T_L-DNA region caused slightly attenuated responses (White *et al.*, 1985).

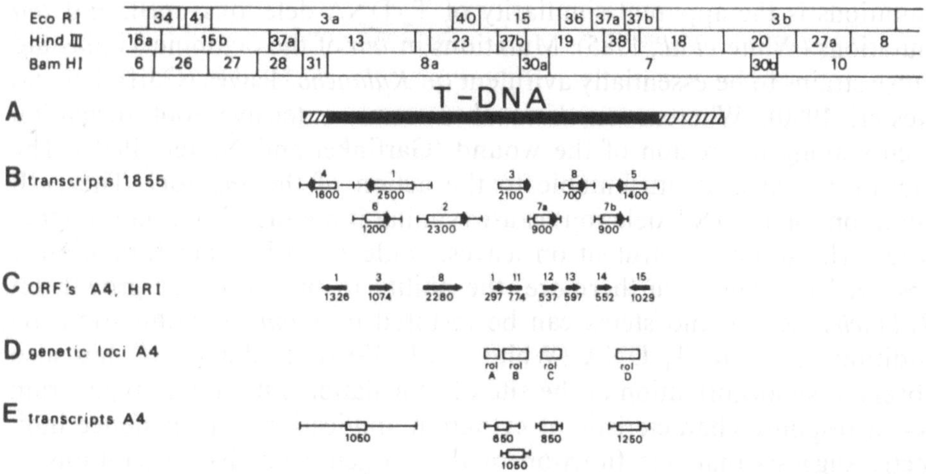

Fig. 4. T_L-DNA of pRiA4. A. Core T-DNA region representative of many T-DNA insertions. B. Transcripts of T_L-DNA as determined by Ooms *et al.* (1986) and Durand-Tardif *et al.* (1985). C. Positions of open reading frames as determined by Slightom *et al.* (1986). D. Genetic loci as determined by White *et al.* (1985). E. Transcripts as determined by Taylor *et al.* (1986). The figure was provided through the courtesy of G. Ooms

The effects of the *rol* mutations are most dramatic on *Kalanchoe* leaves where there is an absolute requirement for both the T_L-DNA locus of *rol*B and the auxin synthesis loci, *tms*1 and *tms*2. However, the genes of the T_L-DNA have effects on other plants as well. Recently, the observation was made that *rol*B is required for root induction on the apical side of carrot root disks, but only in the absence of the auxin genes (Estramareix *et al.*, 1986). The effects of the T_L-DNA can also be seen in the phenotypes of regenerated tobacco plants (discussed further below). Under normal conditions some plant tissues such as carrot disks have an endogenous root induction factor(s) which compensates for the T_L-DNA functions in the presence of adequate auxin levels (Ryder *et al.*, 1985). This is apparently the situation on tobacco and *Kalanchoë* stems, for example, if the *tms* genes are intact. Endogenous auxin levels are adequate for root induction on *Kalanchoe* and apical carrot root disks only in the presence of the T_L-DNA. The most extreme case appears to be *Kalanchoë* leaves where the action of both T_L-DNA and T_R-DNA is required. Thus, two signals may be needed for root induction. One is, of course, auxin. The other is supplied by the T_L-DNA functions and is unknown.

The *rol* mutations present a rather complex picture of T_L-DNA function, and there are few clues as to what the functions might be. At present, the only evidence comes from comparisons of Ri and Ti T-DNA mutations (White *et al.*, 1985), and a morphogenic effect noted on *N. tabacum* leaves (Tepfer, 1984). The most striking characteristic of the *rol* mutations is the apparent similiarity of T_L-DNA deletions, *rol*B, and *tmr* mutations (White *et al.*, 1985). Mutations in *tmr* of the octopine Ti plasmid cause strains to be essentially avirulent on *Kalanchoë* leaves (Garfinkel and Nester, 1980). Whereas on *Kalanchoë* stems, extensive root formation occurs along the region of the wound (Garfinkel and Nester, 1980). The rooting response is attributable to the action of the *tms* loci. The *rol*B mutation (or T_L-DNA deletion) causes similar if not identical tumor phenotypes. The strain is avirulent on leaves, while extensive root formation is observed on stems. Furthermore, the ability to induce callus growth on *Kalanchoë* leaves and stems can be restored to a *tmr* mutation strain by addition of the Ri T_L-DNA (White *et al.*, 1985). Tepfer (1984) has also observed shoot formation at the site of inoculation with a wild type strain A4, a response characteristic of cytokinin application. The evidence indirectly suggests that one function of the *rol* genes involves alterations in cytokinin synthesis or a cytokinin-like effect. Further clarification of the effects of *rol* loci await careful phytohormone analysis of transformed plant tissue.

The T-DNA of the mannopine type Ri plasmid from strain 8196 presents additional unusual features for an *Agrobacterium* virulence plasmid. Only one T-DNA has been identified for pRi8196, although, in the two regenerated carrot plants that have been examined, two different T-DNA contents were detected. One transformant contained 34—42 kb of plasmid DNA representing one contiguous stretch of pRi8196 (Byrne *et al.*, 1983). However, whether the T-DNA was present as one large insert without discontinuities or a series of overlapping sections could not be determined from the data. Most T-DNA regions appear to be in the order of 15—20 kb, thus the T-DNA region of pRi8196 is the largest region yet identified. The second transformed carrot line contained only 18 kb of the T-DNA region (Byrne *et al.*, 1983). The rightward boundary of the T-DNA region from both lines lies within the *Bam*HI-14a fragment suggesting that the right portion represents the core of the T-DNA region (Byrne *et al.*, 1983). The T-DNA region of pRi8196 has not been characterized extensively, and, like the A4 T_L-DNA, there are few clues to the functions of T-DNA genes.

Hybridization data from a comparison of agropine Ri T_L-DNA (pRi1855) and T-DNA of pRi8196 suggest that the two T-DNA regions may be very similar. Clones covering the *Bam*HI-6 and *Bam*HI-19 fragments from pRi8196 (Fig. 2) hybridize with the core T-DNA region from 8196 (Spano *et al.*, 1982). The region of homology apprears to include most of the *rol* loci. Some heterogeneity may exist in the middle of the region since a relatively weak hybridization signal was observed for the central *Eco*RI-15 (*Eco*RI-4.6 of White *et al.*, 1985) fragment which contains three of the *rol* loci (Spano *et al.*, 1982; White *et al.*, 1985). The homology to genes for mannopine synthesis in the extreme right boundary of the pRi8196 indicates that the analogous T_R-DNA region of the agropine Ri plasmid may be fused into one T-DNA with the exclusion of the *tms* genes (Lahners *et al.*, 1984). Two other interesting features of the pRi8196 T-DNA involve the presence of Ti *vir* region sequence homology and possible inclusion of agrocinopine catabolism genes within the T-DNA (Lahners *et al.,* 1984). The function of these regions of homology needs to be further characterized by genetic analysis.

T_R-DNA mutagenesis experiments suggest that agropine type strains which are lacking the auxin synthetic functions behave very similar to strains containing pRi8196 (Cardarelli *et al.,* 1985; White *et al.,* 1985). Restriction fragment heterogeneity between the two T-DNA regions, however, argues aganist a simple deletion of the T_R-DNA of the agropine Ri plasmid to yield the 8196 plasmid. As discussed earlier, perhaps the agropine and mannopine type Ri plasmids are more closely related to each other than to octopine and nopaline Ti plasmids; yet each represents one subgroup of an undetermined number of plasmid types which fall in the Ri category.

V. T-DNA of Ri Transformed Plants

With the characterization of the Ri T-DNA regions, the behavior of hairy root tumors and the transformed plant tissue derived from the tumors can be better understood. Hairy root tumors have a distinct morphological phenotype on many plant species, notably, the pronounced rooting, compared to tumors derived from strains containing Ti plasmids. Ri plasmid transformation of plants does not appear to be as inhibitory to normal differentiation as does the incorporation of phytohormone synthesis genes of the Ti T-DNA region. Plants can be readily regenerated from the callus or roots of infected plant tissue (Ackermann, 1977; Chilton *et al.,* 1982; Spano and Costantino, 1982; Tepfer, 1982, 1984; David *et al.,* 1984; Taylor *et al.,* 1985 a), and the T-DNA can be subsequently transmitted through meiosis to progeny plants (Costantino *et al.,* 1984; Tepfer, 1984; Taylor *et al.,* 1985 a). However, plants containing Ri T-DNA are not always normal in appearance (Fig. 6; Tepfer, 1984; Taylor *et al.,* 1985 a). Transgenic plants often have aberrant features such as wrinkled leaves, shortened internode distances, rounded or indented leaf margins, heterostyly, reduced fertility, reduced apical dominance, increased lateral root initiation, and reduced root geotropic responses (Ackermann, 1977; Spano *et al.,* 1981; Spano and Costantino, 1982; Tepfer, 1982, 1984; Taylor *et al.,* 1985 a). Some of these traits, such as reduced fertility and heterostyly, have been noted in plants regenerated from Ti T-DNA containing tumors (Wullems *et al.,* 1981). The synthesis of otherwise root-specific peroxidase isoenzymes in transformed leaf tissue has also been observed (Benvenuto *et al.,* 1983). Other physiological abnormalities are reduced osmoticum and concomitant reduced K^+ in the cell sap (Ooms *et al.,* 1986 a) and increased transpiration rates (V. Sinkar, unpublished results) in the transformed plants. Different degrees of aberrant morphologies can occur (Tepfer, 1984; V. Sinkar, unpublished results), and abnormal plants can revert to less severe or normal appearance (Fig. 7; Tepfer, 1984; V. Sinkar, unpublished results). Similar

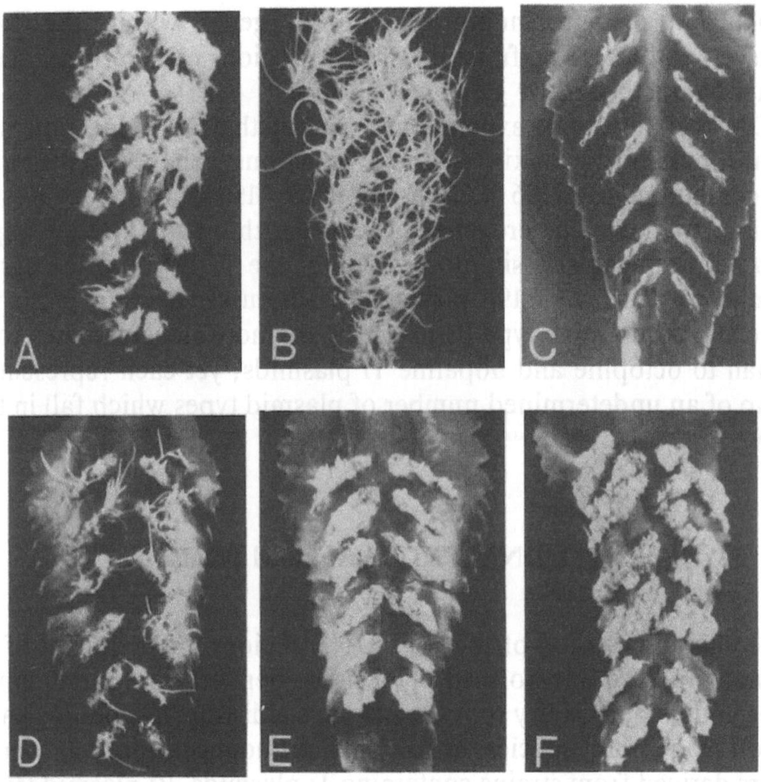

Fig. 5. Effects of *rol* mutations on *Kalanchoë* leaves. A. Wild type; B. *rol*A; C. *rol*B; D. slightly attenuated R1020 insertion; E. *rol*C; F. *rol*D

reversion to normal growth habit has been observed in Ti T-DNA transformed tissue (Turgeon *et al.*, 1976). The suppression of the morphogenic effects has been attributed in some cases to methylation (Amasino *et al.*, 1984) or deletion of the T-DNA sequences (Yang and Simpson, 1981). In other instances, the plants appear to adjust in some unknown way to the effects of the T-DNA genes and establish a relatively normal growth pattern (Barton *et al.*, 1983; Black *et al.*, 1986). The basis of the phenotype switching is not known.

Features of the Ri plasmid suggest a number of possibilities of how the structure and content of the Ri plasmid T-DNA regions might result in the observed morphogenic characteristics and behavior of transformed tissue. In fact, the morphologies may be due to multigenic effects of the T-DNA. Mutagenesis and T-DNA analysis indicate that either T-DNA region of the agropine Ri plasmid can contribute genes which have morphogenic potential (Cardarelli *et al.*, 1985; Durand-Tardif *et al.*, 1985; Taylor *et al.*, 1985a; White *et al.*, 1985). Ultimately, the effect of each locus of both regions needs to be examined. Since the genes are distributed between two T-DNA regions, even the configuration of the Ri T-DNA genes probably plays a role in the characteristics of transformed plants. Different T-DNA

Fig. 6. Morphology of regenerated plants. A. Wrinkled leaf phenotype of regenerated *N. tabacum* (left) and normal plant (right). B. Wrinkled leaf phenotype of regenerated *N. glauca* (left) and normal plant (right)

regions may be transferred to individual cells, and an unequal distribution of these genes to the transformed cells would result in a heterogeneous population of cells with regards to growth characteristics. Some cells may receive only T_R-DNA, others only T_L-DNA, and others may receive both regions. The portions of the population of transformed cells with each type of T-DNA complement will influence both the overall morphology of the initial tumor response and the morphology of individual clonal cell lines derived from the initial tumor. The frequency of Ri T-DNA co-transfer has not been reported. However, separate Ti plasmid T-DNA regions present on the same plasmid (Depicker *et al.*, 1985), or on different vectors in the same bacterium (An *et al.*, 1985), co-transformed cells at a frequency higher than expected for independent transfer (60—70 % vs. 20 %), but not at 100 %. Thus, not every transformed cell receives both T-DNA regions.

The T_R-DNA of the Ri plasmid possesses the gene homologs for auxin synthesis (Huffman *et al.*, 1984; Jouanin, 1984; Cardarelli *et al.*, 1985; White *et al.*, 1985), and integration of T_R-DNA can, therefore, lead to auxin synthesis in transformed cells. Many plants will initiate roots if exposed to high auxin levels. The lack of extensive root initiation on many test plants when inoculated by *Agrobacterium* strains containing Ri *tms* mutations clearly indicates the involvement of T-DNA directed auxin synthesis in hairy root disease (Cardarelli *et al.*, 1985; White *et al.*, 1985). *N. glauca* represents an exceptional case. Phytohormone autonomous tumor lines of *N. glauca* were found to contain the *tms* genes (White *et al.*, 1985). Either the endogenous phytohormone milieu or the sensitivity of the cells is such that autonomous growth is stimulated by auxin alone.

Another factor in the morphological effects of the T_R-DNA is the structure of the region. The left T_R-DNA border of the agropine type Ri

Fig. 7. Reversion of wrinkled-leaf *N. glauca* to normal growth habit. Stem on left spontaneously appeared from regenerated *N. glauca* plant containing T_L-DNA sequences

plasmid appears to be located relatively close to the *tms*2 gene (White *et al.*, 1985). In addition, transformed plants often do not contain the entire portion of the T_R-DNA (De Paolis *et al.*, 1985; Durand-Tardif *et al.*, 1985; Taylor *et al.*, 1985 a). Thus, transfer of the entire region including a complete copy of *tms*2 may be relatively inefficient and a border-like sequence may be located to the right of the *tms* genes. The presence of at least a portion of both T-DNA regions in the transformed plants suggests that cotransfer of the rightward portion of the T_R-DNA occurs at a relatively high frequency.

The analysis of T-DNA of regenerated plants and tumor calli suggests that incorporation of the *tms* genes into *N. glauca* is inhibitory to regeneration. Possibly only those cells which either lost the *tms* genes of T_R-DNA or never received the *tms* genes are capable of regenerating whole plants, since none of the plants examined contained the genes (Taylor *et al.*, 1985 a). On the other hand, regenerated *N. tabacum* plants appear to tolerate the presence of the Ri *tms* loci since in three of four plants these genes appear to be present (Taylor *et al.*, 1985 a). Some of the plants contain the *tms* genes in multiple copies, and the genes appear to be expressed at a low level (Taylor, unpublished results). Durand-Tardif *et al.* (1985) reported that the entire T_R-DNA was not present in several transformed clones of *N. tabacum*. Therefore, regeneration of tobacco may be affected by the presence of the *tms* genes, but not to the same extent as *N. glauca*. The method used in selecting the transgenic plants may also determine the T-DNA content since selection for certain differentiated structures (shoots vs. roots) may favor a particular T-DNA content (Tepfer, 1984).

The results of the mutagenesis and T-DNA analyses implicate the involvement of the T_L-DNA genes in many of the transgenic abnormalities. In particular, the evidence strongly suggests that the wrinkled leaf, dented leaf margin, shortened internode, increased lateral root formation, and partial loss of root geotropism are, in part, due to the T_L-DNA (Durand-Tardif *et al.*, 1985; Taylor *et al.*, 1985 a; White *et al.*, 1985). In wrinkled *N. glauca*, the T_R-DNA pattern indicated that complete copies of either *tms* gene were absent, and one plant did not appear to contain any T_R-DNA sequences (Taylor *et al.*, 1985 a). Similarly, *N. tabacum* plants which have the transformed phenotype also may lack the *tms* genes (Durand-Tardif *et al.*, 1985). Thus, some gene or combination of genes in the T_L-DNA appears to control the phenotype. No direct evidence indicating which genes of the T_L-DNA are responsible exists. However, the effect of T-DNA mutations on root induction on *Kalanchoë* leaves identified four candidates (*rol*A—D). Some *N. tabacum* plants do not contain the region of ORFs 1—7 and still display the transformed phenotypes. Therefore, these potential genes are not likely to be involved (Durand-Tardif *et al.*, 1985; Slightom *et al.*, 1986). The effects of the individual *rol* mutations on the transformed plant phenotype are presently being assessed.

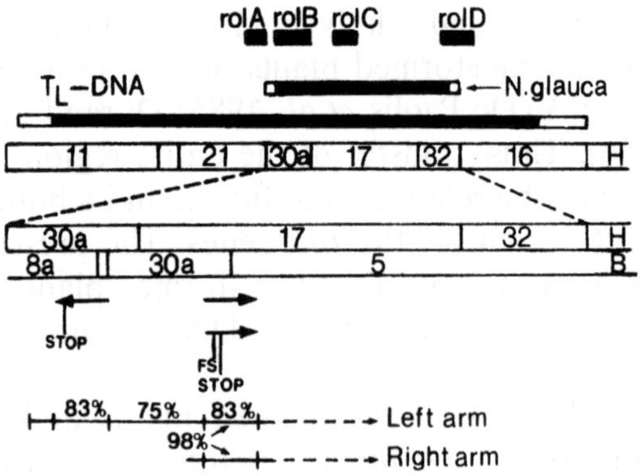

Fig. 8. Cellular T-DNA sequences of *N. glauca*. Arrows below the map represent the left (upper row) and right (lower row) open reading frames of the plant sequences. The percentages indicate the degree of relatedness to the bacterial sequences and the right and left arms. FS, frame shift; STOP, nonsense codon; H, *Hin*dIII; B, *Bam*HI. Data is from Furner *et al.* (1986)

VI. Endogenous T-DNA of Plants

Perhaps one of the most intriguing aspects of Ri T-DNA has been the identification of endogenous homologous T-DNA sequences in different species of plants. The presence of these sequences suggest the exciting possibility that bacterially acquired sequences play an important role in the evolution of some, perhaps all, plant species. Whether in fact the T-DNA regions are contributing evolutionarily significant genes or simply represent "genetic scars" remains to be determined.

Hybridizations of untransformed normal plant genomic DNA with Ti plasmid probes have occasionally been reported (Thomashow *et al.*, 1980; Yang and Simpson, 1981). However, the observations have never been pursued further. Initial T-DNA characterization experiments involving both agropine and mannopine Ri plasmids led to reports of hybridization signals to four different plant species (*D. carota*, Spano *et al.*, 1982; *C. arvensis*, Tepfer, 1982; *N. glauca*, White *et al.*, 1982). In the case of the agropine Ri plasmid, strong hybridization to *N. glauca* DNA was observed with at least one region of the pRiA4b, which was subsequently demonstrated to be the T_L-DNA (White *et al.*, 1983; Fig. 8). The region of homology was cloned and found to contain an approximately 6—8 kb of the central T-DNA region. Further analysis has revealed that the plant copy contains an asymmetrical inverted repeat bringing the total content to approximately 12 kb (Furner *et al.*, 1986). Portions of the same region were also found in other species of *Nicotiana* including *N. tabacum, N. otophora,*

N. tomentosiformis, N. tomentosa, and *N. benavidesii* (Furner *et al.,* 1986). No homology was reported outside the genus *Nicotiana.*

Although it may be difficult to prove whether the important features of the T_L-DNA genes evolved in plants or bacteria, it seems likely that the plant sequences were acquired from an *Agrobacterium* transformation event which occurred early in the evolution of the *Nicotiana* species (Furner *et al.,* 1986). The duplication may have occurred during the integration event similar to the scrambling of T-DNA sequences in present day integrations. The plant T-DNA is closely related but not identical to the A4 T_L-DNA. Both the right and left halves of the plant sequences show approximately 17 % divergence in the coding sequences (*rol*B and C) and 25 % divergence in the intergenic region from the present A4 sequences. Therefore, the inciting strain was somewhat divergent from A4, or the present A4 sequences represent the degree of change during the millions of years since the transformation event. The plant sequences would be expected to change more slowly due to the increased generation time compared to the bacteria. The plant, in a sense, has preserved an ancient bacterial sequence.

The question remains whether the genes have been conserved while in the plant. The region of homology between *N. glauca* and pRiA4b encompasses the *rol*C and *rol*B loci (ORF 11 and 12; Slightom *et al.,* 1986), and therefore potentially represents a functional gene with known plant morphogenic effects (White *et al.,* 1983, 1985). The sequence data has revealed that the left plant copy of the open reading frame that corresponds to the *rol*C locus (ORF 12 of Slightom *et al.,* 1986) contains the complete open reading frame (Furner *et al.,* 1986; Fig. 8). The right copy of the plant *rol*C contains a frame shift and a termination codon suggesting that this copy is not potentially functional. The single ORF which corresponds to the *rol*B locus in the *N. glauca* DNA encodes a polypeptide that is truncated by 48 amino acid residues compared to the Ri copy (Furner *et al.,* 1986). There is approximately 2 % divergance between the right and left halves of the plant sequences. The retention of potentially functional genes may reflect selection and conservation in the plant, or simply insufficient time has passed for mutations to accumulate which disrupt the coding sequences (Furner *et al.,* 1986).

Many questions are posed by the presence of endogenous T-DNA. First of all, how widespread is the occurrence of these sequences in the plant kingdom? Several reports indicated the presence of sequences in plant species outside the *Nicotiana* genus (Spano *et al.,* 1982; Tepfer, 1982). Other instances of homology to plant DNA which were subsequently localized to the cloning vector portion of the hybridization probe in addition to the T-DNA sequences have been reported (Hood *et al.,* 1984). However, these cases have not been examined further. There is no compelling reason not to expect the discovery of T-DNA sequences in other plants. The strains from which T-DNA homology was observed were not originally isolated from *Nicotiana* species, and therefore, there is no special relationship between the *Nicotiana* and *Agrobacterium.*

Experiments designed to test whether the endogenous T-DNA regions can complement bacterial T_L-DNA mutations and, therefore, represent potential plant genes are in progress. However, an even more daunting task will be obtaining evidence that the endogenous sequences might represent a driving force in the evolution of plant species. The capture of bacterial genes, particularly phytohormone genes, may also provide an explanation as to what benefit plants derive from their association with *Agrobacterium.* Perhaps plants do not gain a direct benefit such as in the related *Rhizobium*/legume association for nitrogen fixation, but rather the relationship is part of the process of plant evolution itself.

VII. Conclusions and Further Speculations

Much attention has been given to the pliable nature of Ri plasmid transformed plant tissue, and the relative ease of regenerating whole plants from the tissue. However, the species, for which regeneration from Ri transformation has been characterized (e. g. *N. glauca, N. tabacum, D. carota, Solanum tuberosum,* and *C. arvensis*), are known to be amenable to regeneration techniques. Thus, the Ri plasmid does not necessarily confer on the plant a new regenerative potential, but appears to direct phenotypic changes in plant tissue which allow both identification of transformed cells and regeneration of whole plants from the cells.

The facility of plant regeneration from Ri transformations was also being considered in context of earlier studies concerning regeneration of Ti plasmid transformed tissue (Wullems *et al.*, 1981). Phytohormone synthesis genes for both cytokinin and auxin are tightly linked in the octopine and nopaline Ti plasmid T-DNA. Most cells which are transformed with T-DNA then contain a very potent morphogenic complement of genes. Ti plasmids have since been constructed with the phytohormone synthesis genes of the T-DNA removed (Zambryski *et al.*, 1983; An *et al.*, 1985; Klee *et al.*, 1985), and cells transformed with these altered T-DNA regions can be induced to form whole plants more readily than plant cells containing the wild type T-DNA complement of genes. While genes of Ri T_L-DNA have some cytokinin-like effects, the genes do not appear to have the severe oncogenic effects of the *tmr*/*tms* combination of the Ti plasmid T-DNA, and, therefore, the Ri plasmid can function somewhat like a disarmed Ti plasmid.

The Ri plasmid T-DNA regions can be used for the construction of plant transformation vectors similar to vectors derived from the Ti plasmid T-DNA regions. Plants with either T-DNA region of the Ri plasmids can display growth abnormalities. Therefore, the strategy for vector development is much the same as for the Ti vectors in that genes with morphogenic effects must be removed if the recovery of a normal plant is to be assured. For example, the T_L-DNA of pRiA4 was recently used for the introduction of a herbicide resistance gene (Comai *et al.*, 1985) and a nodule-specific leghaemoglobin gene (Jensen *et al.*, 1986). Promoters for Ri

T-DNA genes can also be identified and used in combination with other coding sequences.

Imaginative ideas have been developed to take advantage of the particular properties of Ri plasmid-directed transformations. One suggestion, which has advanced to the experimental stage, is to devise methods for using *A. rhizogenes* infections in the development of drought resistant plants (Strobel and Nachmias, 1985). *A. rhizogenes* treated almond plants *(Prunus amygdalus)* showed significant differences in the numbers of roots and leaves per plant, stem thickness, and branch length after one season at the Gilat Desert Experiment Station (Strobel and Nachmias, 1985). The apparent benefit of the treated plants may be due to a bacterialization effect either by the production of growth promoting substances or inhibition of the production of deleterious substances by other microorganisms. Whether transformation of the plants by Ri plasmid T-DNA is involved is not known (Strobel and Nachmias, 1985). The benefits may be due to the type of *Agrobacterium* rather than the Ri plasmid. Increases in the nodulation of alfalfa *(Medicago sativa)* by *Rhizobium meliloti* strains which harbor an Ri plasmid have been noted in greenhouse experiments (Strobel *et al.*, 1985). Therefore, transfer of the Ri plasmid to other soil bacteria represents another potential agronomic benefit. On the other hand, the isolation of strains of *A. rhizogenes* was in fact due to severe disease situations where deleterious effects on plant growth and even plant death were recorded (Munnecke *et al.*, 1963). Clearly, further characterization of practical applications of bacteria containing an Ri plasmid needs to be performed, but such approaches offer interesting possibilities.

Other approaches in the potential use of the Ri plasmid involve the use of plants containing the T-DNA genes. Tepfer (1984), for example, has noted the conversion of transformed carrots from a biennial to an annual growth habit, which could potentially improve breeding strategies. The genes of Ri T-DNA may therefore confer agronomically valuable characteristics. However, whether such changes can be separated from deleterious effects (e. g. reduced fertility) remains to be determined. The aberrations in plant growth may also represent important tools in studying the normal process of differentiation in plants (Tepfer, 1984; White *et al.*, 1985; Ooms *et al.*, 1985). Understanding the biochemical alterations due to the T-DNA genes will be essential to realizing the full potential in regards to plant development.

While T-DNA genes of Ri plasmid do not have the deleterious effects of the combination of Ti phytohormone synthesis genes, the question remains regarding what biochemical events Ri genes direct. Clearly, the synthesis of auxin can be ascribed to the T_R-DNA, and the synthesis of auxin is responsible for some of the morphogenic properties of Ri transformed plant tissue. However, even in the absence of T_R-DNA directed auxin synthesis, as in mannopine Ri plasmids which naturally lack the *tms* loci, and agropine Ri *tms* mutants, some root induction is observed. Regenerated plants, which do not contain the *tms* genes, also may have growth abnormalities. Therefore, genes of Ri T_L-DNA direct the synthesis of a substance (or substances) which apparently recruit cells to differentiate into roots under the influence of endogenous auxin synthesis as well as T_R-DNA directed auxin synthesis. The synthesis of these purported substances is likely to be responsible for the other abnormallities as well. Analysis of the phytohormone content of transformed roots and plants should provide clues as to the action of the Ri T-DNA genes. Whether the agropine and mannopine T-DNA equivalents function in a similar manner is not known. More extensive homology comparisons between the two analogous T-DNA regions will undoubtedly resolve the question.

The Ri plasmids have also pointed the way for search of endogenous T-DNA genes in the genomes of the present day plants. This aspect of the research may prove an important contribution to our understanding of plant development and evolution. It should be reiterated that there is no compelling reason to believe that the presence of endogenous T-DNA sequences is not more widespread. Perhaps plants are continually acquiring new genes, partiularly phytohormone synthesis genes, from microbial sources, and these genes may confer new selective advantages to the plant. The application of auxins and cytokinins, for example, are known to affect both developmental and disease resistance processes. Conversely, the prospect that bacteria such as *Agrobacterium* species may shuttle genes between plants, or that T-DNA genes were originally plant-derived remains to be tested. However, the associations of bacteria and plants have existed from the beginning of plant evolution, and attempts to decide which came first, the proverbial chicken or the egg, will likely never be resolved.

VIII. References

Ackermann, C., 1977: Pflanzen aus *Agrobacterium rhizogenes* Tumoren an *Nicotiana tabacum.* Plant Sci. Lett. **8,** 23—30.

Akiyoshi, D. E., Klee, H. J., Amasino, R., Nester, E. W., Gordon, M. P., 1984: T-DNA of *Agrobacterium tumefaciens* codes an enzyme of cytokinin biosynthesis. Proc. Natl. Acad. Sci., U.S.A. **81,** 5994—5998.

Albinger, G., Beiderbeck, R., 1977: Übertragung der Fähigkeit zur Wurzelinduktion von *Agrobacterium rhizogenes* auf *A. tumefaciens.* Phytopathol. Z. **90,** 306—310.

Amasino, R. M., Powell, A. T., Gordon, M. P., 1984: Changes in T-DNA methylation and expression are associated with phenotypic variation and plant regeneration in a crown gall tumor line. Mol. & Gen. Genet. **197,** 437—446.

An, G., Watson, B. D., Stachel, S., Gordon, M. P., Nester, E. W., 1985: New cloning vehicles for transformation of higher plants. EMBO J. **4,** 277—284.

Anderson, A. R., Moore, L. W., 1979: Host specificity in the genus *Agrobacterium.* Phytopathology **69,** 320—323.

Ark, P. A., Thompson, J. P., 1961: Detection of hairy root pathogen, *Agrobacterium rhizogenes,* by use of fleshy roots. Phytopathology **51,** 69—71.

Barker, R. F., Idler, K. B., Thompson, D. V., Kemp, J. D., 1983: Nucleotide sequence of the T-DNA region from the *Agrobacterium tumefaciens* octopine Ti plasmid pTi15955. Plant Mol. Bio. **2,** 335—350.

Barton, K. A., Binns, A. N., Matzke, A. J., Chilton, M.-D., 1983: Regeneration of intact tobacco plants containing full length copies of genetically engineered T-DNA and transmission of T-DNA to R1 progeny. Cell **32,** 1033—1043.

Benvenuto, E., Ancora, G., Spano, L., Costantino, P., 1983: Morphogenesis and isoperoxidase characterization in tobacco "Hairy Root" regenerants. F. Pflanzenphysiol. **110,** 239—245.

Black, R. C., Kuleck, G. A., Binns, A. N., 1986: The initiation of auxin autonomy in tissue from tobacco plants carrying the auxin biosynthesizing genes from the T-DNA of *Agrobacterium tumefaciens.* Plant Physiol. **80,** 145—151.

Brown, N. A, 1929: The tendency of the crown-gall organism to produce roots in conjunction with tumors. J. Agr. Res. **39,** 747—766.

Byrne, M. C., Koplow, J., David, C., Tempe, J., Chilton, M.-D., 1983: Structure of T-DNA in roots transformed by *Agrobacterium rhizogenes.* J. Mol. Appl. Genet. **2,** 201—209.

Cardarelli, M., Spano, L., De Paolis, A., Mauro., M. L., Vitali, G., Costantino, P., 1985: Identification of the genetic locus responsible for non-polar root induction by *Agrobacterium rhizogenes* 1855. Plant Mol. Biol. **5,** 385—391.

Chilton, M.-D., Tepfer, D. A., Petit, A., David, C., Casse-Delbart, F., Tempe, J., 1982: *Agrobacterium rhizogenes* inserts a T-DNA into the genomes of the host plant root cells. Nature **295,** 432—434.

Comai, L., Facciotti, D., Hiatt, W. R., Thompson, G., Rose, R. E., Stalker, D. M., 1985: Expression in plants of a mutant *aroA* gene from *Salmonella typhimurium* confers tolerance to glyphosate. Nature **317,** 741—744.

Costantino, P., Mauro, M. L., Micheli, G., Risuleo, G., Hooykaas, P. J. J., Schilperoort, R., 1981: Fingerprinting and sequence homology of plasmids from different virulent strains by *Agrobacterium rhizogenes.* Plasmid **5,** 170—182.

Costantino, P., Spano, L., Pomponi, M., Benvenuto, E., Ancora, G., 1984: The T-DNA of *Agrobacterium rhizogenes* is transmitted through meiosis to the progeny of hairy root plants. J. Mol. Appl. Genet. **2,** 465—470.

Currier, T. C., Nester, E. W., 1976: Evidence for diverse types of large plasmids in tumor-inducing strains of *Agrobacterium.* J. Bacteriol. **126,** 157—165.

David, C., Chilton, M.-D., Tempe, J., 1984: Conservation of T-DNA in plants regenerated from hairy root cultures. Biotechnology **2,** 73—76.

De Cleene, M. D., De Ley, J., 1981: The host range of infectious hairy root. Bot. Rev. **47,** 147—194.

De Greve, H., Leemans, J., Hernalsteens, J. P., Thia-Toong, L., De Beuekeleer, M., Willmitzer, L., Otten, L., Van Montagu, M., Schell, J., 1982: Regeneration of normal and fertile plants that express octopine synthase, from tobacco crown galls after deletion of tumor-controlling functions. Nature **300,** 752—755.

De Paolis, A., Mauro, M. L., Pomponi, M., Cardarelli, M., Spano, L., Costantino, P., 1985: Localization of Agropine-synthesizing functions in the Tr region of the root-inducing plasmid of *Agrobacterium rhizogenes* 1855. Plasmid **13,** 1—7.

Depicker, A., Herman, L., Jacobs, A., Schell, J., Van Montagu, M., 1985: Frequencies of simultaneous transformation with different T-DNAs and their relevance to the *Agrobacterium*/plant cell interaction. Mol. & Gen. Genet. **201,** 477—484.

Durand-Tardif, M., Broglie, R., Slightom, J., Tepfer, D., 1985: Structure and expression of Ri T-DNA from *Agrobacterium rhizogenes* in *Nicotiana tabacum* an phenotypic specificity. J. Mol. Biol. **186,** 557—564.

Ellis, J. G., Murphy, P. J., 1981: Four new opines from crown gall tumors — their detection and properties. Mol. & Gen. Genet. **181,** 36—43.

Estramareix, C., Ratet, P., Boulanger, F., Richaud, F., 1986: Multiple mutations ir the transferred regions of the *Agrobacterium rhizogenes* root-inducing plasmids. Plasmid **15,** 245—247.

Furner, I. J., Huffman, G. A., Amasino, R. M., Garfinkel, D. J., Gordon, M. P., Nester, E. W., 1986: An *Agrobacterium* transformation in the evolution of the genus *Nicotiana.* Nature **329,** 422—427.

Garfinkel, D. J., Nester, E. W., 1980: *Agrobacterium tumefaciens* mutants affected in crown gall tumorigenesis and octopine catabolism. J. Bacteriol. **144,** 732—743.

Garfinkel, D. J., Simpson, R. B., Ream, L. W., White, F. F., Gordon, M. P., Nester, E. W., 1981: Genetic analysis of crown gall: fine structure map of the T-DNA by site-directed mutagenesis. Cell **27,** 143—153.

Hedgecock, G. G., 1905: The crown-gall hairy root disease of the apple tree. U.S. Dept. Agr. Bur. Plant Indus. Bull. **131,** 21—23.

Hoekema, A., Hooykaas, P. J., Schilperoort, R., A., 1984: Transfer of the octopine T-DNA segment to plant cells mediated by different types of *Agrobacterium* tumor- or root-inducing plasmids. Generality of virulence systems. J. Bacteriol. **158,** 383—385.

Holmes, B., Roberts, P., 1981: The classification, identification and nomenclature of *Agrobacteria.* J. Appl. Bacteriol. **50,** 443—467.

Holsters, M., Silva, V., Van Vliet, F., Genetello, C., DeBlock, M., Dhaese, P., Depicker, A., Inze, D., Engler, G., Villarroel, R., Van Montagu, M., Schell, J., 1980: The functional organization of the nopaline *A. Tumefaciens* plasmid pTi C58. Plasmid **3,** 212—230.

Hood, E. E., Jen, G., Kayes, L., Kramer, J., Fraley, R. T., Chilton, M.-D., 1984: Restriction endonuclease map of pTIBo542, a potential Ti plasmid vector for genetic engineering of plants. Biotechnology **2,** 702—709.

Hooykaas, P. J. J., Hofker, M., den Dulk-Ras, H., Schilperoort, R. A., 1984: A comparison of virulence determinants in an octopine Ti plasmid, a nopaline Ti plasmid, and an Ri plasmid by complementation analysis of *Agrobacterium tumefaciens* mutants. Plasmid **11,** 195—205.

Hooykaas, P. J. J., Klapwijk, P. M., Nuti, M. P., Schilperoort, R. A., Rorsch, A., 1977: Transfer of the *Agrobacterium tumefaciens* Ti plasmid to avirulent *Agrobacterium* and to *Rhizobium ex planta.* J. Gen. Microbiol. **98,** 477—484.

Huffman, G. A., White, F. F., Gordon, M. P., Nester, E. W., 1984: Hairy-root-inducing plasmid: Pysical map and homology to tumor-inducing plasmids. J. Bacteriol. **157,** 269—276.

Inze, D., Follin, A., Van Lijsebettens, M., Simoens, C., Genetello, C., Van Montagu, M., Schell, J., 1984: Genetic analysis of the individual T-DNA genes of *Agrobacterium tumefaciens;* further evidence that two genes are involved in indole-3-acetic acid synthesis. Mol. & Gen. Genet. **194,** 265—274.

Jensen, J. S., Marcker, K. A., Otten, L., Schell, J., 1986: Nodule-specific expression of a chimaeric soybean leghaemoglobin gene in transgenic *Lotus corniculatus.* Nature **321,** 669—675.

Joos, H., Inze, D., Caplan, A., Sormann, M., Van Montagu, M., Schell, J., 1983: Genetic analysis of T-DNA transcripts in nopaline crown galls. Cell **32,** 1057—1067.

Jouanin, L., 1984: Restriction map of an agropine-type Ri plasmid and its homologies with Ti plasmids. Plasmid **12,** 91—102.

Jouanin, L., Vilaine, F., d'Eufert, C., Casse-Delbart, F., 1985: Localization and restriction maps of the replication origin regions of the plasmids of *Agrobacterium rhizogenes* strain A4. Mol. & Gen. Genet. **201,** 370—374.

Keane, P. J., Kerr, A., New, P. B., 1970: Crown gall of stone fruit. Aust. J. Biol. Sci. **23**, 585—595.

Kerr, A., Roberts, W. P, 1976: *Agrobacterium:* Correlation between and transfer of pathogenicity, octopine and nopaline catabolism and bacteriocin 84 sensitivity. Physiol. Plant Pathol. **4**, 37—44.

Kersters, K., De Ley, J., 1973: Numerical taxonomic analysis of *Agrobacterium.* J. Gen. Microbiol. **78**, 227—239.

Klee, H., J., White, F. F., Iyer, V. N., Gordon, M. P., Nester, E. W., 1983: Mutational analysis of the virulence region of an *Agrobacterium tumefaciens* Ti-plasmid. J. Bacteriol. **153**, 878—883.

Klee, H. J., Yanofsky, M. F., Nester, E. W., 1985: Vectors for transformation of higher plants. Biotechnology **3**, 637—642.

Koplow, J., Byrne, M., Jen, Ge., Tempe, J., Chilton, M., 1984: Physical map of the *Agrobacterium rhizogenes* strain 8196 virulence plasmid. Plasmid **11**, 17—27.

Lahners, K. Byrne, M. C., Chilton, M., 1984: T-DNA fragments of hairy root plasmid pRi8196 are distantly related to octopine and nopaline Ti plasmid T-DNA. Plasmid **11**, 130—140.

Moore, L., Warren, G., Strobel, G., 1979: Involvement of a plasmid in the hairy root disease of plants caused by *Agrobacterium rhizogenes.* Plasmid **2**, 617—626.

Munnecke, D. E., Chandler, P. A., Starr, M. P., 1963: Hairy root *(Agrobacterium rhizogenes)* of field roses. Phytopathology **53**, 788—799.

Nester, E. W., Gordon, M. P., Amasino, R. M., Yanofsky, M. F., 1984: Crown gall: a molecular and physiological analysis. Annu. Rev. Plant Physiol. **35**, 387—413.

Ooms, G., Atkinson, J., Bossen, M. E., Leigh, R. A., 1986a: T_L-DNA from *Agrobacterium rhizogenes* plasmid pRi 1855 reduces the osmotic pressure in transformed plants grown *in vitro.* Planta **168**, 106—112.

Ooms, G., Karp, A., Burrell, M. M., Twell, D., Roberts, J., 1985: Genetic modification of potato development using Ri T-DNA. Theor. Appl. Genet. **70**, 440—446.

Ooms, G., Twell, D., Bossen, M. E., Hoge, J. H. C., Burrell, M. M., 1986b: Developmental regulation of Ri T_L-DNA gene expression in roots, shoots and tubers of transformed potato (*Solanum tuberosum* cv. Desiree). Plant Mol. Biol. **6**, 321—330.

Petit, A., David, C., Dahl, G., Ellis, J., Guyon, P., Casse-Delbart, F., Tempe, J., 1983: Further extension of the opine concept: Plasmids in *Agrobacterium rhizogenes* cooperate for opine degradation. Mol. & Gen. Genet. **190**, 204—214.

Pomponi, M., Spano, L., Sabbadini, M. G., Costantino, P., 1983: Restriction endonuclease mapping of the root-inducing plasmid of *Agrobacterium rhizogenes* 1855. Plasmid **10**, 119—129.

Riker, A. J., Banfield, W. M., Wright, W. H., Keitt, G. W., Sagen, H. E., 1930: Studies on infections hairy-root of nursery apple trees. J. Agric. Res. **41**, 507—540.

Risuleo, G., Battistoni, P., Costantino, P., 1982: Regions of homology between tumorigenic plasmids from *Agrobacterium rhizogenes* and *Agrobacterium tumefaciens.* Plasmid **7**, 45—51.

Ryder, M. H., Tate, M. E., Kerr, A., 1985: Virulence properties of strains of *Agrobacterium* on the apical and basal surfaces of carrot root discs. Plant Physiol. **77**, 215—221.

Schell, J., Van Montagu, M., Depicker, A., DeWaele, D., Engler, G., Genetello, C., Hernalsteens, J. O., Holsters, M., Messens, E., Silva, B., Van den Elsacken, S., Van Larabeke, N., Zaenen, I., 1976: *Agrobacterium tumefaciens:* what segment of the plasmid is responsible for the induction of crown gall tissue? In L. Bogorad and J. H. Weil (ed.) *Nucleic Acids and Protein Synthesis in Plants.* New York: Plenum Press, pp. 329—342.

Schröder, G., Waffenschmidt, S., Weiler, E. W., Schröder, J., 1984: The T-region of Ti plasmids codes for an enzyme synthesizing indole-3-acetic acid. Eur. J. Biochem. **138,** 387—391.

Siegler, E. A., 1928: Studies on the etiology of apple crown gall. J. Agr. Res. **37,** 301—313.

Skoog, F., Miller, C. O., 1957: Chemical regulation of growth and organ formation in plant tissues cultured *in vitro.* — The Biological Action of Growth Substances. Symposium XI, Soc. Exp. Biol. Cambridge Univ. Press, Cambridge. p. 118.

Slightom, J. L., Durand-Tardif, M., Jouanin, L„ Tepfer, D., 1986: Nucleotide sequence analysis of T_L-DNA of *Agrobacterium rhizogenes* agropine type plasmid: Identification of open-reading frames. J. Biol. Chem. **261,** 108—121.

Slightom, J. L., Jouanin, L., Leach, F., Drong, R. F., Tepfer, D., 1985: Isolation and identification of T_L-DNA/plant junctions in *Convolvulus arvensis* transformed by *Agrobacterium rhizogenes* strain A4. EMBO J. **4,** 3069—3077.

Spano, L., Costantino, P., 1982: Regeneration of plants from callus cultures of roots induced by *Agrobacterium rhizogenes* on tobacco. Z. Pflanzenphysiol. **106,** 87—92.

Spano, L., Pomponi, N., Costantino, P., Van Slogteren, G. M. S., Tempe, J., 1982: Identification of T-DNA in the root-inducing plasmid of the agropine type *Agrobacterium rhizogenes* 1855. Plant Mol. Biol. **1,** 291—300.

Spano, L., Wullems, G. J., Schilperoort, R. A., Costantino, P., 1981: Hairy root: *In vitro* growth properties of tissues induced by *Agrobacterium rhizogenes* on tobacco. Plant Sci. Lett. **23,** 29—305.

Strobel, G. A., Lam, B., Harrison, L., Hess, B., Lam, S., 1985: Introduction of the hairy root plasmid into *Rhizobium meliloti* results in increased nodulation on its host. J. Gen. Microbiol. **131,** 355—361.

Strobel, G. A., Nachmias, A., 1985: *Agrobacterium rhizogenes* promotes the initial growth of bare root stock almond. J. Gen. Microbiol. **131,** 1245—1249.

Taylor, B. H., Amasino, R. M., White, F. F., Nester, E. W., Gordon, M. P., 1985a: T-DNA analysis of plants regenerated from hairy root tumors. Mol. & Gen. Genet. **201,** 554—557.

Taylor, B. H., White, F. F., Nester, E. W., Gordon, M. P., 1985b: Transcription of *Agrobacterium rhizogenes* A4 T-DNA. Mol. & Gen. Genet. **201,** 546—553.

Tempe, J., Petit, A., Farrand, S. K., 1984: Induction of cell proliferation by *Agrobacterium tumefaciens* and *A. rhizogenes:* A parasite's point of view. In *Plant Gene Research,* eds. D.P.S. Verma and Th. Hohn. Wien - New York: Springer. pp. 271—286.

Tepfer, D., 1981: La transformation genetique de plantes superieures par *Agrobacterium rhizogenes.* 2[e] Colloque sur les Recherches Fruitieres - Bordeaux, 47—59.

Tepfer, D., 1984: Transformation of several species of higher plants by *Agrobacterium rhizogenes:* Sexual transmission of the transformed genotype and phenotype. Cell **37,** 959—967.

Tepfer, D. A., Tempe, J., 1981: Production d'agropine par des racines formees sous l'action d'*Agrobacterium rhizogenes* souche A_4. CR Acad. Sci. Paris, Ser III **292,** 153—156.

Thomashow, L. S., Reeves, S., Thomashow, M. F., 1984: Crown gall oncogenesis: Evidence that a T-DNA gene from the *Agrobacterium* Ti plasmid PTiA6 encodes an enzyme that catalyzes synthesis of indoleacetic acid. Proc. Natl. Acad. Sci, U.S.A. **81,** 5071—5075.

Thomashow, M. F., Hugly, S., Buchholtz, W. G., Thomashow, L. S., 1986: Molecular basis for the auxin-independent phenotype of crown gall tumor tissue. Science **232,** 606—618.

Thomashow, M. F., Nutter, R., Montoya, A. L., Gordon, M. P., Nester, E. W., 1980: Integration and organization of Ti plasmid sequences in crown gall tumors. Cell **19,** 729—739.

Turgeon, R., Wood, H. N., Braun, A., 1976: Studies on the recovery of crown gall tumor cells. Proc. Natl. Acad. Sci., U.S.A. **73,** 3562—3564.

Unger, L., Ziegler, S. F., Huffman, G. A., Knauf, V. C., Peet, R., Moore, L. W., Gordon, M. P., Nester, E. W., 1985: New class of limited-host-range *Agrobacterium* mega-tumor-inducing plasmids lacking homology to the transformed DNA of a wide-host-range, tumor-inducing plasmid. J. Bacteriol. **164,** 723—730.

White, F. F. Garfinkel, D. J., Huffman, G. A., Gordon, M. P., Nester, E. W., 1983: Sequences homologous to *Agrobacterium rhizogenes* T-DNA in the genomes of uninfected plants. Nature **301,** 348—350.

White, F. F., Ghidossi, G., Gordon, M. P., Nester, E. W., 1982: Tumor induction by *Agrobacterium rhizogenes* involves the transfer of plasmid DNA to the plant genome. Proc. Natl. Acad. Sci, U.S.A. **79,** 3193—3197.

White, F. F., Nester, E. W., 1980a: Hairy root: Plasmid encodes virulence traits in *Agrobacterium rhizogenes.* J. Bacteriol. **141,** 1134—1141.

White, F. F., Nester, E. W., 1980b: Relationship of plasmids responsible for hairy root and crown gall tumorigenicity. J. Bacteriol **144,** 710—720.

White, F. F., Taylor, B. H., Huffman, G. A., Gordon, M. P., Nester, E. W., 1985: Molecular and genetic analysis of the transferred DNA regions of the root-inducing plasmid of *Agrobacterium rhizogenes.* J. Bacteriol. **164,** 33—44.

Willmitzer, L., Dhaese, P., Schreier, P. H., Schmalenbach, W., Van Montagu, M., Schell, J., 1983: Size, location and polarity of T-DNA-encoded transcripts in nopaline crown gall tumors; common transcripts in octopine and nopaline tumors. Cell **32,** 1045— 1046.

Willmitzer, L., Sanchez-Serrano, J., Busfeld, E., Schell, J., 1982: DNA from *Agrobacterium rhizogenes* is transferred to and expressed in axenic hairy root plant tissues. Mol. & Gen. Genet. **186,** 16—22.

Wullems, G. J., Molendijk, L., Ooms, G., Schilperoort, R. A., 1981: Retention of tumor markers in F-1 progeny plants from *in vitro* induced octopine and nopaline tumor tissues. Cell **24,** 719—728.

Yang, F., Simpson, R. B., 1981: Revertant seedlings from crown gall tumors retain an portion of the bacterial Ti plasmid DNA sequences. Proc. Natl. Acad. Sci., U.S.A. **78,** 4151—4155.

Yanofsky, M., Montoya, A., Knauf, V., Lowe, B., Gordon, M., Nester, E., 1985: Limited-host-range plasmid of *Agrobacterium tumefaciens:* Molecular and genetic analyses of transferred DNA. J. Bacteriol. **163,** 341—348.

Zambryski, P., Goodman, H., Van Montagu, M., Schell, J., 1983: *Agrobacterium* tumor induction. In *Mobile Genetic Elements,* ed. J. Shapiro. Academic Press. pp. 506—535.

Zambryski, P., Joos, H., Genetello, C., Leemans, J., Van Montagu, M., Schell, J., 1983: Ti plasmid vector for the introduction of DNA into plant cells without alteration of their normal regeneration capacity. EMBO J. **2,** 2143—2150.

Chapter 7

Pathways to Plant Genetic Manipulation Employing *Agrobacterium*

Stephen G. Rogers and Harry Klee

Plant Molecular Biology, Biological Sciences Department, Monsanto Company, 700 Chesterfield Village Parkway, St. Louis, MO 63198, U.S.A

With 8 Figures

Contents

I. Introduction

The adaption of the *Agrobacterium tumefaciens* Ti plasmid as a vector for plant transformation is an enabling technology permitting plant molecular biologists to rapidly expand the frontiers of knowledge about the fundamental biological processes of gene function, expression and regulation (See Fraley *et al.,* 1986 for a recent review). The ability to introduce native or modified genes into plants has moved plant molecular biology from a descriptive mode to a manipulative mode where the effects of specific alterations in gene sequences can be assayed. In the four short years since the first reports of the successful use of the system for the introduction and expression of a foreign gene in plant tissues, researchers have identified specific nucleotide sequences responsible for light regulated expression of certain genes (Broglie *et al.,* 1984; Herrera-Estrella *et al.,* 1984; Morelli *et al.,* 1985). Other groups (Nagy *et al.,* 1985; Beachy *et al.,* 1985; Sengupta-Gopalan *et al.,* 1985) have shown that newly introduced genes exhibit the characteristic tissue specific and/or developmental stage specific expression that they displayed in the plants from which they were isolated. Still others (Schreier *et al.,* 1985) have begun to identify the portions of transit peptides responsible for proper compartmentalization of proteins within a plant cell. The Ti transformation system has become a tool for the study of virus host range, symptom production and replication (Grimsley *et al.,* 1986, Rogers *et al.,* 1986b; Grimsley and Bisaro, this volume) as well as cellular recombination processes. Recently, the Ti system has been used to introduce the maize Ac element into tobacco to examine the process and regulation of transposition and for use as a potential gene tagging system (Baker *et al.,* 1986; Saedler, this volume).

In addition, within this short period of time, the potential of the Ti transformation system for applied plant biology and crop improvement has begun to be realized. Transgenic plants have been created that are resistant to the non-selective herbicide Roundup® (Comai *et al.,* 1985; Shah *et al.,* 1986), that are now resistant to tobacco hornworm because they express the *Bacillus thuringiensis* crystal toxin protein from a chimeric nuclear gene (Adang *et al.,* 1986; J. Vaeck, personal communication; D. Fischhoff, personal communication), and that are now resistant to super-infection by virulent strains of tobacco mosaic virus because they produce the TMV coat protein from a chimeric nuclear gene (Abel *et al.,* 1986). These are just the first of the many exciting basic and applied research discoveries that the transformation system promises.

The following will review alternative strategies used to modify the Ti plasmid for the introduction of homologous and heterologous DNAs and genes into plants and discuss the particular advantages and disadvantages of each so that researchers new to this field may choose a suitable system to apply in making their own contributions to the exciting areas of understanding plant gene function and improvement of plants.

II. Biology of the *Agrobacterium tumefaciens* Ti Plasmid

The following description is based upon the work of many laboratories throughout the world. Their efforts provided the fundamental information that permitted the development of the vector systems. We have attempted to give appropriate credit for these discoveries and apologize for any omissions. More complete, recent reviews of *Agrobacterium* biology and the Ti system may be found in Nester *et al.* (1984) and Gheysen *et al.* (1985).

Agrobacterium tumefaciens and the related *Agrobacterium rhizogenes* (which is discussed in the chapter by White in this volume) are gram-negative, soil bacteria that can infect plants by gaining entry through wounds. Once inside, the bacteria bind to plant cells and transfer a segment of DNA from their large endogenous Ti (tumor inducing) or Ri (root inducing) plasmids into the plant cell where the transferred DNA (T-DNA) becomes covalently inserted into the plant cell genome by a process that is not well understood (Koukolikova *et al.,* this volume for a thorough discussion of this topic). It is known, however, that the products of bacterial genes (*vir* genes) located on the Ti plasmid are essential for this process. In addition to the *vir* gene products, directly repeated sequences of 25 bases called border sequences (Yadav *et al.,* 1982; Zambryski *et al.,* 1981) that flank the T-DNA are essential for the transfer process (Wang *et al.,* 1984; Horsch and Klee, 1986).

Once in the plant cell, this transferred DNA (T-DNA) enters the nucleus and becomes covalently integrated in the plant genome at random sites. The natural T-DNA carries eight to thirteen genes depending on whether it is derived from an octopine or nopaline producing Ti plasmid. These are true plant genes that function in plants but not in the *Agrobacterium* from which they were transferred. Expression of three of these genes in the naturally transformed cells produces enzymes responsible for biosynthesis of the auxin, indoleacetic acid and the cytokinin, isopentenyl-adenosine monophosphate. These growth regulators act on the surrounding tissues resulting in cell proliferation and the neoplastic disease called crown gall. No other genes carried in the T-DNA are involved in the disease production. Therefore, inactivation or deletion of these three genes causes complete loss of gall formation and disease. Production of the phytohormones in transformed cells prevents morphogenesis so that deletion of these genes, thus creating "disarmed" Ti plasmids, is essential to permit analysis of gene regulation in morphologically normal, transgenic plants.

III. Strategies for Inserting Genes into the T-DNA

A. Homogenotization

Knowledge of the biology of the Ti plasmid and T-DNA transfer suggested that foreign DNA located between the T-DNA border sequences would be transferred to plant cells. This was effectively demonstrated by Hernalsteens *et al.* (1980) who showed that a 15 kb Tn*7* transposed into the T-DNA could be introduced into plant cells, become integrated into the genome and be stably maintained. This result led to the design of Ti vector systems that would allow the insertion of foreign DNAs into the T-DNA. The large size of Ti plasmids (200 kb) and consequent lack of unique restriction endonuclease cleavage sites within the T-DNA region for insertion of foreign DNAs as well as the lack of a useful transformation procedure for *A. tumefaciens* prevented the direct use of the entire Ti plasmid as a vector. Other indirect methods to engineer the Ti plasmid were developed.

The initial strategy was to introduce new DNA segments into the T-DNA of the *Agrobacterium* Ti plasmid using microbial genetic techniques developed for study of the *Rhizobium sym* plasmid (Ruvkun and Ausubel, 1981). Matzke and Chilton (1981) modified the procedure for use in *Agrobacterium.* The process involves insertion of the DNA of interest along with a gene encoding antibiotic resistance for *Agrobacterium* into a segment of the T-DNA carried in a broad host-range plasmid in *E. coli.* This "intermediate" plasmid is then transferred by bacterial conjugation into *Agrobacterium.* A double recombination event in which the modified T-DNA segment replaced its homolog in the T-DNA is forced by introducing into the *Agrobacterium* a second broad host range plasmid incompatible with the first and demanding maintenance of the antibiotic resistance carried on the modified T-DNA segment. This process of "homogenotization" may also be scored for by conjugal transfer of the Ti plasmid into another *Agrobacterium,* selecting for maintenance of the antibiotic marker carried in the T-DNA and scoring for the concomittant loss of the antibiotic markers of the broad host range intermediate plasmid (Leemans *et al.,* 1981). Overall, this process, while effective, is quite cumbersome and time consuming. In addition, the frequency of obtaining the double crossover homogenotes can be very low. In spite of these disadvantages, Matzke and Matzke (1986) have recently described a broad host range intermediate plasmid that will create a disarmed T-DNA concommitant with insertion of the new DNA of interest. It is unlikely that general use of such a system will continue due to the comparative ease of use of the co-integrating and binary vector systems.

B. Cointegrating Intermediate Vectors

The rationale for using the homogenote approach was that the resulting modified T-DNA did not contain duplications of flanking DNA segments that could be transferred to the plant genome. It was feared that these duplications might be unstable causing deletion of the introduced DNAs by homologous recombination within the plant cell. Despite this concern several groups (Comai *et al.,* 1983; Fraley *et al.,* 1983; Zambryski *et al.,* 1983) developed intermediate vectors to introduce foreign DNAs into the T-DNA by means of a single homologous recombination event. These cointegrating vectors were assembled in *E. coli* and were much simpler to construct since they required only a single, continuous segment of the T-DNA be present. The plasmids used ColE1-type origins of replication so that they could be made in large quantity for assemblies but these origins of replication do not function in *Agrobacterium*. The intermediate plasmids could only be maintained if they recombine with the resident Ti plasmid T-DNA. This cointegration event is easily obtained by selecting for antibiotic resistance markers carried in the vector. The single crossover or cointegration of these intermediate plasmids occurred in 10^{-3} to 10^{-6} of the *A. tumefaciens* cells receiving the plasmid. More importantly, this procedure did not require the cumbersome manipulations to force the resolution of the cointegrate representing a great improvement over the "homogenotization" method. Because the intermediate vectors cointegrate into the T-DNA and are physically linked to it, these plasmids are referred to as *cis* vectors.

Horsch *et al.* (1984) solved the duplicate flanking DNA segment problem by including a T-DNA border sequence in the intermediate vector so that the duplicated region of homology lay outside the T-DNA borders and only one copy is transferred into the plant genome. The DNA introduced using this system was stably inherited according to Mendelian rules. Subsequently, the concerns about frequent loss of introduced genes by homologous recombination were proven unfounded when DeBlock *et al.* (1984) showed that the resulting T-DNAs flanked by duplicated pBR322 DNAs were stably maintained in the plant genome during meiosis.

C. Binary Vectors

A great simplification in Ti plasmid vector design became possible in 1983 after both Hoekema *et al.* and de Framond *et al.* reported that the T-DNA need not be physically linked to the *vir* genes for transfer of the T-DNA to plant cells to occur. This finding, coupled with the knowledge that the only parts of the T-DNA required for transfer and integration of DNA into a plant genome are the border sequences permitted development of small, binary or *trans* Ti transformation vectors. These plasmids contain the T-DNA border sequences flanking selectable markers and cloning sites joined to broad host range plasmid replication origins. Using binary vectors, the foreign DNA or new gene of interest is directly inserted into the T-DNA during the cloning assembly steps in *E. coli*. The plasmid is then transferred to *A. tumefaciens* by standard bacterial conjugation methods. The great improvement of the binary vectors is their ability to replicate in *Agrobacterium* obviating the need for cointegrate formation and simplifying their introduction into *Agrobacterium* since the frequency of maintenance is essentially the frequency of conjugation. This frequency can be as high as 10 % of the *A. tumefaciens* that are mixed with donor *E. coli* cells. In contrast, the cointegrating vectors which must recombine into the resident Ti plasmid give a frequency of introduction of approximate frequency of 10^{-3} to 10^{-6}. Before describing specific cointegrating and binary vector systems, we will discuss the construction and structures of the disarmed Ti plasmids used with the various cointegrating and binary vectors.

D. Disarming the T-DNA

All of the strategies using the Ti plasmid and T-DNA transfer system to obtain transgenic plants must prevent the phytohormone biosynthetic genes of the natural T-DNA from entering the transformed plant cells. Their expression results in undifferentiated growth and inhibition of morphogenesis. Since the goal of a gene introduction experiment is to study the expression of the new gene in plants and not undifferentiated cells, various methods to "disarm" the T-DNA have been developed. Disarmed Ti plasmids for the cointegrating intermediate vectors (Table 1) retain the *vir* region to provide the functions essential for transfer and integration of the T-DNA and at least the border sequences of the T-DNA. To construct the pTiB6S3-SE disarmed plasmid (Fig. 1), Fraley *et al.* (1985) replaced approximately 80 % of the octopine T_L T-DNA and the entire T_R T-DNA

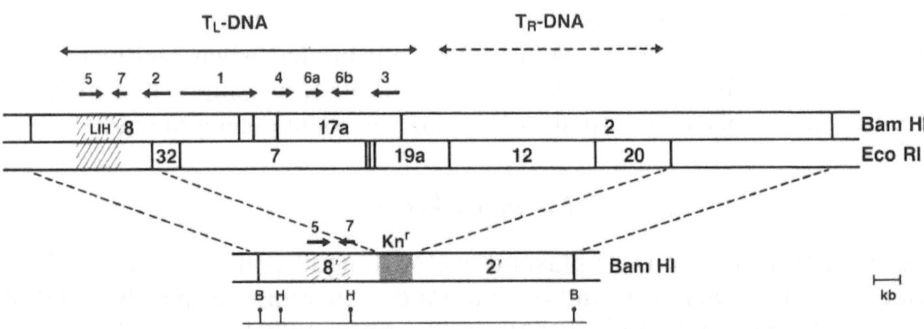

Fig. 1. Method for producing and structure of the disarmed pTiB6S3-SE plasmid. A 4.5 kb BamHI-EcoRI fragment 8′ (from BamHI fragment 8) and a 6.0 kb EcoRI-BamHI fragment 2′ (from BamHI fragment 2) provided homology for replacement of greater than 80 % of the T_L T-DNA and the entire T_R T-DNA with the 1.2 kb Tn*903* segment carrying bacterial kanamycin resistance. The resulting pTiB6S3-SE plasmid carries only the T_L left border and intact transcripts 5 and 7 which are not involved in disease production. Reprinted from Bio/Technology 3, 629—635 (1985) with permission

with a bacterial kanamycin resistance gene. The remaining T_L T-DNA sequences include the left border sequence and two T-DNA genes (transcripts 5 and 7) that are not involved in disease production but whose function is not known. A portion of the T-DNA segment called the LIH for Left Inside Homology, is present in the pMON200 intermediate vector. The LIH segment provides a region of DNA for homologous recombination to form a cointegrate, hybrid T-DNA carrying a gene to be introduced into plants. The disarmed plasmid is called pTiB6S3-SE for *s*plit *e*nd because one of the T-DNA borders is carried on this helper plasmid and the other is carried on the pMON200 plasmid. An analogous nopaline-type pTiT37-SE disarmed plasmid has also been constructed for use with pMON200 (Fischhoff and Rogers, in preparation). In the pTiT37-SE plasmid 95 % of the T-DNA including the right border has been replaced with the pBR322 and LIH regions of pMON200 to provide homology. Following cointegration, a hybrid T-DNA is created in a manner analogous to that described for the pTiB6S3-SE and pMON200. The pTiT37 T-DNA left border which remains functions as the left border of the hybrid T-DNA.

Table 1. Disarmed Helper Ti Plasmids

Helper Plasmid	Parent	Homology	Reference
Cointegrating Helper Plasmids			
pGV3850	pTiC58	pBR322	Zambryski *et al.*, 1983
pTiB6S3-SE	pTiB6S3	T-DNA segment[a]	Fraley *et al.*, 1985
pTiT37-SE	pTiT37	pBR322, T-DNA[b]	Fischhoff and Rogers, in preparation
Binary Helper Plasmids			
pAL4404	pTiAch5	—	Hoekema *et al.*, 1983
pTiB6S3-NE	pTiB6S3	—	Fischhoff and Rogers, in preparation
pMP90RK	pTiC58	—	Koncz and Schell, 1986

[a] This region of homology is derived from an octopine-type T-DNA and is also carried in pMON200.
[b] The pBR322 and T-DNA segments are derived fom pMON200.

Zambryski *et al.* (1983) replaced all of the central region of the T-DNA with a linearized pBR322 DNA. The resulting disarmed Ti plasmid, pGV3850, can now be used as a helper plasmid for all intermediate vectors that contain pBR322 sequences. This elegant strategy allows the assembly of an intermediate vector from any pBR322 plasmid that carries a selectable bacterial marker other than carbenicillin resistance and its ready introduction into *Agrobacterium.*

The restriction that only pMON200 or derivative intermediate vectors must be used with the pTiB6S3-SE plasmid is somewhat limiting compared to the pGV3850-pBR system. However, an advantage of this limitation is that it has permitted direct comparison of constructs and expression experiments performed several years apart without concern that the flanking sequences in the intermediate vector are contributing to transformation or expression differences.

One advantage of binary vectors is that they can be mobilized into any *Agrobacterium* host, not just hosts containing Ti plasmids with homologous DNA. The cointegrating systems suffer from the fact that a specific disarmed Ti plasmid must be used to provide helper functions and thus the host range of plant species that may be transformed is limited by the properties of the plasmid in which the deletions were made. This can be important since various plant species show different degrees of sensitivity to infection by different *Agrobacterium* strains (Owens and Cress, 1985; Byrne *et al.*, 1987).

Furthermore, regenerated plants may be derived from transformants of binary vectors carried in hosts with a non-disarmed Ti plasmid. This "self-disarming" system has been shown to work for binary vectors in nondisarmed *A. tumefaciens* (An *et al.*, 1985) or nondisarmed *A. rhizogenes* (Simpson *et al.*, 1986) hosts. In the latter study the frequency of cotransformation was 50 % suggesting that many of the transformants selected for kanamycin resistance would be free of the *A. rhizogenes* disease genes and able to differentiate. Thus, a single binary plasmid construct can be mobilized into many different *Agrobacterium* hosts to optimize infection and transformation of different plant species.

However, in plant species where transformation or regeneration occurs at lower frequencies, it is best to use *A. tumefaciens* strains carrying disarmed Ti plasmids to avoid possible cotransfer of the phytohormone biosynthetic genes. The binary vectors may be efficiently transferred to plants by *A. tumefaciens* strains carrying the disarmed plasmids created for the cointegrating vectors such as pTiB6S3-SE or pGV3850 which provide all of the *vir* region functions. In addition, special Ti plasmids deleted for the T-DNA region including the border sequences but containing an intact *vir* region have been created (Table 1). Plasmid pAL4404, a large deletion derivative of pTiACH5 carried in strain LBA4404 (Ooms *et al.*, 1980) is the most commonly used. A similar derivative of pTiB6S3 has been created by replacement of the entire T_L and T_R T-DNAs including all of the border sequences with a bacterial Tn*903* kanamycin resistance gene. The resulting plasmid, pTiB6S3-NE (*N*o *E*nds, Fig. 2) retains the octopine catabolism region and its construction will be described in a separate publication (Fischhoff and Rogers, in preparation).

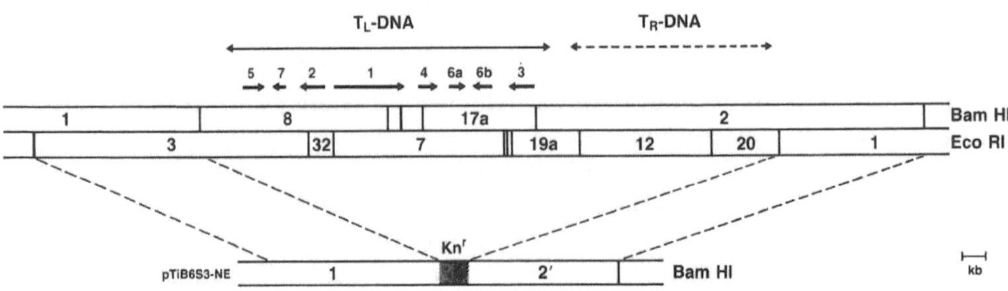

Fig. 2. Structure of the disarmed pTiB6S3-NE helper plasmid for binary vectors. A 7 kb EcoRI-BamHI fragment 1′ (from BamHI fragment 1) and a 6.0 kb EcoRI-BamHI fragment 2′ (from BamHI fragment 2) provided homology for replacement of the entire T_L and T_R T-DNAs with the 1.2 kb Tn*903* segment carrying bacterial kanamycin resistance. All of the T-DNA is removed

Recently, Koncz and Schell (1986) described a disarmed pTiC58 plasmid, pMP90RK, that contains the replication functions of the broad host range RK2 plasmid integrated into the Ti plasmid in place of the T-DNA. This disarmed helper plasmid allowed the authors to develop very small binary vectors containing just the RK2 origin of replication sequences since the pMP90RK plasmid will supply all the proteins required for vector replication.

E. Specific Examples

1. Use of pMON200: A Cointegrating Vector

Plasmid pMON200 is a single crossover intermediate vector that has been used in combination with a disarmed Ti plasmid to obtain over 13000 transgenic plants in our laboratory. The pMON200 plasmid is made up of the well characterized DNA segments shown in Fig. 3. The cloning of additional DNAs into the pMON200 vector is done in *E. coli* as the host using standard recombinant DNA methods. Once clones of the desired structure are identified, these plasmids are transferred to *Agrobacterium* via the procedure called tri-parental mating shown in Fig. 4. Two types of *E. coli*

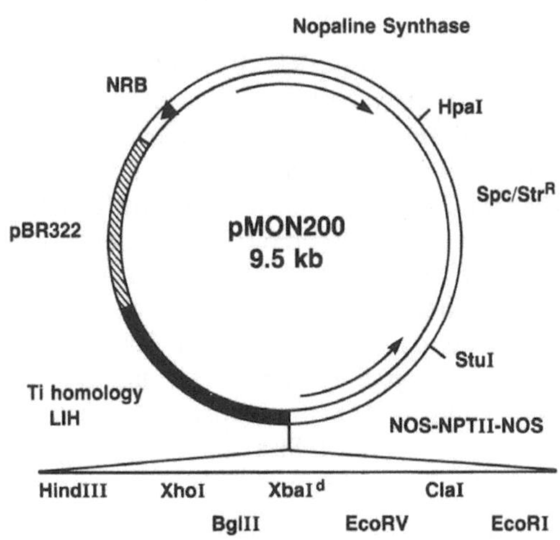

Fig. 3. Map of the cointegrating intermediate transformation vector pMON200. The pMON200 plasmid (Fraley *et al.*, 1985) is made up of a 1.6 kb LIH (left inside homology) segment derived from the octopine-type pTiA6 plasmid which provides a region of homology for recombination with a resident Ti plasmid in *A. tumefaciens*, a 1.6 kb segment carrying the pBR322 origin of replication, a 2.2 kb segment of the nopaline-type pTiT37 plasmid that carries the right border (NRB) of the nopaline T-DNA and intact nopaline synthase (NOS) gene, a 2.7 kb segment of Tn7 carrying the spectinomycin/streptomycin resistance (Spc/Str^R) determinant, a 1.6 kb segment encoding a chimeric NOS-NPTII′-NOS gene that provides selectable kanamycin resistance in transformed plant cells and a synthetic multi-linker containing unique restriction sites for insertion of other DNA segments. All of the restriction endonuclease sites shown are unique. The "d" superscript denotes a site that is cleavable only in DNA prepared from a *dam* methylase deficient *E. coli* host

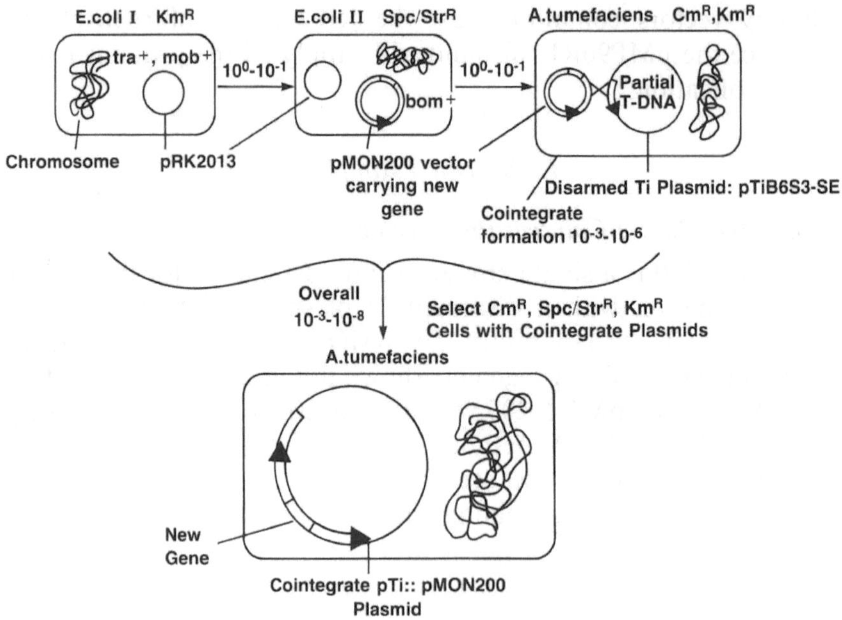

Fig. 4. Tri-parental mating procedure used to introduce pMON200 and derivatives into *Agrobacterium tumefaciens* to produce a cointegrate plasmid carrying a hybrid SEV T-DNA. *E. coli* cells carrying the pRK2013 helper plasmid are mixed with *E. coli* carrying pMON200 or a derivative and *A. tumefaciens* cells carrying the disarmed pTiB6S3-SE helper plasmid are added. The pRK2013 plasmid provides the RK2 transfer functions and the ColE1 *mob* protein which act at the pMON200 *bom* site to mobilize pMON200 into *A. tumefaciens.* Since pMON200 cannot replicate in *A. tumefaciens,* it must recombine with the pTiB6S3-SE helper plasmid for its maintenance. The *Agrobacterium* cells are already resistant to chloramphenicol (Cm^R) and kanamycin (Km^R) and become resistant to spectinomycin and streptomycin (Spc/Str^R) if they carry the pTiB6S3-SE::pMON200 cointegrate plasmid. The numbers next to each of the steps indicate the frequency of transfer or recombination. Further details of the mating procedure may be found in Rogers et al. (1986a)

cells — one carrying a helper plasmid pRK2013 (Ditta *et al.,* 1980) and one carrying the pMON200 derivative — are mixed with the *A. tumefaciens* 3111-SE strain carrying the disarmed pTiB6S3-SE plasmid. Following overnight incubation on a rich medium, an aliquot of the mixture is spread on a plate containing antibiotics to select for the *A. tumefaciens* cells carrying the pMON200 plasmid co-integrated into the disarmed Ti plasmid. The correct structure is confirmed by Southern hybridization methods.

Cointegration results in a hybrid T-DNA made up of the following DNA segments (Fig. 5). First, are the left border sequence and LIH contributed by the disarmed pTiB6S3-SE plasmid followed in order by the synthetic multi-linker where additional genes are inserted, the kanamycin

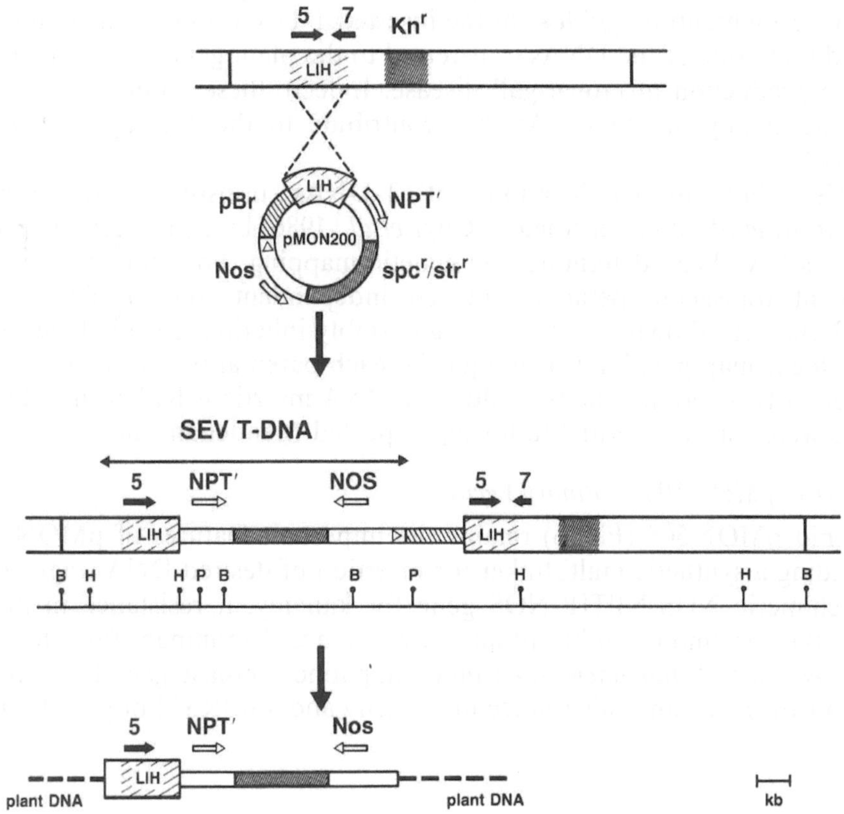

Fig. 5. Transfer and structure of the hybrid, cointegrate T-DNA into plant cells. Cointegration between pMON200 and pTiB6S3-SE at their homologous LIH regions results in the formation of a hybrid SEV T-DNA containing the chimeric NPT′ kanamycin resistance and nopaline synthase (NOS) genes for selection of transformed plant cells and monitoring presence and inheritance of T-DNA in plants. Reprinted from Bio/Technology 3, 629—635 (1985) with permission

resistance gene NPT′ (NOS-NPTII′-NOS), the spectinomycin resistance marker, the scorable nopaline synthase marker (NOS), the right border sequence and then the pBR322 segment. This is then followed by a repeat of the LIH segment. Since each of the border sequences is initially carried on a separate plasmid, we refer to the entire system as the Split End Vector or SEV system.

Several points should be made regarding this vector design. Firstly, only DNA between the border sequences is transferred to the plant cell. This includes any new DNAs inserted between the LIH segment and NPT′ region of the pMON200 plasmid. Secondly, the pBR322 and second copy of the LIH segment are located outside of the borders and will not be transferred into the plant genome. This positioning of the second copy of the LIH region means that a duplicated LIH DNA segment will not be transferred to the plant genome and provide a homologous region for subsequent recombination and loss of the inserted T-DNA. Finally, it should be noted that none of the DNAs transferred to the plant genome are involved in the production of crown gall disease. Indeed, these sequences are not present in any of the DNAs that contribute to the cointegrate hybrid T-DNA.

The chromosomal locations of T-DNAs transferred to petunia (Wallroth *et al.*, 1986) or tomato (Chyi *et al.*, 1986) by cointegrating vector systems have been determined by genetic mapping. For four of six independent transgenic petunia and ten independent tomato plants, the T-DNAs were shown to act as a single, stably inherited genetic locus and each locus map position was unique for each parental transgenic plant. In the other two petunia plants, multiple T-DNA insertions had occurred and these were stably inherited following expected Mendelian rules.

2. Use of pMON 505: A Binary Vector

Plasmid pMON 505 (Fig. 6) retains the important features of pMON 200 including a synthetic multi-linker for insertion of desired DNA fragments, the chimeric NOS-NPTII′-NOS gene for kanamycin resistance in plant cells, the spectinomycin/streptomycin resistance determinant for selection in *E. coli* and *A. tumefaciens,* an intact nopaline synthase gene for scoring of transformants and inheritance in progeny and a pBR 322 origin of repli-

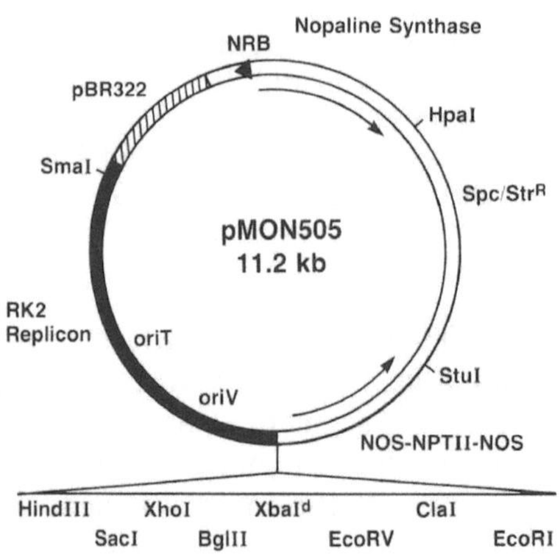

Fig. 6. Map of the binary transformation vector pMON 505. This plasmid is identical to pMON 200 (Fig. 3) except that the LIH region has been replaced by an RK 2 broad host range plasmid replication origin. oriV and oriT are the RK 2 replication and conjugational transfer origins, respectively. The additional segments are described in the legend to Fig. 3. Restriction endonuclease sites shown are unique. The "d" superscript denotes a site that is cleavable only in DNA prepared from a *dam* methylase deficient *E. coli* host

cation for ease in obtaining large amounts of the vector. Unlike most binary vectors, pMON505 contains a single T-DNA border derived from the right end of the pTiT37 nopaline-type T-DNA.

After insertion of the new DNAs into pMON505, the plasmid is transferred to *Agrobacterium* using the triparental mating procedure shown in Fig. 7. Note that because the plasmid is able to replicate in *Agrobacterium,*

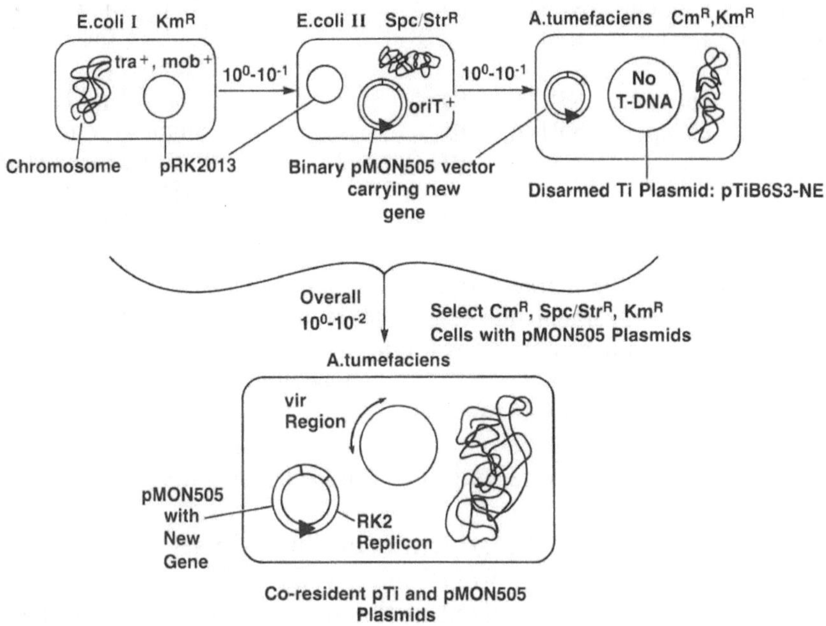

Fig. 7. Triparental mating procedure for the introduction of the pMON505 binary vector into *Agrobacterium.* The frequencies of each of the steps is indicated. Note that mobilization of pMON505 is through the RK2 oriT. The other labels are the same as those in Fig. 4 and 6

the frequency of transfer is essentially the frequency one obtains cells containing the plasmid for use in transformation experiments. This routinely is above 10 % of the starting *Agrobacterium* recipient cells so that one needs merely to streak a loopful of cells from the mating mix on a selection plate to obtain a large number of transconjugants carrying the binary vector.

F. Border Sequences and Binary Vector T-DNA Structure

The most significant difference among the binary vectors listed in Table 2 is the number of border sequences included in the vector. Most of these vectors carry a small T-DNA region flanked by two border sequences. The exceptions are pMON505 and pMON809 which have a single border. Studies by others (Hoekema *et al.,* 1985) and our own (Horsch and Klee, 1986) have shown that for binary vectors a single border is necessary and

Table 2. Sampling of Available Binary Vectors

Vector	*Sel Marker* Bacteria	plant	Replication Origin	Borders	Reference
BIN19	kan[a]	NPTII[b]	RK2	2	Bevan, 1984
pGA471	tc, kan	NPTII	RK2	2	An *et al.*,1985
pEND4K	kan, cm	NPTII	RK2	2	Klee *et al.*, 1985
binary vector	sm, gm	NPTII	incQ	2	Matzke and Matzke, 1986
pMON505	spc	NPTII	RK2	1	Horsch and Klee, 1986
pC22	cb, spc	NPTII	Ri	2	Simoens *et al.*, 1986
pAGS125	tc	hph	RK2	2	van den Elzen *et al.*, 1985b
pMON809	spc	mtx	RK2	1	Eichholtz *et al.*, 1987

[a] The bacterial selectable markers are: kan, kanamycin resistance; spc, spectinomycin resistance; cb, carbenicillin resistance; tc, tetracyline resistance; sm, streptomycin resistance; gm, gentamicin resistance.

[b] The plant selectable markers are NPTII, kanamycin resistance; hph, hygromycin B resistance; mtx, methotrexate resistance.

sufficient for transfer and integration of the DNA into the genome of plant cells.

We have investigated the structure of the T-DNA found in plants transformed with the single border binary vectors. The structure suggests that the binary vector is linearized at the single border and inserted into the plant genome. In a sense, then, the entire pMON505 plasmid is the T-DNA. A diagrammatic representation of the pMON505 binary vector T-DNA structure is shown in Fig. 8. It should be noted that the border sequences are lost during the process of integration of both the cointegrating type or binary T-DNAs into the plant cell genome. This means that the inserted DNA is no longer a functional T-DNA capable of being transferred by the same mechanism that originally inserted the T-DNA into the plant genome. The stability of these T-DNAs has been demonstrated by extensive genetic analysis of transgenic plants and their progeny derived using the cointegrating vectors (Horsch *et al.*, 1984; De Block *et al.*, 1984).

Since most of the binary vectors have a higher copy number in *Agrobacterium* than a Ti plasmid, it was important to determine whether multiply transformed plant cells would arise more frequently than with cointegrating vectors. Data available at this point suggest that binary vectors behave exactly like cointegrating vectors (Simpson *et al.*, 1986). Whether one obtains single or multiple copies of T-DNA is more a matter of how the transformation is done than the vector type. We have examined the T-DNA structure of a number of transgenic plants obtained from transformations with the single border pMON505 and its derivatives (Eichholtz *et al.*, 1987). The entire plasmid usually becomes integrated in the trans-

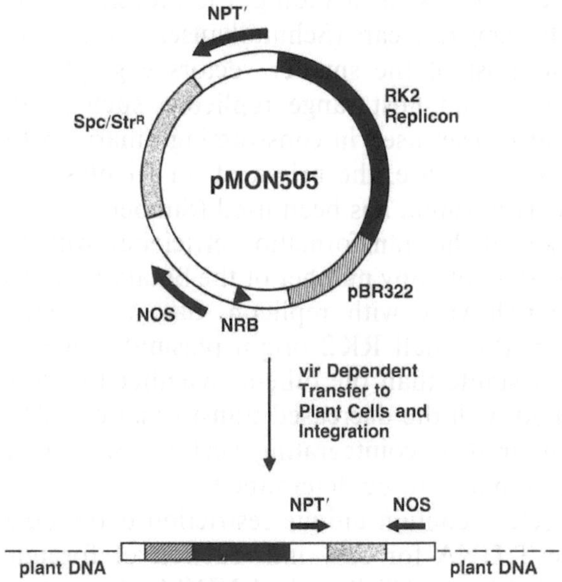

Fig. 8. Schematic representation of transfer of the single border binary vector pMON505 DNA from *A. tumefaciens* and integration into plant genomic DNA. During this process, the entire pMON505 plasmid is linearized by removal of the T-DNA border sequence (NRB) and inserted into the genome by an as yet unknown process

formed plant cells. Most, though not all, of the transgenic plants have a single copy of T-DNA. Plants containing multiple T-DNAs have been observed, but they are less frequent (less than 20% of the transformants).

IV. Practical Catalogue

A. Survey of Binary Vectors

Examples of currently available binary vectors are listed in Table 2. Other binary vectors have been reported (Hoekema *et al.,* 1985; Hepburn *et al.,* 1985; Simpson *et al.,* 1986; Waldron *et al.,* 1985) but we have not included these because they are similar to ones in Table 2, lack multiple unique cloning sites within the T-DNA, do not contain dominant selectable markers or contain phytohormone genes which prohibits their use in obtaining transgenic plants.

The common feature of all binary vectors is the presence of a broad host range origin of replication. Most of the vectors listed in Table 2 utilize the broad host range RK2 replication origin *(incP1)* largely due to the fact that the RK2 (RP4) replicon is very well characterized and has been used successfully in *Agrobacterium* research for a number of years. In particular, a minimal replicon of 6.8 kb including a tetracycline resistance gene has been available for several years (Schmidhauser and Helinski, 1985) and is the replicon for most of the smaller vectors e. g. pGA471, pMON505, pENDK4. Other broad host range replicons such as those from *incQ* plasmids, have also been used in constructing binary vectors (Matzke and Matzke, 1986). In one case, the origin of an Ri plasmid, which is compatible with the Ti plasmid, has been used (Simoens *et al.,* 1986). Since no direct comparison of the transformation efficiency with these vectors has been made, the effect of copy number of the binary vector or stability of its maintenance, which vary with replicon, has not been ascertained. In general, however, the small RK2 origin plasmids such as pMON505 or pGA471 are less stable than the others. Whether this lower plasmid stability is associated with the decreased transformation of tomato seen with pMON505 compared to cointegrating vectors (S. McCormick, personal communication) remains to be determined.

All of the vectors contain unique restriction endonuclease recognition sites within the T-DNA for easy introduction of foreign genes. Several including BIN19 (Bevan, 1984) and pENDK4 (Klee *et al.,* 1985) contain unique restriction sites obtained from the *E. coli* cloning vector pUC19 (Yaenisch-Perron *et al.,* 1985) and retain all the advantage of direct screening for inserts on X-gal indicator dye containing medium.

Several of these vectors (pGA471, pMON505, pMON809) contain a ColE1-type replication origin and thus replicate in *E. coli* at high copy number to permit easy isolation of vector DNA and screening of recombinants. The ColE1 origin becomes part of the T-DNA which may prove useful for reisolation of a T-DNA from the genome of a transformed plant.

All of the vectors contain at least one marker for selection in bacteria and this is often incorporated into the T-DNA which could also facilitate reisolation of the T-DNA from a transformant.

The phage lambda *cos* sequence has been included in several of the vectors (pGA471, pAGS127, pC22 and pEND4K) to permit efficient cloning and packaging of large genomic DNA fragments for "shotgun cloning" experiments.

All of the vectors carry selectable markers for identification of transformed plant tissues. In most cases the marker is NPTII for selection of kanamycin resistant cells although many of the new markers are being constructed in binary vectors such as the methotrexate resistance in pMON809 or hygromycn resistance in pAGS125. In addition to selectable markers, the Matzke binary vector and pMON505 carry a nopaline synthase gene as a scorable marker which allows the T-DNA to be followed through regeneration and subsequent generations. The pMON809 plasmid carries a scorable octopine synthase gene for the same purpose. All the binary vectors are reported to be efficient in transformation and selection and should provide the reader a wide selection of combinations of selectable and scorable markers as well as restriction endonuclease sites from which to choose for obtaining their own transgenic plants.

B. Selectable Markers

One area of Ti plasmid vector development that has received needed attention is the construction of additional selectable marker genes to complement the existing chimeric NPTII kanamycin resistance. These additional dominant selectable markers are useful for species where the selectability of kanamycin resistance is marginal. Such markers also permit multiple, sequential transformations of a single plant to introduce additional copies of the same gene of interest or several different genes for enzymes of a biosynthetic pathway or several different members of the same gene family for the study of their relative expression levels. Alternate selectable markers also provide a means to follow the segregation of several introduced genes in the progeny of transformants. A listing of the current selectable markers and species in which they have been used for obtaining transformants appears in Table 3.

Table 3. Marker Genes for Selection of Transformed Plants

Marker	Selective agent	Species	Reference
NPTII	kanamycin[a]	tobacco	Bevan *et al.*, 1983
			Herrera-Estrella *et al.*, 1983
		petunia	Fraley *et al.*, 1983
		tomato	McCormick *et al.*, 1986
		poplar	Filatti *et al.*, 1986
		B. napus	Fry and Horsch, 1986
NPTI	kanamycin[a]	tobacco	Fraley *et al.*, 1983
dhfr[b]	methotrexate	petunia	Eichholtz *et al.*, 1987
		tobacco	
hph	hygromycin	tobacco	Waldron *et al.*, 1985
		petunia	van den Elzen *et al.*, 1985 a
		Arabidopsis	Lloyd *et al.*, 1986
Ble[c]	bleomycin	*N. plumbaginifolia*	Hille *et al.*, 1986

[a] These marker genes also provide resistance to G418, an aminoglycoside antibiotic with greater toxicity to other eucaryotic cells.
[b] This dhfr coding sequence is derived from the mouse.
[c] Only transformed callus has been obtained with this marker.

Previously described markers for the selection of chloramphenicol or methotrexate resistant plant cells (Herrera-Estrella *et al.*, 1983; DeBlock *et al.*, 1984) are not listed since they provide only marginal selection in tobacco and, in petunia, provide no differential between control and treated cells. New, effective markers include resistance to the aminocyclitol antibiotic, hygromycin B (van den Elzen *et al.*, 1985b; Waldron *et al.*, 1985). This marker provides efficient selection in tobacco and petunia and has been particularly valuable for the development of a coupled transformation-regeneration system for *Arabidopsis thaliana* (Lloyd *et al.*, 1986) where kanamycin resistance is only marginally useful. Another marker gene that has proven effective in petunia and tobacco is derived from a mouse cDNA that encodes a methotrexate resistant dihydrofolate reductase. Plant cells transformed with a chimeric CaMV 35S promoter-mouse dhfr gene exhibit excellent resistance to 1 uM methotrexate which kills nontransformed control cells and regenerated plants display methotrexate insensitive DNA replication (Eichholtz *et al.*, 1987). More importantly, transformed shoots will form roots in the presence of 1 uM methotrexate making this marker essentially equivalent to the kanamycin resistance marker.

Finally, a new marker for resistance to bleomycin, an antitumor agent that causes breaks in DNA, has been reported to be effective in selection of transformed *Nicotiana plumbaginifolia* cells (Hille *et al.*, 1986) and may be of use in other species in which the other markers are ineffective.

C. Expression Cassette Vectors

For those wishing to introduce a new trait into a plant to study its effects, vectors are available that contain plant gene expression signals that allow the insertion of coding sequences adjacent to promoters of varying strengths (Table 4). Cassettes for moderate to high level constitutive expression have been constructed using the mannopine synthase promoter from the octopine type T_R T-DNA (Velten *et al.*, 1984) and from the cauliflower mosaic virus 19S and 35S promoters (Rogers *et al.*, 1987). The latter cassettes have been used to obtain high level expression of the tobacco mosaic virus coat protein (Able *et al.*, 1986) and the petunia EPSP synthase enzyme (Shah *et al.*, 1986) in transgenic plants. Expression cassettes containing regulated promoters to provide leaf, root, stem, flower or seed specific expression are currently being constructed in several laboratories and will be available in the near future.

Table 4. Expression Cassette Vectors

Plasmid	Marker	Promoter	3′ Sequences	Type	Reference
pAS2034	NPTII	T_R 1′ (MAS)[a]	T_L Tr 7[b]	Coint	Velten & Schell, 1985
pMON237	NPTII	CaMV 19S[c]	NOS[d]	Coint	Rogers *et al.*, 1987
pMON316	NPTII	CaMV 35S[e]	NOS	Coint	Abel *et al.*, 1986 Rogers *et al.*, 1987
pMON530	NPTII	CaMV 35S	NOS	Binary	Shah *et al.*, 1986 Rogers *et al.*, 1987

[a] Mannopine synthase
[b] octopine-type T-DNA transcript 7
[c] cauliflower mosaic virus 19S (P66 or gene VI)
[d] nopaline synthase
[e] cauliflower mosaic virus 35S (full length)

V. Getting Genes into Plants

The earliest method for transformation used the co-cultivation procedure of Marton *et al.* (1981). The procedure involves treatment of tobacco or petunia protoplasts with *A. tumefaciens* containing the vector and new gene of interest and plating the treated cells under conditions that select for transformants and kill the residual bacterial cells. Within a few weeks to months, microcalli able to grow in the presence of kanamycin are readily apparent. These calli are then transferred to medium containing phytohormones that stimulate shoot formation and root formation. The time from co-cultivation to plantlets growing in soil was three to four months.

With the proven efficacy of the kanamycin resistance marker, it became possible to directly transform explanted tissues using the leaf disk procedure developed by Horsch *et al.* (1985). This procedure simplifies the method to obtain transgenic plants, avoids the use of protoplasts and permits the researcher to obtain transgenic tobacco or petunia plantlets in soil in as little time as six weeks. This method has been adapted for tomato by McCormick *et al.* (1986) and other species such as lettuce (R. Michelmore, personal communication), *Arabidopsis* (Lloyd *et al.*, 1986), *Brassica napus* (Fry and Horsch, 1986), and poplar (Fillatti *et al.*, 1986). For details of the methods the reader is referred to these publications.

VI. Novel Applications of Ti Transformation

A. Gene Isolation by Complementation or Direct Selection

Methods for gene isolation using binary vectors are currently under development. Gene isolation by complementation of a mutant phenotype or transfer of a dominant phenotype to a new host has been routinely used by bacterial and yeast geneticists for a number of years. Such an approach has not been heretofore possible in plants because of the large genome size of most plants and the lack of a high frequency transformation system to generate the numbers of transgenic plants which must be screened to cover a genome equivalent of a typical plant. The discovery that the *Arabidopsis thaliana* genome is only 7×10^4 kilobase pairs in size and contains little repetitive DNA (Leutweiler *et al.*, 1984; Pruitt and Meyerowitz, 1986) has solved the genome size problem. The development of small, highly efficient, cosmid-type, binary plant transformation vectors and improved transformation procedures has solved the second problem by permitting efficient construction of plant genomic banks containing large inserts and their subsequent mobilization into *Agrobacterium*. Following transfer to *Agrobacterium*, these genomic segments can be transformed directly back into plants using high efficiency transformation systems such as the leaf disc procedure. We now have the means to generate the thousands of independent transformants needed to screen a genomic bank of *Arabidopsis* in a highly transformable species such as petunia or tobacco. There are obvious limitations to this strategy for gene isolation. The gene which is to be rescued must have an easily identified or selected phenotype. The gene must be isolated from a plant with a relatively small genome such as *Arabidopsis* but must be functional in petunia or tobacco. There must be a sufficiently large number of clones and retransformants to screen to insure a reasonable probability of introducing the gene of interest. While these are major limitations, the gene rescue approach should be valuable for cloning a number of genes which have been previously unavailable. Theoretically, a bank constructed from *Arabidopsis* DNA would require only 7,000—15,000 clones, depending in insert size, to have a 99 % probability of containing any sequence. In practice, Simoens *et al.* (1986) have shown that many of the *A. thaliana* clones that they created in a binary cosmid vector in *E. coli* were unstable when transferred by conjugation into *Agrobacterium*. As a result of their studies, they estimate that over 180,000 transconjugants would be required to give a 95 % probability that the genome was represented in the cells used to transform plants. At this time, there has been no successful use of the shotgun system to isolate a plant gene.

We have developed a model system for demonstrating the feasibility of gene rescue in plants using *Arabidopsis* tissue which had been transformed with pMON200 (H. Klee and M. Hayford, in preparation). The single T-DNA in this tissue conferred kanamycin resistance as well as the ability to synthesize nopaline on the tissue. A genomic bank was constructed from this tissue in a binary cosmid vector which could be mobilized into *Agrobacterium* and used to directly retransform plant cells. The entire bank was then used to transform petunia leaf discs for kanamycin resistance. Kanamycin resistant, nopaline positive tissue was obtained at a low frequency (approximately one in twenty leaf discs) effectively demonstrating that a single copy gene can be rescued from a genomic bank by re-transformation. The use of such a procedure to isolate a directly selectable plant gene remains to be demonstrated.

B. T-DNA as a Transposon for Mutation and Promoter Probe

Another application for the Ti plasmid transformation system is use of the T-DNA as an insertional mutagen. This is especially attractive in a species such as *Arabidopsis* where the genome is small and over 60 % of which is comprised of unique sequences. The availability of transformation in *Arabidopsis* (Lloyd *et al.*, 1986) makes the identification of mutants associated with T-DNA insertions quite possible and is an active area of research in our own and several other laboratories.

Teeri and colleagues (1986) have modified the T-DNA to act as a promoter probe transposon similar to those in use in bacterial systems. These researchers have positioned a T-DNA border sequence adjacent to the NPTII coding sequence without a promoter. When this T-DNA inserts into an active transcriptional unit in the plant genome, kanamycin resistant cells and plants result. Both transcriptional and translational fusions have been isolated in this manner as well as mutants (Andre *et al.*, 1986) that may be associated with the insertion event. No genes or promoters have yet been isolated but such work is in progess and will provide a valuable tool for plant gene isolation once the complete strategy is proven successful.

VII. Prospects

It is clear from this brief description of current uses and adaptations of the transformation systems that the *Agrobacterium* Ti plasmid based vectors will continue to provide tools for the further study of plant molecular biology. If the progress during the next four years only keeps pace with the last four, we will undoubtedly be presented with a wealth of new and exciting discoveries about plant gene function and new means for the improvement of crops.

Acknowledgements

The authors wish to thank those colleagues who provided information prior to publication and B. Schiermeyer for tireless preparation of this manuscript. We also thank T. Hohn for his patience.

VIII. References

Abel, P. P., Nelson, R. S., De, B., Hoffmann, N., Rogers, S. G., Fraley, R. T., Beachy, R. N., 1986: Delay of disease development in transgenic plants that express the tobacco mosaic virus coat protein gene. Science **232,** 738—743.

Adang, M. J., De Boer, D., Firoozabady, I., Kemp, J. D., Murray, E., Rocheleau, T. A., Rashka, K. Staffield, G., Stock, C., Sutton, D., Merlo, D. J., 1986: J. Cellular Biochem. Supplement **10 C,** 11.

An, G., Watson, B., Stachel, S., Gordon, M., Nester, E., 1985: New cloning vehicles for transformation of higher plants. EMBO Journal **4,** 277—284.

Andre, D., Colau, D., Schell, J., Van Montagu, M., Hernalsteens, J.-P., 1986: Gene tagging in plants by a T-DNA insertion mutagen that generates APH(3′)II-plant gene fusions. Mol. Gen. Genet. **204,** 512—518.

Baker, B., Schell, J., Lorz, H., Fedoroff, N., 1986: Transposition of the maize controlling element "Activator" in tobacco. Proc. Natl. Acad. Sci., U.S.A. **83,** 4844—4848.

Beachy, R. N., Chen, Z.-L., Horsch, R. B., Rogers, S. G., Hoffmann, N. L., Fraley, R. T., 1985: Accumulation and assembly of soybean beta-conglycinin in seeds of transformed petunia plants. EMBO Journal **4,** 3074—3053.

Bevan, M., Flavell, R., Chilton, M.-D., 1983: A chimaeric antibiotic resistance gene as a selectable marker for plant cell transformatiton. Nature **304,** 184—187.

Bevan, M., 1984: Binary *Agrobacterium* vectors for plant transformation. Nucleic Acids Res. **12,** 8711—8721.

Broglie, R., Coruzzi, G., Fraley, R. T., Rogers, S. G., Horsch, R. B., Niedermeyer, J. G., Fink, C. L., Flick, J. S., Chua, N.-H., 1984: Light-regulated expression of pea ribulose-1,5-biphosphate carboxylase small subunit gene in transformed plant cells. Science **224,** 838—843.

Byrne, M., McDonnell, R., Wright, M., Carnes, M., 1987: Strain and genotype specificity in *Agrobacterium*-soybean interaction. Plant Cell Tissue Organ Culture, in press.

Chyi, Y.-S., Jorgensen, R. A., Goldstein, D., Tanksley, S. D., Loaiza-Figueroa, F., 1986: Locations and stability of *Agrobacterium*-mediated T-DNA insertions in the *Lycopersicon* genome. Mol. Gen. Genet. **204,** 64—69.

Comai, L., Schilling-Cordaro, C., Mergia, A., Houck, C., 1983: A new technique for genetic engineering of *Agrobacterium* Ti plasmid. Plasmid **10,** 21—30.

Comai, L, Facciotti, D., Hiatt, W., Thompson, G., Rose, R., Stalker, D., 1985: Expression in plants of a mutant *aroA* gene from *Salmonella typhimurium* confers tolerance to glyphosate. Nature **317,** 741—743.

De Block, M., Herrera-Estrella, L., Van Montagu, M., Schell, J., Zambryski, P., 1984: Expression of foreign genes in regenerated plants and their progeny. EMBO J. **3,** 1681—1689.

de Framond, A., Barton, K., Chilton, M.-D., 1983: Mini-Ti: a new vector strategy for plant genetic engineering. Bio/Technology **1,** 262—269.

Ditta, G., Stanfield, S., Corbin, D., Helinski, D., 1980: Broad host range DNA cloning system for gram-negative bacteria: Construction of a gene bank of *Rhizobium meliloti.* Proc. Natl. Acad. Sci., U.S.A. **77,** 7347—7351.

Eichholtz, D. A., Rogers, S. G., Horsch, R. B., Klee, H. J., Hayford, M., Hoffmann, N. L., Braford, S. B., Fink, C., Flick, J., O'Connell, K. M., Fraley, R. T., 1987: Expression of a mouse dihydrofolate reductase gene confers methotrexate resistance in transgenic petunia plants. Somatic Cell and Molecular Genetics **13,** 67—76.

Fillatti, J. J., Sellmer, J., McCown, B., Haissig, B., Comai, L., 1987: *Agrobacterium* mediated transformation and regeneration of *Populus.* Mol. Gen. Genet. in press.

Fraley, R. T., Rogers, S. G., Horsch, R. B., 1986: Genetic transformation in higher plants. Critical Reviews in Plant Sciences, **4,** 1, 1—46.

Fraley, R. T., Rogers, S. G., Horsch, R. B., Sanders, P. R., Flick, J. S., Adams, S. P., Bittner, M. L., Brand, L. A., Fink, C. L., Fry, J. S., Galluppi, G. R., Goldberg, S. B., Hoffmann, N. L., Woo, S. C., 1983: Expression of bacterial genes in plant cells. Proc. Natl. Acad. Sci., U.S.A. **80,** 4803—4807.

Fraley, R. T., Rogers, S. G., Horsch, R. B., Eichholtz, D. A., Flick, J. S., Fink, C. L., Hoffmann, N. L., Sanders, P. R., 1985: The SEV system: A new disarmed Ti plasmid vector system for plant transformation. Bio/Technology **3,** 629—635.

Fry, J., Horsch, R., 1986: Transformation of *Brassica napus* by *Agrobacterium tumefaciens* based vectors. Proceedings of IV International Congress of Plant Tissue and Cell Culture.

Gheysen, G., Dhaese, P., Van Montagu, M., Schell, J., 1985: DNA flux across genetic barriers: the crown gall phenomenon. In: Hohn, B., Dennis, E. S. (eds.) Genetic flux in plants. pp. 11—47, Springer, Wien - New York.

Grimsley, N., Hohn, B., Hohn, T., Walden, R., 1986: "Agroinfection", an alternative route for viral infection of plants by using Ti plasmid. Proc. Natl. Acad. Sci., U.S.A. **83,** 3282—3286.

Hepburn, A. G., White, J., Pearson, L., Maunders, M. J., Clarke, L. E., Prescott, A. G., Blundy, K. S., 1985: The use of pNJ5000 as an intermediate vector for the genetic manipulation of *Agrobacterium* Ti-plasmid. J. Gen. Micro. **131,** 2961—2969.

Hernalsteens, J.-P., Van Vliet, F., DeBeuckeleer, M., Depicker, A., Engler, G., Lemmers, M., Holsteers, M., Van Montagu, M., Schell, J., 1980: The *Agrobac-*

terium tumefaciens Ti plasmid as a host vector system for introducing foreign DNA in plant cells. Nature **287,** 654—656.

Herrera-Estrella, L., De Block, M., Messens, E., Hernalsteens, H.-P., Van Montagu, M., Schell, J., 1983: Chimeric genes as dominant selectable markers in plant cells. EMBO Journal **2,** 987—995.

Herrera-Estrella, L., van den Broeck, G., Maenhaut, R., Van Montagu, M., Schell, J., Timko, M., Cashmore, A., 1984: Light-associated and chloroplast-associated expression of a chimeric gene introduced into *Nicotiana tabacum* using a Ti plasmid vector. Nature **310,** 115—120.

Hille, J., Verheggen, F., Roelvink, P., Franssen, H., van Kammen, A., Zabel, P., 1986: Bleomycin resistance: a new dominant selectable marker for plant transformation. Plant Molecular Biology **7,** 171—176.

Hoekema, A., Hirsch, P., Hooykaas, P., Schilperoort, R., 1983: A binary plant vector strategy based on separation of *vir*- and T-region of *Agrobacterium tumefaciens* Ti plasmid. Nature **303,** 179—180.

Hoekema, A., Van Haaren, M. J. J., Fellinger, A. J., Hooykaas, P., Schilperoort, R. A., 1985: Non-oncogenic plant vectors for use in the agrobacterium binary system. Plant Molecular Biology **5,** 85—89.

Horsch, R. B., Fraley, R. T., Rogers, S. G., Sanders, P. R., Lloyd, A., Hoffmann, N. L., 1984: Inheritance of functional foreign genes in plants. Science **223,** 496—498.

Horsch, R. B., Fry, J. E., Hoffmann, N. L., Wallroth, M., Eichholtz, D., Rogers, S. G., Fraley, R. T., 1985: A simple and general method for transferring cloned genes into plants. Science **227,** 1229—1231.

Horsch, R., Klee, H. 1986: Rapid assay of foreign gene expression in leaf discs transformed by *Agrobacterium tumefaciens:* Role of the T-DNA borders in the transfer process. Proc. Natl. Acad. Sci., U.S.A. **83,** 4428—4432.

Klee, H., Yanofsky, M., Nester, E., 1985: Vectors for transformation of higher plants. Bio/Technology **3,** 637—642.

Koncz, C., Schell, J., 1986: The promoter of T_L-DNA gene 5 controls the tissue-specific expression of chimeric genes carried by a novel type of *Agrobacterium* binary vector. Mol. Gen. Genet. **204,** 383—396.

Leemans, J., Shaw, C., Deblaere, R., De Greve, H., Hernalsteens, J. P., Maes, M., Van Montagu, M., Schell, J., 1981: Site-specific mutagenesis of *Agrobacterium* Ti-plasmids and transfer of genes into plant cells. J. Mol. Appl. Genet. **1,** 149—164.

Leutweiler, L., Hough-Evans, B., Meyerowitz, E., 1984: The DNA of *Arabidopsis thaliana.* Mol. Gen. Genet. **194,** 15—23.

Lloyd, A. M., Barnason, A. R., Rogers, S. G., Byrne, M. C., Fraley, R. T., Horsch, R. B., 1986: Transformation of *Arabidopsis thaliana* with *Agrobacterium tumefaciens.* Science **234,** 464—466.

Marton, L., Wullems, G., Molendijk, L., Schilperoort, R., 1981: *In vitro* transformation of cultured cells from *Nicotiana tabacum* by *Agrobacterium tumefaciens.* Nature **277,** 129—131.

Matzke, A. J. M., Chilton, M.-D., 1981: Site-specific insertion of genes into T-DNA of the *Agrobacterium* tumor-inducing plasmid as an approach to genetic engineering of higher plants. J. Mol. Appl. Genet. **1,** 39—49.

Matzke, A. J. M., Matzke, M. A., 1986: A set of novel Ti plasmid-derived vectors for the production of transgenic plants. Plant Molecular Biology **7,** 357—365.

McCormick, S. M., Niedermeyer, J., Fry, J., Barnason, A., Horsch, R., Fraley, R., 1986: Leaf-disc transformation of cultivated tomato *(L. esculentum)* using *Agrobacterium tumefaciens.* Plant Cell Reports **5,** 81—84.

Morelli, G., Nagy F., Fraley, R. T., Rogers, S. G., Chua, N.-H., 1985: A short conserved sequence is involved in the light-inducibility of a gene encoding ribulose 1,5-biphosphate carboxylase small subunit of pea. Nature **315,** 200—204.

Nagy, F., Morelli, G., Fraley, R. T., Rogers, S. G., Chua, N.-H., 1985: Photoregulated expression of a pea rbcS gene in leaves of transgenic plants. EMBO J. **4,** 3063—3068.

Nester, E. W., Gordon, M. P., Amasino, R. M., Yanofsky, M. F., 1984: Crown gall: a molecular and physiological analysis. Ann. Rev. Plant Physiol. **35,** 387—413.

Ooms, G., Klapwijk, P., Poulis, J., Schilperoort, R., 1980: Characterization of Tn 904 insertions in octopine Ti plasmid mutants of *Agrobacterium tumefaciens.* J. Bacteriol. **144,** 82—91.

Owens, L., Cress, D., 1985: Genotypic variability of soybean response to *Agrobacterium* strains harboring Ti or Ri plasmids. Plant Physiol. **77,** 87—94.

Pruitt, R., Meyerowitz, E. 1986: Characterization of the genome of *Arabidopsis thaliana.* J. Mol. Biol. **187,** 169—183.

Rogers, S. G., Horsch, R. B., Fraley, R. T., 1986a: Gene transfer in plants: Production of transformed plants using Ti plasmid vectors. In Weissbach, H., Weissbach, A. (eds.), Plant Molecular Biology, Methods in Enzymology, **118,** p. 627—640, New York, Academic Press.

Rogers, S. G., Bisaro, D. M., Horsch, R. B., Fraley, R. T., Hoffmann, N. L., Brand L., Elmer, J. S., Lloyd, A., 1986b: Tomato Golden Mosaic Virus A component DNA replicates autonomously in transgenic plants. Cell **45,** 593—600.

Rogers, S. G., Klee, H. J., Horsch, R. B., Fraley, R. T., 1987: Improved vectors for plant transformation: expression cassette vectors and new selectable markers. In: Wu, R., Grossman, L. (eds.), Methods in Enzymology Recombinant DNA, Parts D and E, New York, Academic Press, in Press.

Ruvkun, G., Ausubel, F., 1981: A general method for site directed mutagenesis in prokaryotes. Nature **289,** 85—88.

Sengupta-Gopalan, C., Reichert, N. A., Barker, R. F., Hall, T. C., Kemp, J. C., 1985: Developmentally regulated expression of the bean beta-phaseolin gene in tobacco seed. Proc. Natl. Acad. Sci., U.S.A. **82,** 3320—3324.

Schreier, P., Seftor, E., Schell, J., Bohnert, H., 1985: The use of nuclear-encoded sequences to direct the light regulated synthesis and transport of a foreign protein into plant chloroplasts. EMBO J. **4,** 25—32.

Schmidhauser, T., Helinski, D., 1985: Regions of broad-host-range plasmid RK2 involved in replication and stable maintenance in nine species of gram-negative bacteria. J. Bacteriology **164,** 446—455.

Shah, D. M., Horsch, R. B., Klee, H. J., Kishore, G. M., Winter, J. A., Tumer, N. E., Hironaka, C. M., Sanders, P. R., Gasser, C. S., Aykent, S., Siegel, N. R., Rogers, S. C., Fraley, R. T., 1986: Engineering herbicide tolerance in plants. Science **233,** 478—481.

Simoens, C., Alliotte, Th., Mendel, R., Muller, A., Schiemann, J., Van Lijsebettens, Schell, J., Van Montagu, M., Inze, D., A binary vector for transferring genomic libraries to plants. Nucleic Acids Res. **14,** 8073—8090.

Simpson, R. B., Spielmann, A., Margossian, L., McKnight, T. D., 1986: A disarmed binary vector from *Agrobacterium tumefaciens* functions in *Agrobacterium rhizogenes.* Plant Molecular Biology **6,** 403—415.

Teeri, T., Herrera-Estrella, L., Depicker, A., Van Montagu, M., Plava, E., 1986: Identification of plant promoters in situ by T-DNA-mediated transcriptional fusions to the npt-II gene. EMBO J. **5,** 1755—1760.

van den Elzen, P., Lee, K., Townsend, J., Bedbrook, J., 1985a: Simple binary vectors for DNA transfer to plants. Plant Molecular Biology **5,** 149—154.

van den Elzen, P., Townsend, J., Lee, K., Bedbrook, J., 1985b: A chimaeric hygromycin resistance gene as a selectable marker in plant cells. Plant Molecular Biology **5**, 299—302.

Velten, J., Schell, J., 1985: Selection-expression plasmid vectors for use in genetic transformation of higher plants. Nucl. Acids. Res. **13**, 6981—6998.

Waldron, C., Murphy, E., Roberts, J. Gustafson, G. Armour, S., Malcolm, S., 1985: Resistance to hygromycin B. Plant Molecular Biology **5**, 103—108.

Wallroth, M., Gerats, A. G. M., Rogers, S. G., Fraley R. T., Horsch, R. B., 1986: Chromosomal location of foreign genes in *Petunia hybrida*. Mol. Gen. Genet. **202**, 6—15.

Wang, K., Herrera-Estrella, L., Van Montagu, M., Zambryski, P., 1984: Right 25 bp terminus of the nopaline T-DNA is essential for and determines direction of DNA transfer from *Agrobacterium* to the plant genome. Cell **38**, 455—462.

Yadav, N., Vanderleyden, J., Bennett, D., Barnes, W., Chilton, M.-D., 1982: Short direct repeats flank the T-DNA on a nopaline Ti plasmid. Proc. Natl. Acad. Sci., U.S.A. **79**, 6322—6326.

Yaenisch-Perron, C., Vieira, J., Messing, J., 1985: Improved M13 phage cloning vectors and host strains: nucleotide sequences of the M13mp18 and pUC19 vectors. Gene **33**, 103—119.

Zambryski, P., Depicker, A., Kruger, D., Goodman, H., 1982: Tumor induction by *Agrobacterium tumefaciens:* analysis of the boundaries of T-DNA. J. Mol. Applied Genet. **1**, 361—370.

Zambryski, P., Joos, H., Genetello, C., Leemans, J., Van Montagu, M., Schell, J., 1983: Ti plasmid vector for the introduction of DNA into plant cells without alteration of their normal regeneration capacity. EMBO J. **2**, 2143—2150.

Chapter 8

Plant Transposable Elements: Unique Structures for Gene Tagging and Gene Cloning

U. Wienand and H. Saedler

Max-Planck-Institut für Züchtungsforschung, D-5000 Köln 30, Federal Republic of Germany

With 4 Figures

Contents

I. Introduction

Much of the recent data on genome organization and gene expression are the result of improved and new techniques in recombinant DNA research. Among these the cloning and isolation of genes is one of the most technically difficult subjects.

In comparison to procaryotes, molecular cloning of eucaryotic genes is still rather complicated. This difference in ease of cloning can be explained by the greater complexity of eucaryotic genomes which generally include a high number of genes and large quantities of repetitive DNA (Davidson and Britten, 1973; Flavell *et al.,* 1974; Hake and Walbot, 1980). As a result only a few plant genes have so far been isolated. Of those that have been isolated most are present in high copy numbers (Baulcombe and Verma, 1978; Wienand *et al.*, 1979; Evans *et al.*, 1980) or expressed at a high level in specific tissues (Geiser *et al.*, 1980). This is because a routine procedure used for the isolation of eucaryotic genes is based on the synthesis and isolation of a cDNA (complementary DNA) probe from mRNA which is homologous to the codogenic region of the desired gene (for review see Land *et al.*, 1983). Genes, which are expressed at a low level and thus encode non-abundant proteins, are therefore very difficult to isolate via this method. However, a gene of this nature can be cloned, if the amino acid sequence of the gene product is known. A DNA oligonucleotide can then be synthesized (Szostak *et al.*, 1979) and used as a gene-specific probe in the isolation procedure.

If, however, the product from the gene of interest is unknown, as is the case for most structural and regulatory genes, the two methods previously mentioned cannot be applied.

Nevertheless, it is possible to isolate some of these genes using transposable elements as specific probes in an isolation procedure called transposon tagging (Bingham *et al.*, 1981).

Transposable elements have been characterized at the genetic and molecular level in procaryotes (for reviews see Starlinger, 1980; Kleckner, 1981; Derbyshire and Grindley, 1986) and eucaryotes (for review see Flavell, 1983). In plants, they have been studied extensively in *Zea mays* (for reviews see Fincham and Sastry, 1974; Fedoroff, 1983; Freeling, 1984; Döring and Starlinger, 1984; Nevers *et al.*, 1985; Dellaporta and Chomet, 1985; Döring, 1985; Peterson, 1986; Döring and Starlinger, 1986) and *Antirrhinum majus* (for reviews see Saedler *et al.*, 1984; Sommer *et al.*, 1985a). Detailed genetical analysis of these elements carried out by McClintock (1953, 1958, 1965), Peterson (1953, 1978) and others in *Zea mays* and Fincham, Carpenter and Harrison in *Antirrhinum majus* (Fincham and Harrison, 1967; Harrison, 1971; Harrison and Carpenter, 1973) led to a basic knowledge of their function. In each species a few of the genetically characterized transposable elements have been cloned molecularly, a breakthrough with two important consequences. First, it allowed a detailed molecular analysis of structure and function of transposable elements. Second, these elements have been available for use in cloning genes which otherwise would have been difficult to isolate.

Through the combination of both genetic and molecular techniques it is now possible to either transpose a yet unidentified transposable element to a known and already cloned gene or to transpose a known and cloned transposable element to an unknown and desired gene.

Use of gene-specific sequences as a probe allows the isolation of unknown transposable elements while transposable-element-specific sequences are used in the isolation of new genes.

Some basic aspects of this technique, called gene tagging, will be described in this article.

II. Isolation and Characterization of a Transposable Element

A. Recognition of a Transposable Element

Transposition — insertion and excision — is the unique and remarkable feature of transposable elements. It was this behavior which first led to their detection. However, transposition *per se* is often not detectable. Rather it is certain consequences of transposition that can be scored. Transposable elements can be detected and studied when they integrate into a gene and thereby induce a mutation. If this gene affects, for example, the pigmentation, or the structure or the development of a plant, the mutation may lead to a recognizable change in phenotype. The best known examples of transposable-element-induced mutations are certainly those affecting pigmentation. Such mutants have been found in many plants (for review see Nevers *et al.,* 1985) and have been extensively studied in *Zea mays* and *Antirrhinum majus.* Pigmentation is the result of a co-ordinated interaction of regulatory and structural genes. In maize, about a dozen loci are known to be involved in the anthocyanin pathway among these are *c 1, vp 1, r 1* (regulatory genes) and *c 2, a 1, a 2, bz 1, bz 2, pr* (structural genes) (Styles and Ceska, 1977; Coe and Neuffer, 1977; Reddy and Peterson, 1977; Dooner and Nelson, 1977; Dooner and Nelson, 1979; Coe, 1979; Dooner, 1983; Soave and Salamini, 1984). Transposition of an element into one of these loci interrupts or alters gene expression, which depending on the locus involved, may result in a colorless phenotype. Excision of the inserted transposable element often (but not always) restores gene function. The restoration of gene function (via excision) in some of the somatic cells and in their clonal progeny results in a variegated phenotype. Hence, one of the characteristic features of transposable-element-induced mutations is that they are somatically unstable. Somatic variegation may appear as spots or stripes in kernels (Fig. 1), anthers, leaves or flowers, the frequency varying from many to only a few, indicating that transposition can be frequent or rare. Occasionally reversion affects the germline of the plant as well, which then leads to the isolation of revertants, (usually) phenotypically indistinguishable from wildtype phenotype.

B. Genes Suitable for the Isolation of Transposable Elements

The molecular cloning of a transposable element by using standard procedures for gene isolation is not possible, since data concerning transcription and translation of element-encoded functions are usually unknown. Nevertheless transposable elements can be isolated.

Transposon mutagenesis is used on a gene which can be (or already has been) cloned. This procedure results in the integration of a transposable element into the gene. The gene affected by the integration of a transposable element is changed in its structure, due to the integration of a stretch of DNA (the transposable element) in either its regulatory or

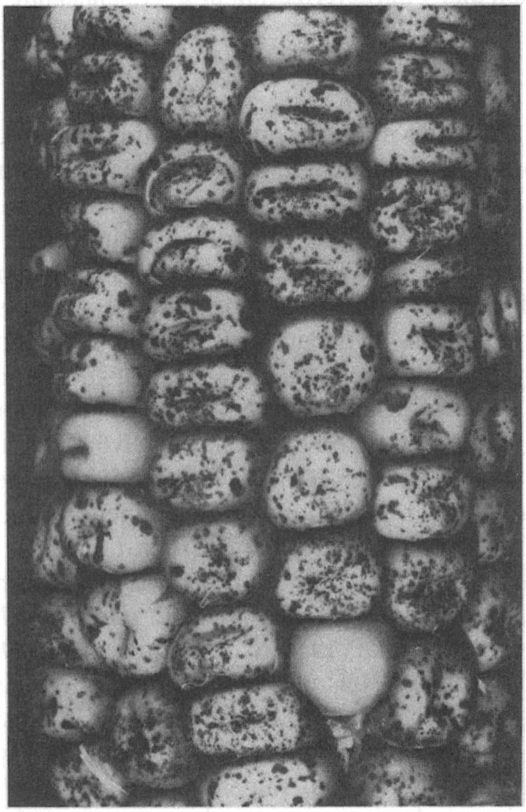

Fig. 1. Variegated pattern of maize kernels due to the presence of a transposable element at a gene involved in pigmentation

codogenic parts. The comparison of gene-specific sequences from mutant and wildtype (either by heteroduplex or restriction enzyme analysis) therefore reveals additional sequences present in the mutated gene which are transposable-element specific (Fig. 2).

This strategy requires genes which first are not too difficult to isolate and second, at which mutations are easily scored phenotypically. In maize at least three such genes are known and have been extremely valuable in the isolation of transposable elements. Two of the genes are expressed in the endosperm at a rather high level and thus were amenable to isolation via cDNA cloning. The first of these is the *shrunken-1 (Sh1)* gene (Geiser *et al.*, 1982; Werr *et al.*, 1985; McCarty *et al.*, 1986), coding for sucrose synthase (Chourey and Nelson, 1976) and the second is the *waxy (Wx)* gene (Shure *et al.*, 1983; Schwarz-Sommer *et al.*, 1984; Klösgen *et al.*, 1986), coding for UDP-glucose-starch transferase (Nelson and Rines, 1962). The recessive mutation of the *Sh1* gene *(sh1)* changes the structure of the maize endosperm, leading to a collapsed kernel, whereas *wx* affects the amylose content and thereby turns the endosperm opaque with a waxy texture, also

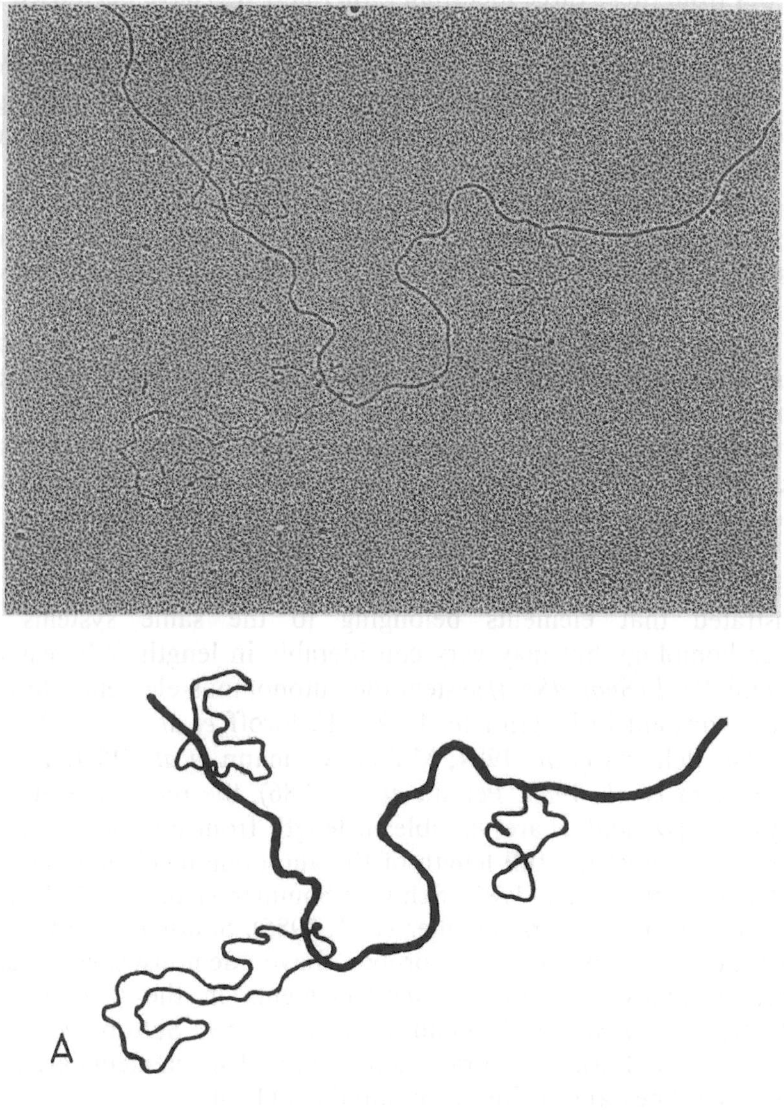

Fig. 2. Heteroduplex of recombinant lambda clones containing the wildtype and an En element-induced mutant allele of the waxy gene. The loop (A) represents the controlling element En present at the waxy gene in the mutant *wx-844* (Pereira *et al.*, 1985). The two other loops represent non homologous sequences of both clones

easily detectable in this case by staining the endosperm with iodine (KI/I2). The *alcohol dehydrogenase* gene *(Adh1)* in maize is the third gene which has been extensively analyzed (Gerlach *et al.*, 1982; Dennis *et al.*, 1984) and used to trap and subsequently to clone transposable elements (Peacock *et al.*, 1983; Döring *et al.*, 1984b). This gene is inducible under anaerobic conditions. A mutation in pollen for this gene confers allyl alcohol resistance to the *adh1* pollen grains and makes them directly selectable from wildtype pollen (Schwartz and Ostermann, 1976).

Probes from these three loci *(Adh1, Sh1* and *Wx)* have been used in the isolation of several genetically distinct transposable elements. These include the *Ac* (Activator; Fedoroff *et al.*, 1983), the *En* (Enhancer; Pereira *et al.*, 1985), the *Spm* (Suppressor-mutator; Wienand *et al.*, 1986; genetically equivalent to *En;* Peterson, 1965) and the *Mu1* (Mutator; Bennetzen *et al.*, 1984) elements.

Both the *Ac* and the *En (Spm)* element belong to the so-called two-component systems in maize, each of which is composed of an autonomous (or regulatory) element *(Ac* or *En [Spm])* and a nonautonomous (or receptor) element (McClintock, 1953, 1958, 1961; Peterson, 1953). These receptor elements called *Ds* (Dissociation) in the *Ac/Ds* system and *I/dSpm* (Inhibitor/defective Spm) elements in the *En/I (Spm/dSpm)* system are (usually) deletion derivatives of the autonomous elements and are transposition defective. Autonomous and receptor elements can interact within a system because autonomous elements can transactivate receptor elements to move within the genome. However there is no interaction between different systems. In fact, the presence of these interactions is the definitional basis of the systems. Molecular studies have demonstrated that elements belonging to the same systems share sequence homology but may vary considerably in length. Whereas in the *Ac/Ds* and *En/I (Spm/dSpm)* system the autonomous elements *Ac* and *En (Spm)* are constant in length (*Ac:* 4.5 kb; Fedoroff *et al.*, 1983; Pohlmann *et al.*, 1984; Behrens *et al.*, 1984; Müller-Neumann *et al.*, 1984; *En (Spm):* 8.3 kb; Pereira *et al.*, 1985; Pereira *et al.*, 1986), the receptor elements of both systems *(Ds* and *I)* are variable in length from only a few hundred base pairs to almost the full length of the autonomous element (Fedoroff *et al.*, 1983; Sachs *et al.*, 1983; Schwarz-Sommer *et al.*, 1984; Döring *et al.*, 1984a, 1984b; Schwarz-Sommer *et al.*, 1985). Southern analyses using element-specific sequences as probes have demonstrated that the presence of many *Ds* and *I (dSpm)* elements in the genome (about 30—60) (Geiser *et al.*, 1982; Fedoroff *et al.*, 1983; Sutton *et al.*, 1984; O'Reilly *et al.*, 1985a). In contrast, active *En* and *Ac* elements are present in only a few lines and at low copy numbers (1—4).

The *Mu1* element in *Zea mays* (Robertson, 1978; Bennetzen *et al.*, 1984) is not related to any of the two component systems known so far (Robertson and Mascia, 1981), although some data suggest, that *Mu1* might be part of a (not yet identified) more complex system (Schnable and Peterson, 1986). The *Mu1* element (Bennetzen *et al.*, 1984) was first detected in a line showing an exceptionally high mutation rate (Robertson, 1978) as compared to other maize lines. This line and those descended from it have been designated "Mutator lines". The *Mu1* element as known from molecular analysis is present in Mutator lines with a copy number varying between 20 to 50 (Bennetzen, 1984; Alleman and Freeling, 1986). The majority of *Mu1* elements are 1.4 kb in length (Barker *et al.*, 1984), although recently *Mu* elements have been cloned which are larger (1.7 kb) or smaller (1.1 and 0.8 kb) (Taylor and Walbot, 1985; Taylor *et al.*, 1986).

The *Antirrhinum majus* elements also differ in size. The largest transposable element isolated so far (form any species) is the *Tam1* element (*t*ransposon *a*ntirrhinum *m*ajus) which is 15 kb in length (Bonas *et al.*, 1984; Piotrowiak, 1986). The *Tam3* transposable element (3.5 kb in length; Sommer *et al.*, 1985b) has been successfully used in the isolation of the *pallida* locus of *Antirrhinum majus.* Both Tam elements are present in several copies (between 5 and 15) in the *Antirrhinum majus* genome (Martin *et al.*, 1985).

It is important to know the copy number of transposable elements used for gene tagging, because mutations caused by an element present in low copy number are much easier to molecularly identify than those induced by a high-copy-number element (Ralston and Dooner, 1986).

An alternative approach for cloning genes interacting with transposable elements utilizes heterologous probes in the gene isolation procedure. This can be extremely helpful, if the desired gene from the species of interest is too difficult to isolate but can be detected and cloned via cross hybridization with the heterologous probe.

The *chalcone synthase* cDNA from *Petroselinum hortense* (Kreuzaler *et al.*, 1983; Reimold *et al.*, 1983), for example, was successfully utilized as a probe to isolate the *nivea* locus of *Antirrhinum majus* (Wienand *et al.*, 1982; Sommer and Saedler, 1986). This locus is involved in pigmentation and is affected by the integration of the transposable element *Tam1* in the variegated mutant *niv-rec-53* and *Tam3* in the variegated *niv-rec-98* mutant.

However, gene cloning using heterologous probes, although very helpful in specific cases, cannot be used as a general strategy, because cross hybridization often is limited to a small number of related species. This is illustrated by the fact that cloning of the chalcone synthase gene with the heterologous probe from parsley was only possible from other dicotyledonous plants such as *Antirrhinum majus* (Wienand *et al.*, 1982) and *Petunia hybrida* (Reif *et al.*, 1985) and was not possible from the monocotyledonous species *Zea mays* (Wienand *et al.*, 1982). Therefore, although heterologous probes are useful in specific instances gene tagging is still required in many cases.

With a battery of cloned transposable elements at hand in *Zea mays* and in *Antirrhinum majus* the predominant problem in the isolation of a gene is and will continue to be the tagging of a gene with one of the already cloned elements.

C. Transposon Tagging of a Gene in Zea mays

A tagging experiment in principle is rather easy to perform, provided a recessive mutation of the desired gene affects the phenotype of the plant and is already available, because in this case the tagged gene will be visible in the F1 of an appropriate cross (see Fig. 3). The principle structure of such crosses is:

Transposon (TE) tagging of a gene (G)

Initial cross

$$\frac{a^r\ G\ TE}{a^r\ G\ TE} \times \frac{a\ g}{a\ g}$$

Phenotype: variegated at a, since a^r responds to TE × not variegated

F1 majority

$$\frac{a^r\ G\ TE}{a\ \ g\ \ -}$$

Phenotype: variegated at a
wildtype at G

F1 exceptions

Phenotype: variegated at a
variegated at g

alternative I

TE is integrated into G $= g^{TE}$

alternative II

receptor element of TE is integrated into $G = g^r$

Testcrosses

Alternative I:

$$\frac{a^r\ g^{TE}}{a\ \ g} \times \frac{a^r\ g}{a^r\ g}$$

		a^r	g
a^r	g^{TE}	+	+
a^r	g	–	–
a	g^{TE}	+	+
a	g	–	–

Alternative II:

$$\frac{a^r\ g^r\ TE}{a\ \ g\ \ -} \times \frac{a^r\ g}{a^r\ g}$$

			a^r	g
a^r	g^r	TE	+	+
a^r	g^r	–	–	–
a^r	g	TE	+	–
a^r	g	–	–	–
a	g^r	TE	+	+
a	g^r	–	–	–
a	g	TE	+	–
a	g	–	–	–

+ = variegated
– = not variegated

Co-segregation of G-variegated and TE

Independent segregation of G-variegated and TE

The female plant carries one or a few copies of one or a few families of active transposable elements and in addition the wild-type allele of the gene to be tagged. These female plants are pollinated by a line (lacking active transposable elements) carrying a recessive stable mutation of the gene of interest. In most instances, the progeny of this cross exhibit a wild-type phenotype. If, however, an element is transposed into the gene of interest a mutant phenotype results. Such exceptional progeny are isolated and analyzed further (Peterson, 1978; Reddy and Peterson, 1983).

D. Genetical Analysis of the Tagged Mutants

The analysis of a newly isolated and hopefully tagged mutant, is very laborious and usually takes several maize generations and years. Assuming the new mutation exhibits the characteristic instability associated with transposable element inserts the following questions must be answered via classical genetical analysis:

1. Is the exceptional phenotype heritable?
2. Is the mutation affecting the target gene due to the integration of a transposable autonomous element or a receptor element?
3. To which system does that element belong?

An exceptional progeny arising in a gene tagging experiment is first tested for heritability by testcrossing it to a line carrying the stable recessive

---

Fig. 3. The crosses performed for the analysis of tagged mutants are given in a very simplified and shortened form in order to demonstrate the principle method of analysis. A detailed description of the sequences of crosses performed in practice is given by Peterson (Peterson, 1978; Peterson, 1986). In the initial cross a line containing the gene to be tagged (G; homozygous dominant), a transposable element (TE) and a receptor element of TE (present at gene a [a^r]) is crossed by a line containing the homozygous recessive alleles a and g. The majority of kernels in the F1 generation will have the genotype outlined and would therefore be phenotypically identical to the female parent. The variegation of a^r in the F1 is the result of presence and activity of the transposable element TE and therefore defines a specific system relationship between the transposable element and the receptor element present at a. If variegation in the desired gene (G) is observed, the transposable element (TE) either integrated in gene G (g^{TE}; alternative I) or a receptor element is present at G (g^r) and the transposable element is located elsewhere (alternative II). The two alternatives can be distinguished by a cross of the mutant to the tester line, a^r, g. If in this cross a ratio of 1 : 1 is observed for the segregation of variegated a and g phenotype versus non variegated kernels, the element is present at g (g^{TE}) and cosegregates with g. If variegation in the desired gene G is due to the integration of a receptor element into G (g^r) and the autonomous element is located elsewhere in the genome (alternative II) variegation observed at a and g does not cosegregate. Instead of that a ratio of 2 : 1 : 1 is observed comparing non variegated kernels versus those variegated at a and g and those variegated only at a. It should be noted that the analysis of a mutant that generates germinal nonresponsive and revertant alleles at a high rate is much more complicated allele. This cross tests not only the heritability of the mutant phenotype but also confirms that the mutant is allelic to the locus of interest. If heritability is confirmed, further crosses are performed to clarify the nature of the new mutant.

If the exceptional isolate is variegated two alternatives (as shown in Fig. 3) are of interest. In alternative I the autonomous transposable element (TE) is integrated into the desired gene (G). In alternative II a receptor element is integrated into G generating a responsive allele, g^r, while the autonomous element (TE) is located somewhere else in the genome and activates the receptor element present at the locus. The cross diagrammed in Fig. 3 distinguishes between these two alternatives.

If somatic (germinal) instability at the new mutant always cosegregates with the mutant (G) phenotype (a 1 : 1 segregation in the appropriate backcross, see Fig. 3) it is reasonable to assume that an insertion of an autonomous element is the cause of the mutation in G. However, detailed analysis in this respect is necessary to unequivocally identify the nature of the mutation (Peterson, 1978; Pereira *et al.*, 1985). For example, the segregation of many independent autonomous elements (belonging to the same system) in a line can complicate this analysis as can alleles which generate non responsive derivatives (Peterson, 1978).

If the variegated phenotype does not co-segregate with the autonomous element the mutation is due to the integration of a receptor element into the gene (G). This insertion generates a g^r allele. The presence of variegation demonstrates that an autonomous element (TE) is present somewhere else in the genome which in trans triggers reversion of the mutant g^r to G. Such a situation is indicated if the variegated G-phenotype is observed and a 2 : 1 : 1 segregation is observed in the backcross of the mutant to so called tester lines. These lines (MERLs = mobile element receptor lines; Peterson and Salamini, 1986) allow the identification of the transposable element system which is active in the mutant. This is important to know because the female parent in the tagging experiment might have carried more than one system of autonomous transposable elements. Without knowing which system is responsible for the mutation the proper molecular probe for isolation of the mutated gene cannot be chosen. The MERLs contain a receptor element of a defined two component system at a phenotypically easily detectable gene, e. g., the *A1* gene which is involved in pigmentation. This insertion inactivates the *A* gene (a^r), and hence leads to a colorless kernel phenotype. The receptor element (defective in transposition) can only be (trans) activated by an autonomous element belonging to the same system. This activation often leads to a restoration of the *A1* gene (from a^r to A) and hence somatically to colored spots on a colorless background (variegation). This a^r allele is said to be responding to one of the trans functions (transposase) of the transposable element. This interaction between autonomous and receptor elements are system specific because an autonomous *Ac* for example can only complement in trans the otherwise transposition defective *Ds* (a deletion derivative of *Ac*) and not the *I* element, a deletion derivative of the *En*

(Spm) system. In maize MERLs are available for the following systems: *Ac (Mp), En (Spm), Fcu (Spf), Dt, Bg, Uq, Cy, Mut* and *Mrh* (McClintock, 1951, 1955; Peterson, 1953; Rhoades, 1938; Gonella and Peterson, 1977; Salamini, 1981; Friedemann and Peterson, 1982; Schnable and Peterson, 1986; Rhoades and Dempsey, 1982; Rhoades and Dempsey, 1983).

Progeny from the cross of a newly isolated unstable mutant with a MERL results in variegation provided the line carrying the mutant allele also carries an active copy of the appropriate autonomous element. If none of the receptor lines are triggered by the cross to the mutant line, the autonomous element responsible for instability at the new mutant (either in trans or cis) represents a new system (Friedemann and Peterson, 1982), which although present in the female line used in the tagging experiment was only uncovered by its insertion into a scorable gene.

E. Frequency of Mutation

The frequency of mutations caused by the integration of a transposable element in maize varies.

This frequency may depend on the element used, the original position of the autonomous element in the genome and the number of elements present in the lines. Transposition in maize has been analyzed by many geneticists. The capacity of an element to undergo transposition and reinsertion seems not to be limited to specific target sequences (Peterson, 1970). Short intra-chromosomal transpositions occur at a higher rate than long intra-chromosomal transpositions which occur at higher rates than intra-chromosomal transpositions which occur at higher rates than inter-chromosomal transpositions (van Schaik and Brink, 1958; Peterson, 1970; Nowick and Peterson, 1981). Data obtained by detailed analysis of transposition of *Mp* (Modulator of pericarp, equivalent to *Ac;* Barclay and Brink, 1954) (Greenblatt and Brink, 1963; Greenblatt, 1984) and *En* (Peterson, 1978) indicate that elements probably differ in reinsertion capabilities. Internal environmental factors may influence transposition as well, as has been shown for the induction of mutations using the *Mu1* element (Robertson, 1980; Alleman and Freeling, 1986). Increased *Mu* activity is observed in outcrosses as opposed to inbreeding (Robertson, 1983). Also the mutation rate may be higher when mutator lines are used as males as compared to when they serve as females (Robertson, 1985). In addition, transposition can be influenced by environmental parameters such as temperature, for example, as has been described for several transposable element induced mutants in *Antirrhinum majus* (Harrison and Fincham, 1964; Harrison and Carpenter, 1973) and in *Zea mays* (Rhoades, 1941; Peterson, 1958).

The frequency of mutation induction at certain loci in maize has been tested. For the *En (Spm)* element a number of data are available indicating that this element induces mutations at a frequency of about 1×10^{-5} (Peterson, 1983a, 1983b; Peterson and Cormack, 1984). Differences between mutation frequencies of different genes seem to exist as for example observed when the *a2* and *c1* loci of maize were targeted using the *En* system (Peterson, 1978). Such differences have also been described for the *Mu1* activity in maize where the frequency of recovered mutations was found to differ among several loci. The frequency of *Mu1*-induced mutations at the *Y1, Yg2, Bz1, Sh2* and *Wx* genes range between 7×10^{-5} and 4×10^{-6} and in some experiments have been shown to be as high as $1-4 \times 10^{-4}$ (Robertson, 1985). Such high mutation rates have been also reported for *Bg* induced mutations (Marotta *et al.*, 1986). Mutants isolated from *Cy* and *Uq* element containing lines have been induced with frequencies of 1×10^{-4} (*c2* mutants in a *Cy* containing line; Schnable and Peterson, 1986) and 2×10^{-6} (*Uq* induced mutations at the *c1* locus; Oberthur and Peterson, 1984).

III. Transposable Elements as Molecular Probes for Gene Isolation

A. General Aspects

Cloned transposable elements can be used as specific probes to isolate unknown genes (Bingham *et al.*, 1981).

The strategy of such an experiment is to construct a genomic library from DNA of the transposable-element-induced mutant and to identify sequences (clones) homologous to the probe (transposable element). The second step of analysis is the identification of gene-specific sequences on one (or several) of them. The whole procedure is very laborious and needs detailed molecular analysis of each of the clones isolated.

The number of clones obtained in such cloning experiments is proportional to the number of elements present in the mutant line, hence it is much easier to clone genes from mutants induced by an autonomous element *(Ac* or *En/Spm)* than from non-autonomous receptor element induced mutants *(Ds* or *I/dSpm)*. This is because, in a given line, autonomous elements are present in lower copy number than are receptor elements. Even if the copy number of the element used in the screening procedure is rather high as for example *Ds, I (dSpm)* or *Mu1* elements, certain strategies can occasionally be used to reduce the number of clones to be analyzed. An important result from the molecular analysis of transposable elements, which facilitates their application in gene cloning, is that in Southern experiments no cross hybridization is observed between elements belonging to distinct systems even though, as indicated by detailed sequence analysis, some of the systems seem to be related (Schwarz-Sommer and Saedler, 1985; Sommer *et al.*, 1985b).

B. Isolation of Mutants Induced by Autonomous Elements (Ac and En [Spm])

Most sequences from autonomous *Ac* and *En (Spm)* elements cannot be used to molecularly identify active copies of these elements, because they cross hybridize to the many deletion derivatives *Ds* and *I (dSpm)* respectively, present in the maize genome. However, certain short fragments of both elements, for example an internal sequence of the open reading frame 1 in *Ac* (Pohlman *et al.*, 1984; Müller-Neumann *et al.*, 1984) and a sequence including the first exon of gene 1 in *En (Spm)* (Pereira *et al.*, 1986) are almost unique and present only in genetically active elements. Hence these sequences can be used to clone genes from mutants induced by the integration of an autonomous *Ac* or *En* element.

The molecular isolation of an autonomous *Ac* or *En (Spm)* element induced mutant is done by screening a lambda library prepared from partially digested DNA (Maniatis *et al.*, 1982) of the (homozygous) mutant, with either the *Ac*- or *En*-specific internal probe. Depending on the number of recombinant phages analyzed (normally between 5×10^5 and 2×10^6) about 10—30 clones will light up, which are further analyzed for presence of a full length *Ac* or *En* element. The comparison of those clones with the autonomous element includes restriction site mapping and Southern hybridization experiments. This analysis must be precise because a large number of elements almost identical to the autonomous element in size and sequence might be present in the mutant line analyzed (O'Reilly *et al.*, 1985b). Only the isolation of an element which is structurally identical to the *Ac* or *En* element unequivocally proves the isolation of an autonomous element from the mutant. All clones containing a full size element should be considered as putative candidates to also contain at least part of the desired gene. However, it can happen that due to cloning of partially digested DNA the restriction enzyme cuts within the element, hence only part of an autonomous element is present on a clone, simulating the presence of a defective element (Wienand *et al.*, 1986). Cloning of genes using autonomous elements as probes has been successfully applied in the isolation of the *bz1* (Fedoroff *et al.*, 1984; Dooner *et al.*, 1985) and *p* (Lechelt *et al.*, 1986; Peterson and Schwartz, 1986) loci from *Ac*-induced mutants as well as the *a1* (O'Reilly *et al.*, 1985a), *c2* (Wienand *et al.*, 1986), *c1* (Paz-Ares *et al.*, 1986) locus from *En* induced mutants. The analyses of these mutants revealed only one (in case of *a1*, *c2* and *c1*) or two (in case of *bz1;* Fedoroff *et al.*, 1984) active elements in the genome of these lines.

An alternative strategy to isolate active elements (used for cloning the *c1* locus; Cone *et al.*, 1986) has been developed recently based on the observation, that active elements are molecularly distinguishable from inactive elements (Dellaporta and Chomet, 1985). Active elements have been found to be less or nonmethylated and therefore can be cleaved by methyl-sensitive restriction enzymes. Digestion of total genomic DNA with such an enzyme will preferentially cleave undermethylated sites, whereas the majority of DNA (and inactive elements) will remain uncut. A transposable-element-specific probe in Southern experiments might light up a low number of distinct fragments containing sequences of an active element. If one of the bands co-segregates with the mutant phenotype, isolation and cloning of this band might lead to the isolation of a fragment containing element and gene specific sequences.

C. *Isolation of Clones Carrying Receptor Elements*

The isolation of genes from mutants induced by the integration of a receptor element (or any other highly repetitive element, for example *Mu1*) is much more difficult compared to the situation mentioned above (Ralston and Dooner, 1986). This is the case, whenever there are more than about ten copies of the element present in the genome, as is true for all receptor elements of the maize two component systems as well as for the *Mu1* element and the *Tam3* element of *Antirrhinum majus*.

The difficulty in the analysis of receptor-element-induced mutants is that either the size of the element present at the gene is unknown (in case of *Ds* and *I* elements) or that all (or most) of the elements are identical *(Mu1; Tam3)*. These difficulties seem at first of such magnitudes as to prevent cloning of a gene from a mutant induced by a high-copy-number transposable element. Nevertheless cloning of some of these genes has been possible.

One cloning strategy that overcomes these difficulties is to perform detailed restriction pattern analysis and by Southern hybridization experiments to follow the segregation of element-specific sequences in the DNA of plants segregating for the specific mutant phenotype. If a band co-segregating with the phenotype can be observed this considerably reduces the number of clones to be analyzed as was demonstrated in the cloning of both, the *bz2* (Theres *et al.*, 1987) locus in maize and the *pallida* locus in *Antirrhinum majus* (Martin *et al.*, 1985).

The probability of cloning a gene from a high-copy-number transposable-element induced mutant can be increased if several independently isolated mutants of the same gene exist. Such independent mutants are especially valuable if they were induced by transposable elements belonging to different systems. The underlying idea is simple. If two mutants are induced by two independent elements (genetically and molecularly unrelated) then cross hybridization between clones obtained from the two mutant libraries would be expected only, if sequences outside the transposable elements are identical. These, if unique, most likely represent part of the gene to be isolated, since it can be proven via genetic tests that both transposable elements affect the same gene. This method is extremely helpful, since a relatively large number of clones can be tested simultaneously. For example such a procedure led to the isolation of the *a1* locus (O'Reilly *et al.*, 1985a) in maize by cross screening two libraries derived from *En* and *Mu1* induced mutants.

However, a word of caution is necessary. A *Ds1* element exists (Sutton *et al.*, 1984), which only shares the terminal 11 base pair inverted repeat with *Ac* or other *Ds* elements. In such a situation (*Ds* and *Ds1* cannot be distinguished by genetical tests), both a *Ds* and an internal *Ds1* probe have to be used. This makes the experiment more cumbersome.

D. Identification of Gene-Specific Sequences

Once a clone has been isolated from a transposable-element-induced mutant containing either an autonomous or a receptor element further tests are necessary to prove the presence of gene-specific sequences on that clone. From the nature of the transposable-element-induced mutations (reduction or blocking of gene activity) it must be assumed that gene-specific sequences are in close proximity to the element. Hence the first step in further analysis is to determine whether unique or low repetitive sequences are located next to the element. This is done by cutting the recombinant phage DNA with various restriction enzymes and using

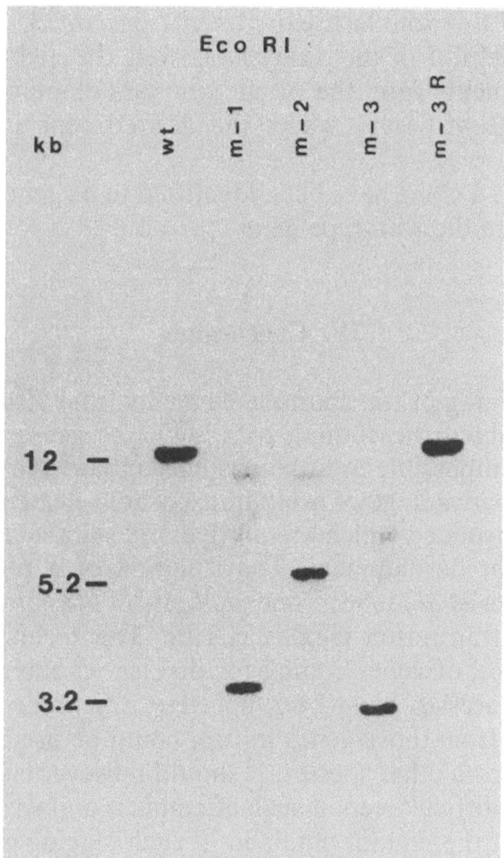

Fig. 4. Hybridization pattern of a gene specific sequence to genomic DNA isolated from wildtype, mutant and revertant lines. The pattern shows hybridization of a fragment, specific for the *c2* locus of *Zea mays* to EcoRI digested wildtype (wt), mutant (transposable element induced mutants: *c2-m1; c2-m2; c2-m3*) and revertant *(c2-m3R)* DNAs (Wienand *et al.*, 1986). The alteration in size of the fragments hybridizing in the mutant DNAs is due to the presence of various transposable elements at the *c2* locus in the mutant alleles

labelled total genomic DNA as a probe (Gupta *et al.*, 1984). Fragments of the clone which after a certain time of exposure do not light up are unique and may contain gene-specific sequences. These fragments can be used in Southern experiments to analyze various alleles of the gene of interest and to see whether differences in the hybridization pattern can be obtained. Such analysis includes probing of DNA isolated from wildtype (progenitor), transposable-element-induced mutant(s), and germinal revertant(s) of the gene of interest. If the fragment used as probe contains part of the gene, the pattern obtained in such a hybridization experiment is predictable. The size of the fragments hybridizing in wildtype and revertant DNA should be identical, whereas the size of fragments hybridizing in the mutants should be different (Fig. 4). This kind of hybridization experiment in many cases has been successfully used to prove the presence of gene specified sequences on the cloned DNA.

In addition to the Southern experiments described, Northern experiments might be helpful in the analysis as well. In such experiments the same unique fragments from the isolated mutant clone are used to probe RNAs extracted from tissues where the desired gene is expected to be expressed.

If sequences on a clone have been identified to be gene specific they are then used to isolate the wild-type gene.

IV. Conclusions

The molecular cloning of transposable elements from *Zea mays* and *Antirrhinum majus* enabled the cloning of a series of genes which otherwise would have been impossible to isolate using standard gene isolation procedures. The unique advantage of using transposable elements as gene tags is that genes can be isolated which are particularly suitable for studying gene regulation or plant development. The isolation of a regulatory gene in maize (*c1;* Paz-Ares *et al.*, 1986; Cone *et al.*, 1986) are convincing examples of how powerful transposon tagging can be. This technique may further lead to the isolation of genes coding for disease resistance or involved in developmental processes from both species, *Zea mays* and *Antirrhinum majus*. Sequences from those genes in turn could be used to clone homologous sequences from other species. It should however be stressed that the limiting and most difficult step in such attempts is and will be the genetical characterization of the mutant obtained in such tagging experiments.

A second very interesting approach for gene tagging of unknown genes in plants other than *Zea mays* and *Antirrhinum majus* has already been started in several laboratories. It uses recently developed methods in gene transformation to transfer transposable elements from *Zea mays* or *Antirrhinum majus* to other (dicotyledonous) plants, not only to study the function of the elements but also to eventually transpose these elements into interesting genes. In most of these plants no transposable elements have yet been isolated, though elements have been described (genetically) in many plant species. Transfer of the *Ac* element to *Nicotiana tabacum* (Baker *et al.,* 1986) has demonstrated that *Ac* is capable of transposition in this heterologous system. Such analyses has been performed with other transposable element systems in other species as well (Lillis *et al.,* 1985). The availability of transgenic plants containing already cloned and characterized elements from other plants facilitates the cloning of genes in these species.

Acknowledgements

We are grateful to P. Schnable for critically reading the manuscript.

V. References

Alleman, M., Freeling, M., 1986: The *Mu* transposable elements of Maize: Evidence for transposition and copy number regulation during development. Genetics **112,** 107—119.

Baker, B., Schell, J. Lörz, H., Fedoroff, N., 1986: Transposition of the maize controlling element "Activator" in tobacco. Proc. Natl. Acad. Sci., U.S.A. **83,** 4844—4848.

Barclay, P. C., Brink, R. A., 1954: The relation between modulator and activator in maize. Proc. Natl. Acad. Sci., U.S.A. **40,** 1118—1126.

Barker, R. F., Thompson, D. V., Talbot, D., R., Swanson, J., Bennetzen, J. L., 1984: Nucleotide sequence of the maize transposable element *Mul.* Nucl. Acid Res. **12,** 5955—5967.

Baulcombe, D., Verma, D. P. S., 1978: Preparation of a complementary DNA for leghaemoglobin and direct demonstration that leghaemoglobin is encoded by the soybean genome. Nucl. Acid Res. **6,** 2707—2715.

Behrens, U., Fedoroff, N., Laird, A., Müller-Neumann, M., Starlinger, P., Yoda, J., 1984: Cloning of the *Zea mays* controlling element *Ac* from the *wx-m7* allele. Mol. Gen. Genet. **194;** 346—349.

Bennetzen, J. L., 1984: Transposable Element *Mul* is found in multiple Copies only in Robertson's *mutator* maize lines. J. of Mol. and Appl. Genetics **2,** 519—524.

Bennetzen, J. L., Swanson, J., Taylor, C., Freeling, M., 1984: DNA insertion in the first intron of maize Adh1 affects message levels: cloning of projenitor and mutant Adh1 alleles. Proc. Natl. Acad. Sci., U.S.A. **13,** 4125—4129.

Bingham, P. M., Leavis, R., Rubin, G., 1981: Cloning of DNA sequences from the *white* locus of *Drosophila melanogaster* by a novel and general method. Cell **25,** 693—704.

Bonas, U., Sommer, H., Harrison, B. J., Saedler, H., 1984: The transposable element *Tam1* of *Antirrhinum majus* is 17 kb long. Mol. Gen. Genet. **194,** 138—143.

Coe, E. H., Jr., 1979: Specification of the anthocyanin biosynthetic function by *B* and *R* in maize. Maydica XXIV, 49—58.

Coe, E. H., Jr., Neuffer, M. G., 1977: The genetics of corn. In: Sprague, G. F. (ed.) Corn and Corn improvement. American Society of Agronomy, Inc. Madison Wisconsin, pp. 111—223.

Cone, K. C., Burr, F. A., Burr, B., 1986: Molecular Analysis of the Maize Regulatory Locus Cl. Proc. Natl. Acad. Sci., U.S.A., **83,** 9631—9635.

Chourey, P. S., Nelson, O. E., 1976: The enzymatic deficiency conditioned by the shrunken-1 mutation in maize. Biochem. Genet. **14,** 1041—1055.

Davidson, E. H., Britten, R. J., 1973: Organization, transcription and regulation in the animal genome. Quart. Rev. Biol. **48,** 565—613.

Dellaporta, S. L., Chomet, P. S., 1985: The Activation of Maize Controlling Elements. In Hohn, B., Dennis, E. S. (eds.): Genetic Flux in Plants. pp. 169—216. Springer-Verlag, Wien.

Dennis, E. S., Gerlach, W. L., Pryor, A. J., Bennetzen, J. L., Inglis, A., Llewellyn, D., Sachs, M. M., Ferl, R. J., Peacock, W. J., 1984: Molecular analysis of alcohol dehydrogenase (Adh1) gene of maize. Nucl. Acid Res. **12,** 3983—4000.

Derbyshire, K. M., Grindley, N. D. F., 1986: Replicative and Conservative Transposition in Bacteria. Cell **47,** 325—327.

Döring, H. P., 1985: Plant transposable elements. Bio Essays **3,** 164—171.

Döring, H. P., Starlinger, P., 1984: Barbara McClintocks controlling elements: now at the DNA level. Cell **39,** 253—259.

Döring, H. P., Starlinger, P., 1986: Molecular Genetics of Transposable elements in Plants. Ann. Rev. Genet. **20,** 175—200.

Döring, H. P., Tillmann, E., Starlinger, P., 1984a: DNA sequence of the maize transposable element Dissociation. Nature **307,** 127—130.

Döring, H. P., Freeling, M., Hake, S., Johns, A., Kunze, R., Merkelbach, A., Salamini, F., Starlinger, P., 1984b: A *Ds*-mutation of the Adh1 gene in *Zea mays* L. Mol. Gen. Genet. **193,** 199—204.

Dooner H. K., 1983: Coordinate genetic regulation of flavanoid biosynthetic enzymes in maize. Mol. Gen. Genet. **189,** 136—141.

Dooner, H., K., Nelson, O. E., 1977: Genetic Control of UDP glucose: Flavonol 3-O-Glucosyltransferase in the Endosperm of Maize. Biochem. Genetics **15,** 509—515.

Dooner, H. K., Nelson, O. E., 1979: Interaction among C, R and VP in the control of the Bz Glycosyltransferase during endosperm development in maize. Genetics **91,** 209—315.

Dooner, H. K., Weck, E., Adams, S., Ralston, E., Favreau, M., English, J., 1985: A molecular genetic analysis of insertions in the *bronze* locus in maize. Mol. Gen. Genet. **200,** 240—246.

Evans, M., Cray, R. R. D., Brown, P., Baulter, D., 1980: Synthesis of complementary DNAs to partially purified mRNAs coding for the storage proteins of *Pisum sativum L,* Biochim. Biophys. Acta **610,** 81—95.

Fedoroff, N., 1983: Controlling elements in maize. In: Shapiro, J. (ed.) Mobile Genetic Elements, pp. 1—63. New York: Academic.

Fedoroff, N. Wessler, S., Shure, M., 1983: Isolation of the Transposable Maize controlling Elements *Ac* and *Ds*. Cell **35,** 235—242.

Fedoroff, N., Furtek, D., Nelson, O. E., 1984: Cloning of the *bronze* locus in maize by a simple and generalizable procedure using the transposable controlling element *Ac*. Proc. Natl. Acad. Sci., U.S.A. **81,** 3825—3829.

Fincham, J. R. S., Harrison, B. H., 1967: Instability at the *pal* locus in *Antirrhinum majus*. 2. Multiple alleles produced by mutation of one original mutable allele. Heredity **22,** 211—224.

Fincham, J. R. S., Sastry, G. R. K., 1974: Controlling elements in maize. Ann. Rev. Genet. **8,** 15—50.

Flavell, A., 1983: Mobile genetic elements in eukaryotes. In: MacLean, N., Gregory, S. P., Flavell, R. A. (eds.) Eukaryotic Genes, their Structure, Activity and Regulation, pp. 236—276; Butterworth, London - Boston.

Flavell, R. B., Bennett, M. D., Smith, J. B., Smith, D. B., 1974: Genome size and proportion of repeated nucleotide sequence DNA in plants. Biochem. Genet. **4,** 257—269.

Freeling, M., 1984: Plant transposable elements and insertion sequences. Ann. Rev. Plant Physiol. **35,** 277—298.

Friedemann, P., Peterson, P. A., 1982: The *Uq* controlling element system in maize. Mol. Gen. Genet. **187,** 19—29.

Geiser, M., Döring, H. P., Wöstemeyer, J., Behrens, U., Tillmann, E., Starlinger, P., 1980: A cDNA clone from *Zea mays* endosperm sucrose synthase mRNA. Nucl. Acid Res. **8,** 6175—6188.

Geiser, M., Weck, E., Döring, H. P., Courage-Tebbe, U., Tillmann, E., Starlinger, P., 1982: Genomic clones of wild type allele and a transposable element induced allele of the sucrose synthase gene of *Zea mays* L. EMBO J. **1,** 1455—1460.

Gerlach, W. L., Pryor, A. J., Dennis, E. S., Ferl, R. J., Sachs, M. M., Peacock, W. J., 1982: cDNA cloning and induction of the alcohol dehydrogenase gene (Adh 1) of maize. Proc. Natl. Acad. Sci., U.S.A. **79,** 2981—2985.

Gonella, J. A., Peterson, P. A., 1977: Controlling elements in a trible maize from Columbia: *Fcu,* a two unit system. Genetics **85,** 629—645.

Greenblatt, J. M., Brink, R. A., 1963: Transposition of Modulator in maize into divided and undivided chromosome segments. Nature **197,** 412—443.

Greenblatt, I. M., 1984: A possible dose-dependent inactivation of mutator *(Mu)* in maize. Mol. Gen. Genet. **191,** 86—90.

Gupta, M., Shepherd, N. S., Bertram, I., Saedler, H., 1984: Repetitive sequences and their organization on genomic clones of *Zea mays.* EMBO J. **3,** 133—139.

Hake, S., Walbot, V., 1980: The genome of *Zea mays,* its organization and homology to related grasses. Chromosoma **79,** 251—270.

Harrison, B. J., 1971: Association of three highly mutable genes involving pigmentation in *Antirrhinum majus.* Mutation Research **12,** 381—390.

Harrison, B. J., Carpenter, R., 1973: A comparison of the instabilities at the *nivea* and *pallida* loci in *Antirrhinum majus.* Heredity **31,** 309—323.

Harrison, B. J., Fincham, J. R. S., 1964: Instability at the *pal* locus in *Antirrhinum majus.* 1. Effects of environment on frequencies of somatic and germinal mutation. Heredity **19,** 237—258.

Kleckner, N., 1981: Transposable elements in prokaryotes. Ann. Rev. Genet. **15,** 341—404.

Klösgen, R. B., Gierl, A., Schwarz-Sommer, Zs., Saedler, H., 1986: Molecular analysis of the *waxy* locus of *Zea mays.* Mol. Gen. Genet. **203,** 237—244.

Kreuzaler, F., Ragg, H., Fautz, E., Kuhn, D. N., Hahlbrock, K., 1983: UV induction of chalcone synthase mRNA in cell suspension cultures of *Petroselinum hortense.* Proc. Natl. Acad. Sci., U.S.A. **80,** 2591—2593.

Land, H., Grez, M., Hauser, H., Lindenmaier, W., Schütz, G., 1983: Synthesis of ds cDNA involving addition of dCMP tails to allow cloning of 5′-Terminal mRNA sequences. In: Wu, R. (ed.) Methods in Enzymology. 100, 285—292.

Lechelt, C., Laird, A., Starlinger, P., 1986: Cloning of DNA from the P locus. Maize Gen. Coop. News Lett. **60,** 40.

Lillis, M., Spielmann, A., Simpson, R. B., 1985: Transfer of a Robertson's Mutator element to tomato and tobacco cells. In Freeling, M. (ed.) Plant Genetics pp. 213—224; New York: Alan R. Liss, Inc.

Maniatis, T., Fritsch, E. F., Sambrook, J., 1982: Molecular cloning. A laboratory manual. Cold Spring Harbor Laboratory Press, New York.

Marotta, R., Ponziani, G., Motto, M., Hartings, H., Gierl, A., DiFonzo, N. D.,

Soave, C., Bianchi, A., Salamini, F., 1986: Genetic instability of the *shrunken* and *waxy* loci in the o2-m(r)-Bg strain of maize. Maydica XXXI, 131—151.

Martin, C., Carpenter, R., Sommer, H., Saedler, H., Coen, E. S., 1985: Molecular analysis of instability in flower pigmentation of *Antirrhinum majus,* following isolation of the *pallida* locus by transposon tagging. EMBO J. **4,** 1625—1630.

McCarty, D. R., Shaw, J. R., Hannah, L. C., 1986: The cloning, genetic mapping and expression of the constitutive sucrose synthase locus of maize. Proc. Natl. Acad. Sci. U.S.A. **83,** 9099—9103.

McClintock, B., 1951: Chromosome organization and genetic expression. Cold Spring Harb. Symp. Quant. Biol. **16,** 13—47.

McClintock, B., 1953: Induction of Instability at selected Loci in maize. Genetics **38,** 579—599.

McClintock, B., 1955: Controlled mutation in maize. Carnegie Instit. Washington Yearbook **54,** 245—255.

McClintock, B., 1958: The suppressor-mutator system of control of gene action in maize. Carnegie Inst. Washington, Yearbook **58,** 452—456.

McClintock, B., 1961: Further studies of the suppressor-mutator system of control of gene action in maize. Carnegie Inst. Washington Yearbook **60,** 469—476.

McClintock, B., 1965: The control of gene action in maize. Brookhaven Symp. Biol. **18,** 162—184.

Müller-Neumann, M., Yoder, J., Starlinger, P., 1984: The DNA sequence of the transposable element *Ac* of *Zea mays* L. Mol. Gen. Genet. **198,** 19—24.

Nelson, O. E., Rines, H. W., 1962: The enzymatic deficiency in the waxy mutant of maize. Biochem. Biophys. Res. Commun. 9, 297—300.

Nevers, P., Shepherd, N., Saedler, H., 1985: Plant transposable elements. Adv. Bot. Res. **12,** 102—203.

Nowick, E. M., Peterson, P. A., 1981: Transposition of the Enhancer Controlling Element System in maize. Mol. Gen. Genet. **183,** 440—448.

Oberthur, E., Peterson, P. A., 1984: New c mutants: test of system relationship. Maize Gen. Coop. News Lett. **58,** 5.

O'Reilly, C., Shepherd, N. S., Pereira, A., Schwarz-Sommer, Zs., Bertram, I., Robertson, D. S., Peterson, P. A., Saedler., H. 1985a: Molecular cloning of the *a1* locus of *Zea mays* using the transposable elements *En* and *Mu1*. EMBO J. **4,** 877—882.

O'Reilly, C. Shepherd, N. S., Pereira, A., Schwarz-Sommer, Zs., Niesbach, U., Weydemann, U., Wienand, U., Robertson, D. S., Peterson, P. A., Saedler, H., 1985b: The use of *Spm (En)* and *Mu* transposable elements for molecular analysis of the A1 and C2 loci. Maize Gen. Coop. News lett. **59,** 35.

Paz-Ares, J., Wienand, U., Peterson, P. A., Saedler, H., 1986: Molecular cloning of the *c* locus of *Zea mays:* a locus regulating the anthocyanin pathway. EMBO J. **5,** 829—833.

Peacock, W. J., Dennis, E. S., Gerlach, W. L., Llewellyn, D., Lörz, H., 1983: Gene transfer in maize: Controlling elements and the alcohol dehydrogenase genes. In: Downry, K. Vollmy, R. W., Ahmad, F., Schultz, J. (eds.): Proc. 15th Miami Winter Symp. 20, pp. 311—326. New York, Academic.

Pereira, A., Schwarz-Sommer, Zs., Gierl, A., Bertram, I., Peterson, P. A., Saedler, H., 1985: Genetic and molecular analysis of the Enhancer *(En)* transposable element system in *Zea mays.* EMBO J. **4,** 17—23.

Pereira, A., Cuypers, H., Gierl, A., Schwarz-Sommer, Zs., Saedler, H., 1986: Molecular analysis of the *En/Spm* transposable element system of *Zea mays.* EMBO J. **5,** 835—841.

Peterson, P. A., 1953: A mutable pale green locus in maize. Genetics **38,** 682—683.

Peterson, P. A., 1958: The effect of temperature on the mutation rate of a mutable locus in maize. J. Heredity **49,** 120—124.

Peterson, P. A., 1965: A relationship between the *Spm* and *En* control systems in maize. Am. Nat. **99,** 391—398.

Peterson, P. A., 1970: The *En* Mutable System in Maize III. Transposition associated with Mutational events. Theor. Appl. Genet. **40,** 367—377.

Peterson, P. A., 1978: Controlling elements: The induction of mutability at the A2 and C loci in maize. In: Walden, D. B. (ed.): Maize Breedings and Genetics pp. 601—631.

Peterson, P. A., 1983a: Newly originated mutable alleles at the C2 locus. Maize Gen. Coop. News Lett. **57,** 2.

Peterson, P. A., 1983b: Newly originated mutable alleles at the bz locus. Maize Gen. Coop. News Lett. **57,** 2.

Peterson, P. A., 1986: Mobile elements in maize, Plant Breeding Review, pp. 81—122.

Peterson, P. A., Cormack, J., 1984: Wx alleles newly originated from an *En*-containing plot. Maize Gen. Coop. News Lett. **58,** 3.

Peterson, P. A., Salamini, F., 1986: A search for active mobile elements in the iowa stiff stalk synthetic maize population and some derivatives. Maydica XXXI, 163—172.

Peterson, T., Schwarz, D., 1986: Isolation of a candidate clone of the maize P locus. Maize Gen. Coop. News Lett. **60,** 36.

Pohlmann, R. F., Fedoroff, N. V., Messing, J., 1984: The Nucleotide Sequence of the Maize Controlling Element Activator. Cell **37,** 635—643.

Piotrowiak, R., 1986: Die molekulare Struktur des transponierbaren Elementes *Tam 1* aus *Antirrhinum majus.,* Dissertation, Universität Köln.

Ralston, E., Dooner, H., 1986: *Mu 1* as transposon tag. Maize Genet. Coop. News Lett. **60,** 107—109.

Reddy, A. R., Peterson, P. A., 1977: Controlling element alleles in the analysis of the genetic control of the anthocyanin pathway in maize. Can. J. Genet. Cytol. **19,** 111—117.

Reddy, A. R., Peterson, P. A., 1983: Transposable elements of maize: Genetic basis of pattern differentiation of some mutable *c* alleles of the Enhancer system. Mol. Gen. Genet. **192,** 21—31.

Reif, H. J., Niesbach, U., Deumling, B., Saedler, H., 1985: Cloning and analysis of two genes for chalcone synthase from *Petunia hybrida.* Mol. Gen. Genet. **199,** 208—215.

Reimold, U., Kröger, M., Kreuzaler, F., Hahlbrock, K., 1983: Cloning and 3′ non coding nucleotide sequence of chalcone synthase mRNA and assignment of amino acid sequence of the enzyme. EMBO J. **2,** 1801—1805.

Rhoades, M. M., 1938: Effect of the Dt gene on the mutability of the a1 allele in maize. Genetics **23,** 377—395.

Rhoades, M. M., 1941: The genetic control of mutability in maize. Cold Spring Harb. Symp. Quant. Biology **9,** 138—144.

Rhoades, M., Dempsey E., 1982: The induction of mutable systems in plants with the high-loss mechanism. Maize Genet. Coop. News Lett. **56,** 21—26.

Rhoades, N., Dembsey, E., 1983: Further studies on two unit mutable systems found in our high-loss studies and on the specificity of insertion of responding and controlling elements. Maize Gen. Coop. News Lett. **57,** 14—17.

Robertson, D. S., 1978: Characterization of a mutator system in maize. Mutation Res. **51,** 21—28.

Robertson, D. S., 1980: The timing of *Mu* activity in maize. Genetics **94,** 969—978.

Robertson, D. S., 1985: Differential activity of the maize mutator *Mu* at different loci and in different cell lineages. Mol. Gen. Genet. **200,** 9—13.

Robertson, D. S., Mascia, P. N., 1981: Test of four controlling element systems of maize for Mutator activity and their interaction with *(Mu)* Mutator. Mutation Res. **84,** 283—289.

Robertson, D. S., 1983: A possible dose-dependent inactivation of mutator *(Mu)* in maize. Mol. Gen. Genet. **191,** 86—90.

Sachs, M. M., Peacock, W. J., Dennis, E. S., Gerlach, 1983: Maize Ac/Ds Controlling elements, a molecular viewpoint. Maydica XXVIII, 289—302.

Saedler, H., Bonas, U., Gierl, A., Harrison, B. J., Klösgen, R. B., Krebbers, E., Nevers, P., Peterson, P. A., Schwarz-Sommer, Zs., Sommer, H., Upadhyaya, K., Wienand, U., 1984: Transposable elements in *Antirrhinum majus* and *Zea mays.* Cold Spring Harb. Symp. on Quant. Biology, XLIX, Cold Spring Harbour Lab, pp. 335—361.

Salamini, F., 1981: Controlling elements at the Opaque-2 locus of maize: their involvement in the origin of spontaneous mutation. Cold Spring Harbour Symp. Quant. Biol. **45,** 467—476.

Schnable, P. S., Peterson, P. A., 1986: Distribution of genetically active Cy transposable elements among diverse maize lines. Maydica XXXI, 59—81.

Schwartz, D., Osterman, J., 1976: A pollen selection system for alcohol dehydrogenase-negative mutants in plants. Genetics **83,** 63—65.

Schwarz-Sommer, Zs., Gierl, A., Klösgen, R. B., Wienand, U., Peterson, P. A., Saedler, H., 1984: The *Spm(En)* transposable element controls the excision of a 2 kb DNA insert at the *wx-m8* allele of *Zea mays.* EMBO J. **3,** 1021—1028.

Schwarz-Sommer, Zs., Gierl, A., Berntgen, R., Saedler, H., 1985: Sequence comparison of 'states' of *al-ml* sugest a model of *Spm(En)* action. EMBO J. **4,** 2439—2443.

Schwarz-Sommer, Zs., Saedler, H., 1985: Plant transposable elements 1984. Oxford surveys of Plant Molecular + Cell Biology **2,** 353—360.

Shure, M., Wessler, S., Fedoroff, N., 1983: Molecular Identification and isolation of the Waxy Locus in Maize. Cell **35,** 225—233.

Soave, C., Salamini, F., 1984: The role of Structural and Regulatory Genes in the Development of Maize Endosperm. Developmental Genetics **5,** 1—25.

Sommer, H., Krebbers, E., Piotrowiak, R., Bonas, U., Hehl, R., Saedler, H., 1985a: Transposable elements of *Antirrhinum majus.* In: Liss, A. L. (ed.) Plant Genetics pp. 433—444.

Sommer, H., Carpenter, R., Harrison, B. J., Saedler, H., 1985b: The transposable element *Tam3* of *Antirrhinum majus* generates a novel type of sequence alterations upon excision. Mol. Gen. Genet. **199,** 225—231.

Sommer, H., Saedler, H., 1986: Structure of the chalcone synthase gene of *Antirrhinum majus.* Mol. Gen. Genet. **202,** 429—434.

Starlinger, P., 1980: IS elements and transposons. Plasmid **3,** 241—259.

Styles, E. D., Ceska, O., 1977: The genetic control of flavanoid synthesis in maize. Can. J. Genet. Cytol. **19,** 289—302.

Sutton, W. D., Gerlach, W. L., Schwartz, D., Peacock, W. J., 1984: Molecular analysis of *Ds* controlling element mutations at the Adh1 locus of maize. Science **223,** 1265—1268.

Szostak, J. W., Stiles, J. I., Tye, B.-K., Chiu, P., Sherman, F., Wu, R., 1979: Hybridization with Synthetic oligonucleotides. In: Methods in Enzymology, Wu, R. (ed.), **68,** pp. 419—428.

Taylor, L. P., Walbot, V., 1985: Variability among Mutator elements. Maize Gen. Coop. News Lett. **59,** 96.

Taylor, L. P., Chandler, V. L., Walbot, V., 1986: Insertion of 1.4 kb and 1.7 kb Mu elements into the bronze 1 gene of *Zea mays* L. Maydica XXXI, 31—45.

Theres, N., Scheele, T., Starlinger, P., 1987: Cloning of the *bz2* locus of *Zea mays* using the transposable element *Ds* as a gene tag. Mol. Gen. Genet. in press.

van Schaik, N. W., Brink, R. A., 1958: Transposition of Modulator, a component of the variegated pericarp allele in maize. Genetics **44,** 725—738.

Werr, W., Frommer, W. B., Mass, C., Starlinger, P., 1985: Structure of the sucrose synthase gene on chromosome 9 of *Zea mays* L. EMBO J. **4,** 1373—1380.

Wienand, U., Brüschke, C., Feix, G., 1979: Cloning of double stranded DNAs derived from polysomal mRNA of maize endosperm: isolation and characterization of zein clones. Nucl. Acids Res. **6,** 2707—2715.

Wienand, U., Sommer, H., Schwarz, Zs., Shepherd, N., Saedler, H., Kreuzaler, F., Ragg, H., Fautz, E. Hahlbrock, K., Harrison, B. J., Peterson, P. A., 1982: A General Method to identify Plant Structural Genes Among Genomic DNA Clones Using Transposable Element induced Mutations. Mol. Gen. Genet. **187,** 195—201.

Wienand, U., Weydemann, U., Niesbach-Klösgen, U., Peterson, P. A., Saedler, H., 1986: Molecular cloning of the *c2* locus of *Zea mays,* the gene coding for chalcone synthase. Mol. Gen. Genet. **203,** 202—207.

Chapter 9

Direct Gene Transfer to Plants

I. Potrykus, J. Paszkowski, R. D. Shillito, and M. W. Saul

Friedrich Miescher-Institut, P. O. Box 2543, CH-4002 Basel, Switzerland

With 1 Figure

Contents

Recent progress in cell biology and molecular biology has made available a variety of techniques enabling gene transfer across the borders set by the biology of pollination and fertilisation. These novel techniques include "somatic hybridisation" leading to the combination of entire genomes (Gleba and Sytnik, 1984), "subprotoplast fusion" yielding the transfer of a partial genome (Maliga *et al.,* 1982), X-irradiation prior to protoplast fusion to achieve organelle transfer (Aviv *et al.,* 1980), transfer of isolated nuclei (King and Saxena, Saxena and King, 1985) and metaphase chromosomes (DeLaat and Blaas, 1985). Besides these methods which often cause unpredictable and severe alterations of the host genome, there are others which introduce not more than small stretches of foreign DNA, often not more than just a single foreign gene. For years this has been possible only via the bacterial vector *Agrobacterium tumefaciens.* At present more methods of gene transfer to plants are available. Gene transfer via *A. tumefaciens* has already been discussed on many occasions (for a recent review see Gheysen *et al.,* 1985 and chapter 7 in this volume). This volume contains also articles which deal with the use of viral vectors (see chapters 1 and 2) and the combination of *A. tumefaciens* and viral vectors — Agroinfection (chapter 4) therefore we shall concentrate only on direct, vectorless gene transfer to plants.

The use of biological vectors for gene transfer was essential at the onset of genetic engineering with plants and has been developed with *A. tumefaciens* to a convenient and very efficient technique. There are also problems, however, which led to attempts to develop vectorless transformation: biological vectors are limited by a hostrange. The most serious limitation of the *A. tumefaciens* system is its inability to transfer genes to graminaceous monocots especially the cereals. This limitation may or may not be overcome by future developments. Independent of the hostrange problem it was a challenge to achieve gene transfer directly. The approaches which have been successful so far include "direct gene transfer" (including electroporation), "liposome fusion", "spheroplast fusion" and "microinjection". "Direct gene transfer" has been developed to a routine and efficient technique which has been reproduced in several laboratories.

I. Direct Gene Transfer

A. Introduction

From work with isolated protoplasts it was known that naked plant cells take up foreign DNA (e. g. Lurquin and Kado, 1977). As it was also well documented that fertile plants can be regenerated from isolated protoplasts (e. g. Potrykus and Shillito, 1985), the obvious approach to vectorless gene transfer was via protoplasts. Davey *et al.* (1980), Draper *et al.* (1982) and Krens *et al.* (1982) had shown that the isolated Ti-plasmid of *A. tumefaciens* could be integrated into protoplast-derived clones. On this basis a routine and efficient protocol for hostrange independent gene transfer to plants was developed. The basic prerequisites for an experiment yielding clear answers were a) a selectable marker gene which could be expected to produce a clear phenotype in protoplast-derived cell clones, b) expression signals which could be expected to be active independent of the physiological state of the host cell, c) precise linkage of the protein-coding region of the selectable gene to the expression signals guaranteeing expression of the hybrid gene in the host genome d) DNA stretches on both sides of the hybrid gene which would allow for integration into the host genome without disruption of the gene construct itself, e) a protoplast system with routine high plating efficiency and plant regeneration capacity, and f) an absolutely tight selection protocol for presumptive transformants.

B. A Representative Experiment

A hybrid selectable marker gene was constructed by placing the protein coding region of the bacterial gene for aminoglycoside-phosphotransferase (APH[3′])II from Tn5 under the control of the expression signals from CaMV (Cauliflower Mosaic Virus) gene VI. For control experiments a construction was made containing the protein coding part of the gene in inverse orientation versus the expression signals. Protoplasts from a shoot culture of *N. tabacum* "Petit Havana SR1" were then incubated with the gene, calf thymus DNA and polyethylene glycol, and cultured without selective pressure for one week. Plating the protoplasts into agarose gel segments and further incubation in larger volumes of liquid culture medium ("bead-type culture", Shillito *et al.,* 1984) containing 50 mg/l Kanamycin sulphate with weekly replacement of the liquid phase assured tight selection. In five different control treatments including the gene with the inverse orientation versus the expression signals not a single Kanamycin-resistant clone was found from 10^{10} protoplasts treated. Kanamycin-resistant clones were recovered exclusively from the treatment with the gene in the correct orientation versus the expression signals. Southern blot hybridisation with radioactively labelled probe covering the protein coding region of the gene, as well as enzyme analysis for kanamycin-dependent phosphorylation proved that the novel phenotype of the clones surviving selection was indeed based on the integration and activity of functional copies of the foreign gene into the host genome. Plants were regenerated and the phenotype shown to be transmitted to sexual offspring (Paszkowski *et al.,* 1984).

C. *Protocol and Transformation Frequency*

The frequency for stable integration of functional copies of the foreign gene was initially in the range of one per one million protoplasts treated. On this basis a variety of parameters of possible importance was studied systematically to optimise gene transfer. Those which contributed to an increase in transformation frequency included a) electroporation, b) heat-shock, c) concentration and form of the gene, d) concentration and form of the carrier DNA, e) concentration of polyethylene glycol, f) quality of the protoplasts, g) handling and culture of protoplasts after the transformation treatment, h) sequence of addition of DNA and PEG in the protocol. A protocol giving routinely high frequencies of stable integrative transformation of genes with leaf protoplasts of *N. tabacum* is briefly as follows: Protoplasts are isolated as appropriate for subsequent culture (e. g. Potrykus and Shillito, 1985). Following washing and purification they are resuspended in 0.4 M mannitol containing 6 mM $MgCl_2$ and 0.1 % MES buffer (Morpholinoethansulfonic acid) at pH 5.6 to a population density of 1.5×10^6/ml. $MgCl_2$ is added (3—5 % of a 0.3 M solution) to bring the electric resistance of the suspension into the range of 1.2 Kilo Ohm (for later electroporation). This is followed by incubation at 45° C for five minutes (heat shock treatment) and cooling to room temperature on ice. The suspension is subdivided into aliquots and the gene added to a final concentration of 10 µg/ml and carrier DNA added (calf thymus) to 40 µg/ml. Thereafter polyethylene glycol (MW 6000) is added to a final concentration of 8 % and carefully mixed. For electroporation aliquots are transferred to the chamber of an electroporator (DIA-LOG GmbH, D-4000 Düsseldorf, FRG) and three high voltage pulses applied at ten second intervals to give 50 % survival (1 to 1.5 kV/cm from a capacitor of 10nF). Thereafter the protoplast suspension is transferred into petri dishes, a tenfold volume of culture medium containing 0.6 % agarose (Sea Plaque) is added without prior washing of the protoplasts. The protoplasts are carefully distributed and the gel allowed to solidify. After incubation under optimal growing conditions for seven days the protoplastagarose gel is cut into segments and transferred into a tenfold volume of liquid culture medium containing, if desired, a selective agent.

The transformation frequency is established by a) screening without prior selection a few hundred randomly picked protoplast-derived clones for the physical presence and activity of the gene, or, b) by relating the number of clones surviving selection to the number of clones growing in an unselected aliquot of the treatment. With both methods of calculation frequencies of 1—2 % were achieved with *N. tabacum* "Petit Havana SR1" (Shillito *et al.*, 1985) and with other partially optimised protoplast systems (e. g *Petunia hybrida, Hyoscyamus muticus, N. plumbaginifolia, N. tabacum* cell lines) frequencies of ca. 0.1 % were obtained.

D. Electroporation

Electroporation has been used to transfer genes into animal cells (Naumann *et al.,* 1982), bacteria (Shivarova *et al.,* 1983), and yeasts (Karube *et al.,* 1985). There has therefore been a great deal of interest in applying this technique to plants.

Most electroporation for the purpose of gene transfer has been carried out using a very short discharge pulse of high voltage, in contrast to the longer pulses normally applied to fuse protoplasts. In our case we have used carefully controlled pulses to stably transform leaf protoplasts of *N. tabacum* and other species, as well as cell culture protoplasts, in order to obtain, reproducibly, transformation frequencies in the range of 1—3 % of the number of colonies formed after treatment (Shillito *et al.,* 1985)

However, it is possible to reach frequencies of transformation in the range of 0.1 % by a simple procedure consisting of a heat shock, addition of a suitable selectable gene on a linearised plasmid followed by a treatment with 13—25 % w/v PEG (dependant on the species). Special procedures involving electroporation need to be used only to obtain frequencies in the 1 % range.

E. No Hostrange Limitations

In analogy to the leaf protoplasts used in experiments with *N. tabacum,* leaf protoplasts from *Brassica rapa* (Paszkowski *et al.,* 1986), *Petunia hybrida* and *N. plumbaginifolia* were transformed. Likewise direct gene transfer to cell line protoplasts did not pose any problem in the case of *Hyoscyamus muticus* and *N. tabacum.*

Graminaceous monocots, including the cereals, do not fall within the host range of *A. tumefaciens* and may, therefore, not be accessible to *A. tumefaciens*-mediated gene transfer. The transformation reported for non-graminaceous monocots (Hooykaas-Van Slogteren *et al.,* 1984; Hernalsteens *et al.,* 1984) unfortunately has no impact in this respect. Direct gene transfer into protoplasts of the graminaceous monocots *Lolium multiforum* (Potrykus *et al.,* 1985b) and *Triticum monococcum* (Loerz *et al.,* 1985) and *Zea mays* (Fromm *et al.,* 1986) demonstrate the general applicability of the method of direct gene transfer.

The remaining problem with genetic engineering of graminaceous plant species is thus in plant regeneration from protoplasts (Potrykus *et al.,* 1979) which has, to date, not been demonstrated convincingly.

F. Foreign Gene Mendelian Inheritance

From transformation experiments with the hybrid gene for resistance to kanamycin, ca. 150 independent primary transformants of *N. tabacum* have been analysed. So far, several of these have been studied through six plant sexual generations. The large number of seeds produced per flower (up to 3000), in the majority of the plants regenerated, being morphologically normal and fully fertile, as well as the simple phenotype of the gene (green versus white seedlings on kanamycin) facilitated a detailed genetic analysis. In the great majority of cases the inheritance of the novel phenotype follows Mendelian rules. In all cases analysed, the inheritance of the phenotype was strictly correlated with the transmission of the foreign DNA and the gene activity (Potrykus *et al.*, 1985 a). In the majority (77 %) the gene was inherited as single, dominant Mendelian factor. There were, however, also cases where the foreign gene was integrated in two (6 %), three (3 %) and multiple (7 %) independent chromosomal loci (Potrykus and Shillito, in preparation). There is also a group of primary transformants (7 %) which might be of specific interest in as far as the transmission of the phenotype shows clear reciprocal differences in crosses with the untransformed wildtype. As, so far, clear molecular data on a possible integration of the foreign gene into the plastome or the chondriome are missing, one has to also consider the possibility that these cases represent effects of pollen lethality and polyploidy. Further molecular analysis will allow for a definite interpretation.

G. Stability of the Foreign Gene

A foreign stretch of DNA integrated into chromosomal DNA, of no obvious advantage for the genome under non-selective conditions, should not *a priori* be expected to be stably maintained. The fate of the foreign gene has been studied, so far, through several sexual generations and for 24 months without selective pressure, and both stability and instability have been found, Stability predominates. This may be illustrated by a few results: a) Subpopulations from one experiment yield the same transformation frequency when grown directly under selective conditions as when grown for 5—6 weeks without selection and thereafter transferred to selective conditions. b) Of 60 subclones regenerated from 2 transformed clones, 57 maintained the gene under non-selective conditions. c) Regenerated plants are neither chimeras nor mosaics. d) The gene is regularly transmitted to sexual offspring. e) The inheritance of the gene follows strictly mendelian rule. f) The gene has been maintained over six plant generations. g) The gene is not only stably maintained when present in a homozygous situation, but also when in a heterozygous state. These data demonstrate a high degree of mitotic and meiotic stability. The genetic data are supported by the inheritance of a transformant-specific hybridization pattern in southern blot analysis (Potrykus *et al.*, 1985 c and unpublished data.

H. Instability of the Foreign Gene

Besides clones which expressed a high degree of stability of the foreign gene others were found which showed considerable instability. This may be illustrated by a few cases. A selfcross of a homozygous R1 transformant identified in the offspring of a primary transformant yielded an R2 population consisting of 422 resistant and 4 sensitive seedlings. Pollen contamination and seed contamination as possible cause have been excluded. To study the possibility of sectoring-out of tissue which had lost the gene, 60 further R2 populations of the same plant were produced by selfing 60 flowers from primary, secondary and tertiary inflorescences. All 60 seedling populations contained ca. 0.5 % sensitive individuals. One of the resistant R2 plants tested further produced R3 populations with again ca. 0.5 % sensitive seedlings in resistant populations. Three resistant R3 plants were taken to produce R4 populations. Two of them again had ca. 0.5% sensitive seedlings. One R3 plant, however, produced a population segregating 840 : 98 for resistant versus sensitive. It is assumed that these data are indicative of a meiotic instability of the gene. There are also data to demonstrate an instability of the foreign DNA on the molecular level: Among numerous cases where the transformant specific Southern hybridization pattern was maintained and inherited unaltered, one case was found where all bands had been lost except one of 1.2 kb, representing the functional copy of the gene, and this plant was resistant. A complementary case was also detected, where all bands except the 1.2 kb fragment were maintained, and this plant was sensitive (Potrykus *et al.*, 1985 c and unpublished data).

These examples indicate that it might be dangerous to rely on stable maintenance of an established foreign gene, although instability appears to be, so far, not a too frequent event. It would be interesting to compare these results with comparable studies on *A. tumefaciens* transformed material.

I. Molecular Proof for Transformation

The fact that specific treatment with a specific gene produced the expected phenotype which is inherited according to Mendelian laws provides strong evidence for gene transfer, however, no definite proof. Although it would be very difficult to formulate a hypothesis which could explain all the biological and genetic data without gene transfer, it was important to add the molecular evidence. This was done by demonstrating a strict correlation between the physical presence of the gene (by Southern blot analysis and by *in-situ* hybridization to metaphase chromosomes, as it was performed before for animal cells) (Gerhard *et al.*, 1981) and the activity of the gene (by enzyme assay for neomycin-phosphotransferase activity). For the Southern blot analysis radioactively labelled DNA probes covering the protein coding region with or without expression signals were hybridised to DNA isolated from the presumptive transformants. The most extensive studies have been done with kanamycin-resistant clones recovered from treatment with the plasmid pABDI (Paszkowski *et al.*, 1984) carrying the bacterial gene for neomycin-phosphotransferase. There was no hybridization to plant material recovered from wildtype protoplasts. Kanamycin-resistant clones recovered from treatment with the gene, yielded without exception DNA hybridising to the probe, expressed a transformant-specific hybridization pattern, and contained, after EcoRV digestion, the expected 1.2 kb band representing the functional gene, in 1—30 copies per haploid genome. The transformant-specific hybridization pattern was inherited, together with the phenotype, to sexual offspring. Rare cases of alterations of the hybridization pattern were also detected (see I. H.). Neomycin-phosphotransferase inactivates kanamycin (and some other aminoglycosidic antibiotics) by transfer of a phosphate from ATP. The enzyme assay (Reiss *et al.*, 1984) makes use of the transfer of a radioactively labelled phosphate from ATP to kanamycin. For this purpose protein extracts from tissue are separated on a non-denaturing gel, the gel is overlayed with an agarose gel containing radioactive ATP and Kanamycin. Following incubation to allow for the reaction to proceed, the reaction products are transferred onto phosphocellulose paper placed on top of the gel. The phosphorylated kanamycin can be visualised via its radioactively labelled phosphate by autoradiography. Kanamycin-dependent phosphorylation was found exclusively in kanamycin-resistant material recovered from protoplasts treated with pABDI and containing at least one copy of the 1.2 kb fragment. The enzyme activity is inherited coordinately with the 1.2 kb fragment and the phenotype.

J. Gene Localization by in situ Hybridization

The precise localization of a gene on a specific chromosome and at a specific position on the chromosome is of interest for various reasons. It would allow to study whether or not there are preferential sites in the genome for integration of foreign genes, whether the position in the chromosome has an effect on the expression of the gene and on the intensity of expression, whether it plays a role with respect to the stability or instability of the gene, whether a gene changes its position in the genome, etc. Localization via linkage studies with defined marker genes is rather complicated and laborious in *N. tabacum* for which the majority of transformants have been produced so far. It was, therefore, attempted to develop *in-situ* hybridization to metaphase chromosomes as an alternative method. Homozygous transformants carrying one genetic locus for kanamycin resistance on different chromosomes, selected from the offspring of two independent transformants, were induced to produce roots on cuttings. Protoplasts isolated from the root meristems were subjected to *in-situ* hybridization with a radioactive probe of the foreign gene. Following exposure for twelve weeks the position of the foreign gene was visible on homologous chromosomes which were different in the independent transformants (Mouras *et al.*, 1986). It can be envisaged that the same procedure will also allow the localization of original plant genes for which DNA probes are available.

K. Arrangement of Foreign DNA in the Host Genome

Integration of plasmid molecules into the nuclear genome occurs in a fashion similar to that described for DNA transfection of animal cells. We

(Paszkowski *et al.*, 1984; Potrykus *et al.*, 1985a, 1985b; Shillito *et al.*, 1985) and others (Hain *et al.*, 1985) have found integration of several copies of the transforming plasmid DNA. Both functional and partial copies of the APH(3')II hybrid gene and sequences from the bacterial plasmid were found. Normally the integrated DNA copies are not randomly distributed on several chromosomes, but integrated in only one physical location in the genome. Although integration occurs at a single location, this locus is on different chromosomes in different transformed lines. The mechanism of such integration of foreign DNA is currently under investigation. It has been observed that the pattern of integration and the degree of rearrangement is dependent upon the species and the developmental state of the recipient protoplast and the transformation method applied (Paszkowski *et al.*, 1984; Potrykus *et al.*, 1985a, 1985b; Shillito *et al.*, 1985). These conclusions based on molecular and genetic studies have been confirmed by the findings from *in-situ* hybridization.

L. Co-Transformation with Non-Selectable Genes

Transformation with a gene which is not selectable would be considerably facilitated if the phenomenon of "cotransformation", known from transformation experiments with animal cells (Wigler *et al.*, 1979) also existed for plant cells. To study the question whether or not those plant protoplasts which take up and integrate a selectable foreign gene also take up and integrate other non-selectable genes, the plasmid pMSI carrying a Zein gene (Weinand *et al.*, 1981) was mixed with plasmid pABDI (Paszkowski *et al.*, 1984) carrying the hybrid gene for kanamycin-resistance, in ratios of ca. 1+1 and 3+1, and introduced into protoplasts using our standard protocol (see I. C.) without addition of carrier DNA. Kanamycin-resistant clones were recovered and randomly picked clones from each treatment were grown up and analysed by Southern blot analysis, using a probe covering the entire Zein gene. 72% of the kanamycin-resistant clones recovered from the 1+1 mixture, and 88% of the resistant clones from the 3+1 mixture contained DNA hybridizing to the Zein probe. There was no hybridization to DNA from control clones (Schocher *et al.*, 1986).

A full sized fragment containing the entire gene was found in 28% of the clones. In co-transformation experiments with a nopaline synthase gene (from Ti-plasmid) up to 47% of the kanamycin-resistant clones contained the full copy of the non-selected gene and all those clones expressed the gene as shown by nopaline assay (Schocher *et al.*, 1986). The method of co-transformation has then been used in our laboratory to transfer a series of further non-selectable genes into *N. tabacum*.

The high transformation frequency available with *N. tabacum* makes it feasible to transfer non-selectable genes without the help of co-transformation: screening of protoplast-derived clones for the presence and activity of a non-selectable gene after standard direct gene transfer yielded positive clones.

a
b
c
d
e
f
g
h
SalI PstI EcoRV HindIII HindIII EcoRV BglII SalI
ATG pABDI (5.3kb)TGA

M. Gene Transfer from Total Genomic DNA

So far, cases have been discussed where isolated, cloned genes were transferred, of which the molecular structure and expression signals were known. It is, however, desirable to transfer genes directly from one genome to a host plant without having to go through the very laborious procedure of gene isolation and molecular analysis. This will be especially true for genes for which isolation procedures are difficult: e. g. for single copy genes which do not produce large amounts of m-RNA during specific developmental stages of the plant. To study whether transfer of single copy genes from a total plant genome via direct gene transfer might be possible, two series of experiments have been performed: In reconstruction experiments protoplasts have been treated with a mixture of plant DNA and a kanamycin resistance gene at ratios down to those corresponding to a single copy gene in a tobacco genome. Kanamycin-resistant clones could be recovered reproducibly at ratios as low as 10 copies of the gene per amphihaploid genome from populations of ca. 300,000 protoplasts treated. This result demonstrated that it was feasible to test total genomic DNA: DNA isolated from plants containing one copy of the kanamycin resistance gene stably integrated into the genome was isolated, partially digested and used to treat tobacco protoplasts with the standard direct gene transfer protocol. Transformants with this gene were recovered at the expected low frequency of ca. one per 3 million protoplast-derived clones (Mueller *et al.*, 1986). These results indicate that it will be possible to transfer original plant genes which have not previously been isolated, by treatment of protoplasts with partially digested genomic DNA and subsequent selection for the expected phenotype. As gene transfer with this method works at a low frequency (as to be expected because the gene in question will represent only a minute fraction of the total genome) this will be feasible only with genes which produce a phenotype which allows for tight selection at the protoplast culture level.

Fig. 1. Direct gene transfer — a representative experiment. Freshly isolated protoplasts, e. g. from shoot cultures of N. tabacum "Petit havana SR1" (a) are incubated with a selectable marker gene, e. g. the bacterial gene for aminoglycoside-phosphotransferase under control of the expression signals from CaMV gene VI (b), polyethylene glycol and carrier DNA. Culture in the presence of lethal levels of kanamycin allows for the development of the expected transformed and kanamycin-resistant clones (c). These develop to shoots on kanamycin-containing regeneration medium (d) and continue to grow to complete plantlets with roots (e) whereas shoots regenerated from wildtype protoplasts under non-selective conditions and thereafter transferred to kanamycin medium bleach and die (f). Following transfer to potting compost plants grow to maturity and develop fertile flowers and seed pots (g) which can be used to analyse the transmission of the novel phenotype and the foreign gene into the next sexual generation. Both phenotype (h) and foreign gene are inherited according to Mendelian laws

N. Limitations for Direct Gene Transfer

Direct gene transfer, so far, requires isolated protoplasts. This in itself does not pose any problems. However, sustained development from isolated protoplasts and especially plant regeneration from protoplast-derived clones is not available for every plant species, and especially not for the cereals and other important crop plants. It is still open whether or not there is a size limitation for the stretch of foreign DNA which can be integrated without rearrangement. For transient expression experiments protoplasts from virtually every cell type appear to be open for studies.

II. Other Vectorless Gene Transfer Systems

A. Liposome Fusion

Liposomes are artificial lipid vesicles. They can be used to encapsulate DNA. Fusion of liposomes with protoplasts transfers the DNA into the cell. This approach has been used to show convincingly uptake, integration, and inheritance of a foreign gene in tobacco (Caboche and Deshayes, 1985). As for direct gene transfer, the method requires protoplasts as acceptor cells. The efficiency is not as high as with direct gene transfer, and DNA uptake into the protoplast does not appear to be the limiting step in this technique in any case. There might be an advantage for gene transfer into organelles, although there are so far no data to support this idea.

B. Spheroplast Fusion

Spheroplasts are, in analogy to protoplasts, naked bacteria. They can be fused with protoplasts. If they contain plasmids carrying a gene of interest, this gene can be transferred into the plant genome after release from the spheroplast. This has been achieved in one case (Hain *et al.,* 1984). Advantages and disadvantages are as for liposome fusion.

C. Microinjection

Mechanical injection of DNA into nuclei of protoplasts via microscopic injection needles has been developed very recently to an efficient method (Steinbiss *et al.,* 1985; Miki, 1985; Reich *et al.,* 1985; Goodman, 1985) (see also chapter 10 in this volume).

III. Direct Gene Transfer in Theoretical and Applied Genetics

The experimental possibility of transferring DNA directly and with high efficiency into naked plant cells from which clones and fertile plants can be regenerated opens numerous experimental routes to the study of the plant genome. Many of the experimental approaches discussed below are, of course, not unique for this technique, but may also be possible using *A. tumefaciens* bacterial, and viral vector systems and the other vectorless systems discussed earlier.

A. Gene Isolation

It can be envisaged that DNA clones carrying single copy genes could be identified via feeding pools and subpools of bacteriophage lambda or cosmid libraries to protoplasts and screening for a certain phenotype. If successful, this would constitute an important method for the isolation from libraries of genes which cannot be identified e. g. by differential hybridization to RNY populations or by immunological reactions (see e. g. Glover, 1984).

B. Gene Identification

Insertion of DNA into the plant genome can be expected to also lead to inactivation of genes. The question whether direct gene transfer can lead to insertion mutagenesis is under study. If DNA is being used which carries sequences which allow for subsequent reisolation this might not only allow for the identification but also isolation of genes. The difficulty with this approach could be a high level of spontaneous mutations and/or somaclonal variation. Therefore to limit the background in these experiments, particular conditions with known low level of variations have to be first established.

C. Replication

Replication origins are probably distributed throughout the genome at 50 kb intervals. It might be possible to isolate these by integration of fragments of the genome into a plasmid carrying a selectable gene and studying the effect of such plasmids on the transformation frequency or the level of expression following standard gene transfer. Once replication origins have been isolated they could be used to construct non-integrating vectors for the study of replication including studies of their stability. The low stability of such vectors was significantly improved in other systems by including into their structure centromeric and telomeric chromosomal sequences (e. g. Murray A. W., 1985), however, this approach in plants will be probably more complicated but could be considered as a long term goal.

D. Gene Replacement

If homologous recombination works in higher plants as in microbial and mammalian cells (Smithies O. *et al.*, 1985; Thomas, K. R. *et al.*, 1986), it can be envisaged that it will be possible to insert DNA at specific positions. Thus, directed mutagenesis, gene replacement and the study of position effects would become possible. However, many basic experiments have to be performed in order to assay frequency of homologous recombination in plants versus frequency of illegitimate recombination in order to find optimal experimental conditions for efficient gene replacement.

E. Gene Regulation

This can be studied by combining upstream sequences of the genes in question with coding sequences of the reporter genes. This approach has already been used for a number of genes using *A. tumefaciens* mediated plant transformation. As genes may not only be regulated by adjacent sequences, but also by their presence on a specific chromosome and in a specific chromosomal location, it will be interesting to make use of the *in-situ* hybridization method (see I. J.) in studying the effect of gene position on gene expression. As integration after gene transfer is probably random it will not be too difficult to identify a series of clones having the foreign gene integrated at different loci.

F. Stability and Instability

As mentioned in I. G. and I. H., transgenic plants have been identified which show absolute stability of the foreign gene in several sexual generations and without any selective pressure, and others, which show some kind of instability. An analysis of the chromosomal position and of the adjacent host DNA as well as the integration pattern might yield data which will help in understanding this phenomenon.

G. Gene Transfer to Chloroplasts and Mitochondria

The study of the plastome and the chondriome would be facilitated if it were possible to introduce foreign genes or to re-introduce organelle genes following their modification. There is one report in the literature (DeBlock *et al.*, 1985) indicating that T-DNA could be found in chloroplast DNA isolated from *A. tumefaciens* transformed plants. This would indicate that foreign DNA can pass the membrane system of chloroplasts once it is inside the cell. As direct gene transfer delivers considerable quantities of foreign DNA molecules into protoplasts, genes offered to plant cells via this route might have a chance of also ending up in the organelles.

H. Gene Transfer into Cereals

Direct gene transfer requires, to date, isolated protoplasts. Despite several publications on plant regeneration from protoplasts of graminaceous monocots and recent information on plant regeneration from rice protoplasts from several laboratories in Japan (e. g. Fujimura *et al.*, 1985), plant regeneration of graminaceous species is still a problem. As long as plant regeneration from cereal protoplasts is not routine and efficient, and as long as gene transfer through cell walls is not available, gene transfer into cereals will be difficult. It will also be a question of future developments whether gene transfer via bacterial or viral vectors into cereals will become a reality. To date there is no well documented gene transfer into cereal plants.

I. Gene Transfer into Potentially Totipotent Cells

Because of this limitation direct gene transfer into totipotent cells or tissues would constitute a major improvement. Three approaches are under study: If it were possible to transfer foreign genes to the egg cell with the germinating pollen via the natural fertilisation process (pollen transformation) this would solve the problems mentioned above. This idea goes back to the early 1970's (e. g. Hess, 1975) and numerous experiments have produced interesting results, but no conclusive evidence. The reproduction of these attempts with a clearly and molecularly defined selectable marker gene, e. g. the gene for kanamycin resistance described above, which allows application of rigorous molecular techniques to provide tight evidence for, or against, gene transfer, should allow verification or falsification of the hypothesis that genes can be transferred via this route. There is, probably, more potential in the approach to microspore transformation. Competent microspores (e. g. Heberle-Bors, 1986) can be grown to haploid embryos and plants, which can be diploidised. Genes passing the microspore cell wall would have a similar chance for integration into the host genome as in protoplast systems. Attempts in our laboratory to transfer the gene for kanamycin-resistance into *N. tabacum* microspores have produced, so far, false positives only, and this is, probably the general experience so far. The third approach is transformation of totipotent tissues, either in the form of morphogenic cell cultures or of organ explants. Numerous different experimental approaches have yielded kanamycin-resistant clones; for none of those, however, complete molecular proof of the gene transfer was shown so far.

J. Gene Transfer Without Pre Cloning in Bacteria

For crop plants for which routine protoplast culture and plant regeneration is at hand, transfer and stable integration of any gene available in the cloned form is no major problem and, probably, easily performed via cotransformation (and of course *A. tumefaciens*). The major problem is the availability of genes with applied importance. On the basis of the results from gene transfer with total genomic DNA it appears to be feasible, to approach the transfer of genes, independent of their isolation, by incubation of protoplasts with total genomic DNA isolated from plants which contain the desired gene. This might be possible for cases where the phenotype produced by the gene allows for selection at the protoplast culture level, where a dominant phenotype allows for an easy screening at the cell-culture level; or where large populations of plants can be easily regenerated from protoplasts and the expected phenotype can be screened for in the field either by the experimentor, or selected for by application of biotic or physical stress.

K. Gene Transfer into Organelles

Gene transfer into chloroplasts (to introduce for example herbicide resistance) or into mitochondria (to manipulate male sterility) may be achieved in the near future. DeBlock *et al.* (1985) present strong indicative evidence that DNA from the Ti-plasmid can be transported into chloroplasts. Final proof provided by an altered chl-DNA restriction pattern and hybrid DNA fragments of vector and chl-DNA is not yet available. As mentioned earlier (paragraph I. F.) a collection of clones which are indicative of organelle transformation by clear reciprocal differences in the transmission of the foreign gene in reciprocal crosses with the wildtype have been identified. A possible explanation for these cases of non-Mendelian inheritance might, however, come from causes like e. g. "pollen lethality" and "polyploidy".

L. Tagging of Chromosomes

Some benefit for conventional breeding programs may develop from a combination of gene transfer and *in-situ* hybridization (as mentioned in paragraph I. J.). Visibly labelling chromosomes with a selectable marker gene might help in identifying chromosomes, correlating chromosomes with linkage groups, following the fate of specific chromosomes in pedigrees etc. It can be envisaged that an entire set of chromosomes could be labelled with specific molecular probes, although at present not representing independently selectable marker genes.

M. Modulation of Expression

High levels of expression of genes could be achieved by providing strong and constitutive promoters e. g. 35 S promoter of Cauliflower Mosaic Virus (Pietrzak M. *et al.*) and by integration of multiple copies of the gene; regulated expression of the genes will be possible in combination with organ-specific (Rosahl *et al.*) or of environment-specific expression signals.

N. Conclusions

Direct gene transfer system should be a useful method of studying several basic questions of plant molecular biology. However many aspects of the system should be studied further, but the simplicity, efficiency and reproducibility of the method can help in understanding gene organisation, structure, function and regulation as well as in studies of architecture and of dynamic changes in plant genome like DNA replication and recombination.

Gene transfer has also some potential for applied genetics in the, hopefully, not too distant future. Before the techniques demonstrated with, and applied to a small group of specific model plants can be used in breeding programs, very difficult problems have to be solved. These include the identification, molecular analysis, and isolation of genes responsible for important characters of crop plants and the development of efficient gene transfer techniques for the most important crop plants.

IV. References

Aviv, d., Fluhr, R., Edelmann, M. Galun, E., 1980: Progeny Analysis of Interspecific Somatic Hybrids: *Nicotiana tabacum* (cms) + *Nicotiana sylvestris* with respect to nuclear and chloroplast markers. Theor. Appl. Genet. **56,** 145—150.

Brisson, N., Paszkowski, J., Penswick, J. R., Gronenborn, B., Potrykus I., Hohn T., 1984: Expression of a Bacterial Gene in Plants Using a Viral Vector. Nature **310,** 511—514.

Brisson, N., Hohn T., 1985: Plant virus vectors: Cauliflower Mosaic Virus. In: Weissbach A., Weissbach H. (eds.), Methods in Enzymology **118,** 659—667.

Chilton, M. D., Tepfer, D. A., Petit, A., David, C., Casse-Delbert, F., Tempe, J., 1982: Agrobacterium rhizogenes inserts T-DNA into the host plant root cells. Nature **295,** 432—434.

Davey, M. R., Cocking, E. C., Freeman, J., Pearce, N., Tudor, I., 1980: Transformation of Petunia Protoplasts by Isolated *Agrobacterium* plasmids. Pl. Sci. Lett. **18,** 307—313.

De Laat, A. M. M., Blaas, J., 1985: Potential Applications for Microinjection of Plant Protoplasts. Abstracts 1st Intl. Congress Plant Mol. Biol., G. A. Galun (ed.), University of Georgia Center for Continuing Education, Athens, GA, U.S.A. p. 111.

Deshayes, A., Herrera-Estrella, L. Caboche, M., 1985: Liposome Mediated Transformation of Tobacco Mesophyll Protoplasts by an *Eschericia coli* Plasmid. EMBO J. **4,** 2731—2739.

Draper, J., Davey, M. R., Freeman, J. P., Cocking E. C., Cox, B. G., 1982: Ti plasmid homologous sequences present in tissues from *Agrobacterium tumefaciens* transformed Petunia tissues. Pl. & Cell Physiol. **23,** 451—458.

Fromm, M., Taylor, L. P., Walbot, V., 1986: Stable transformation of maize after gene transfer by electroportation. Nature **319,** 791—793.

Fujimura, T., Sakurai, M., Akagi, H., Negeshi, T., Hirose, A., 1985: Regeneration of rice plants from protoplasts. Plant Tissue Culture Lett. **2,** 74—75.

Gerhard, D. S., Kawasaki, E. S., Carter Bancroft, F.,, Szabo P., 1981: Localisation of a unique gene by direct hybridisation *in situ.* Proc. Natl. Acad. Sci., U.S.A. **78,** 4458—4460.

Gheysen, G. Dhaese, P., Van Montagu, M., Schell, J., 1985: DNA flux Across Genetic Barriers: The Crown Gall Phenomenon. In: Genetic Flux in Plants, Hohn, B., Dennis, E. S. (eds.), Plant Gene Research, Wien - New York: Springer, 1985, pp. 12—47.

Gleba, Y. Y., Sytnik, K. M., 1984: Protoplast Fusion. In: Genetic Engineering in Higher Plants, Shoeman, R. (ed.), Berlin - Heidelberg - New York: Springer.

Glover, D. M., 1984: Gene cloning — The Mechanics of DNA mainpulation. London - New York: Chapman and Hall.

Goodman, R. D., 1985: The geminiviruses: past, present, future. In: Galau, G. A., (ed.), Abstracts First International Congress Plant Molecular Biology. University of Georgia Center for Continuing Education, Athens, GA, U.S.A., pp. 12.

Goodman, R. D., 1985: Personal communication.

Grimsley, N., Hohn, B., Hohn, T., Walden, R., 1986: Agroinfection, a Novel Route for Plant Viral Infection. Proc. Natl. Acat. Sci, U.S.A. **83,** 3282—3286.

Hain, R., Steinbiss, H. H., Schell, J., 1984: Fusion of *Agrobacterium* and *E. coli* Spheroplasts with *Nicotiana* Protoplasts — Direct Gene Transfer from Microorganism to Higher Plant. Pl. Cell. Rep. 60—64.

Hain, R., Stabel, P., Czernilofsky, A. P., Steinbiss, H. H., Herrera-Estrella, L., Schell, J., 1985: Uptake, integration, expression and genetic transmission of a selectable chimeric gene by plant protoplasts. Mol. Gen. Genet. **199,** 161—168.

Heberle-Bors, E., 1986: *In vitro* haploid formation from pollen: a critical review. Theor. Appl. Genet. **71,** 361—374.

Hess, D., 1975: Uptake of DNA and bacteriophage into pollen and genetic manipulation. In: Ledoux, L. (ed.), Genetic manipulation with plant material. New York: Plenum Press, pp. 519—537.

Karube, I., Tamiya, E., Matsuoka, H., 1985: Transformation of *Saccharomyces cerevisiae* spheroplasts by high electric pulse. FEBS Lett. **182,** 90—94.

King, J., Saxena, P., 1985: Genetic transformation of plant cells by isolated nuclei. 1. Preliminary conditions for uptake of nuclei. In: Galau, G. A. (ed.), Abstracts First International Congress Plant Molecular Biology. University of Georgia Center for Continuing Education, Athens, U.S.A., p. 28.

Krens, F. A., Molendijk, L. Wullems, G. J., Schilperoort, R. A., 1982: In Vitro Transformation of Plant Protoplasts with Ti-plasmid DNA. Nature **296,** 72—74.

Lörz, H., Baker, B., Schell, J., 1985: Gene Transfer to Cereal Cells Mediated by Protoplast Transformation. Mol. Gen. Genet. **199,** 178—182.

Lurquin, P. F., Kado, C. I., 1977: *Escherichia coli* plasmid pBR313 insertion into plant protoplasts and into their nuclei. Mol. Gen. Genet. **154,** 113—121.

Maliga, P., Lörz, H., Lazar, G. B., Ferenczy, L., 1982: Cytoplast-protoplast Fusion for Interspecific Chloroplast Transfer in *Nicotiana.* Mol. Gen. Genet. **185,** 211—215.

Miki, B., 1985: The development of microinjection technology. In: Galau, G. A. (ed.), Abstracts First International Congress Plant Molecular Biology. University of Georgia Center for Continuing Education, Athens, U.S.A., p. 28.

Mouras, A., Saul, M. W., Potrykus, I., 1987: *In situ* Hybridisation of a Transformed Gene in *Nicotiana tabacum.* Mol. Gen. Genet. **207,** 204—209.

Mueller, T., Saul, M. W., Paszkowski, J., Shillito, R. D., Potrykus, I., 1986: Direct gene transfer from a total *N. Tabacum* genome, in preparation.

Mullineaux, P. M., Donson, J., Morris-Krsinich, B. A. M., Boulton, M. I., Davies, J. W., 1984: The nucleotide sequence fo maize streak virus. EMBO J. **3,** 3063—3068.

Murray, A. W., 1985: Chromosome structure and behaviour. TIBS March 112—115.

Neumann, E., Scaefer-Ridder, M., Wang, Y., Hofschneider, P. H., 1982: Gene transfer into mouse lyoma cells by electroporation in high electric fields: EMBO J. **1,** 841—845.

Paszkowski, J., Shillito, R. D., Saul, M. W., Mandák, V., Hohn, T., Hohn, B., Potrykus, I., 1984: Direct Gene Transfer to Plants. EMBO J. **3,** 2717—2722.

Paszkowski, J., Pisan, B., Shillito, R. D., Hohn T., Hohn B., Potrykus, I., 1986: Genetic tranformation of *Brassica campestris* var. rapa protoplasts with an engineered Cauliflower Mosaic Virus genome. Plant Mol. Biol., in press.

Pietrzak, M., Shillito R. D., Hohn, T., Potrykus, I., 1986: Expression in plants of two bacterial antibiotic resistance genes after protoplast transformation with new plant expression vector. Nucl. Acid Res., in press.

Potrykus, I., 1980: The Old Problem of Protoplast Culture-Cereals. In: Advances in Protoplast Research, Ferenczy, L., Farkas, G. L. (eds.), Budapest: Hungarian Academy of Sciences, p. 243.

Potrykus, I., Shillito, R. D., 1985: Protoplasts: isolation, culture and plant regeneration. Methods in Enzymology **118,** 549—577.

Potrykus, I., Saul, M., Petruska, J., Paszkowski, J., Shillito, R. D., 1985: Direct Gene Transfer to Cells of a Graminaceous Monocot. Mol. Gen. Genet. **199,** 183—188.

Potrykus, I., Paszkowski, J., Saul, M., Petruska, J., Shillito, R. D., 1985: Molecular and General Genetics of a Hybrid Foreign Gene Introduced into Tobacco by Direct Gene Transfer. Mol. Gen. Genet. **199,** 169—177.

Potrykus, I. Shillito, R. D., Saul, M., Paszkowski, J., 1985: Direct Gene Transfer: State of the Art and Future Potential. Plant Mol. Biol. Rep. **3,** 117—128.

Reich, T., Iyer, V. N., Miki, B., 1985: Transformation of Alfalfa Protoplasts by Intranuclear Microinjection of Ti-plasmids. Abstracts 1st Intl. Congress Plant Mol. Biol., Galun, G. A. (ed.), University of Georgia Center for Continuing Education, Athens, GA, U.S.A., p. 28.

Rosahl, R., Schmidt, R., Schell, J., Willmitzer, L., 1986: Isolation and characterisation of a gene from *Solanum tuberosum* encoding patatin, the major storage protein of potato tubers. Mol. Gen. Genet. **203,** 214—220.

Saxena, P., King, J., 1985: Genetic Transformation of Plant Cells by isolated Nuclei. II. Rescue of transformed clones and evidence for transfer of foreign genome. Abstracts 1st Intl. Congress Plant Mol. Biol., Galun, G. A. (ed.), University of Georgia Center for Continuing Education, Athens, GA, U.S.A., p. 28.

Schocher, R. J., Shillito, R. D., Saul, M. W., Paszkowski, J., Potrykus, I., 1987: Cotransformation of foreign genes into plants. Biotechnology **4,** 1093—1096.

Shillito, R. D., Saul, M. W., Paszkowski, J., Müller, M., Potrykus, I., 1985: High Efficiency Direct Gene Transfer to Plants. Biotechnology **3,** 1099—1103.

Shivarova, N., Forster, W., Jacob, H.-E., Grigorava, R., 1983: Microbiological implications of electric field effects VII. Stimulation of plasmid transformation of *Bacillus cereus* protoplasts by electric field pulses. Z. Allg. Mikrobiol. **23,** 595—599.

Smithies O., Gregg, R. C., Boggs, S. S., Koralewski, M. A., Kucherlapati, R. S., 1985: Insertion of DNA sequences into the chromosomal B-globin locus by homologous recombination. Nature **317,** 230—234.

Steinbiss, H. H., Stabel, P., Töpfer, R., Hirtz, R. D., Schell, J., 1985: Transformation of plant cells by microinjection of DNA. Proceedings Wye International Symposium on Experimental Manipulation of Ovule Tissue, Wye College, London, in press.

Thomas, K. R., Folger, K. R., Capecchi, M. R., 1986: High frequency targeting of genes to specific sites in the mammalian genome. Cell **44,** 419—428.

Wienand, U., Langridge, P., Feix, G., 1981: Isolation and characterisation of genomic sequence of maize coding for a zein gene. Mol. Gen. Genet. **182,** 440—444.

Wigler, M., Sweet, R., Sim, G. K., Wold, B., Pellicer, A., Lacy, E., Maniatis, T., Silverstein, S., Axel, R., 1979: Transformation of a mammalian cells with genes from prokaryotes and eukaryotes. Cell **16,** 777—785.

Chapter 10

Microinjection: An Experimental Tool for Studying and Modifying Plant Cells

Brian L. A. Miki[1], Terry J. Reich[2], and V. N. Iyer[2]

[1] Genetic Engineering Section, Plant Research Centre, Agriculture Canada, Ottawa, Ontario K1A OC6 (Canada)

[2] Department of Biology, Carleton University, Ottawa, Ontario K1S 5B6 (Canada)

With 5 Figures

Contents

I. Introduction

The conviction that DNA is the genetic material of cells has its origins in a set of experiments in which purified DNA was transferred and expressed in bacterial cells. It is therefore no surprise that since these initial experiments with bacteria, methods of transferring DNA and other molecules into eukaryotic cells have been explored with varying degrees of success and usefulness. The most direct way of delivering a population of molecules into a cell or cell compartment is to microinject it mechanically (Celis, 1984; Celis *et al.,* 1986). Techniques to accomplish this have been developed for frog oocytes (Gurdon *et al.,* 1971; Kressman *et al.,* 1978; Rusconi and Schaffner, 1981), insect embryos (Rubin and Spradling, 1982), animal egg cells (Brinster *et al.,* 1985; Gordon *et al.,* 1980; Hammer *et al.,* 1985; Wagner *et al.,* 1981) and mammalian somatic cells (Capecchi, 1980; Diacumakos, 1973; Graessman *et al.,* 1980a, b). The anatomical organization of plant cells and the experimental difficulties associated with their manipulation in culture prevented a facile transposition of such techniques from animal to plant cells (Steinbiss *et al.,* 1984). Fortunately, this situation is changing rapidly. It is now possible to microinject some plant cells and various protoplast types successfully (Crossway *et al.,* 1986; Reich *et al.,* 1986a, b, c). It is the intent of this contribution to review these methodologies so that they can be extended, improved upon and become easily accessible to the community of plant cell and molecular biologists. Through this discussion we hope that the experimental advantages offered by microinjection become apparent.

II. Recipient Cell Systems

A. Cell Types

By microinjection, DNA can be introduced into small numbers of single cells. For genetic analyses, it is important that a large proportion of them regenerate into normal fertile plants. Egg cells would seem to be the obvious targets for microinjection given the success of gamete transformation in generating transgenic mammals (Gordon *et al.,* 1980; Wagner *et al.,* 1981; Hammer *et al.,* 1985), amphibians (Rusconi and Schaffner, 1981) and insects (Rubin and Spradling, 1982); however, methods have not been developed for the isolation and culture of egg cells or embryo sacs in plants. Attempts to introduce DNA into ovules by injection (Zhou *et al.,* 1983; Flavell and Mathias, 1984) have not yielded transgenic plants. Plant regeneration from single cells in culture can be achieved with protoplasts (Evans and Bravo, 1983; Jones, 1985) and isolated microspores (Dunwell, 1985; Heberle-Bors, 1985). Such alternative cell types are preferable because they are simpler to manipulate, but the range of plants for which these technologies exist is severely restricted. Multicellular recipients including the small cell clumps of regenerable suspension cultures of proembryos (Ammirato, 1983; Maddock, 1985) must also be considered for microinjection; however, efficient selection systems may be needed to avoid the regeneration of plants as mosaics of transformed and normal cells. The application of microinjection as a method for genetic transformation is therefore governed by the technologies available for culturing single cells and regenerating plants from them.

The plant cell wall is a major obstacle to the physical and chemical methods for DNA delivery, including direct DNA uptake (Krens *et al.*, 1982, 1985; Paszkowski *et al.*, 1984; Shillito *et al.*, 1986), liposome fusion (Deshayes *et al.*, 1985) and microinjection (Steinbiss *et al.*, 1984; Crossway *et al.*, 1986; Reich *et al.*, 1986a, b, c). In Suspension culture, cells have brittle walls that do not seal after typical micropipettes are withdrawn. Many years ago, it was found that a plug formed around the tip of the micropipette if it was left in the cell for 38—48 h (Nims *et al.*, 1976b) and this allowed its removal without killing the cell. Although this approach was used successfully for studying the introduction of virus particles into tobacco *(Nicotiana tabacum)* cells (Nims *et al.*, 1967a; Halliwell and Gazaway, 1975), the small number of cells that could be manipulated has prohibited the general application of this technique for genetic transformation. Experimental approaches to overcome this barrier such as altering the cell wall properties in culture or modifying the micropipette tip and DNA delivery system have not been fully explored. The cell walls of various cell types differ in their properties and it is likely that certain types will be naturally amenable to microinjection. For instance, embryogenic microspores of rapeseed *(Brassica napus)* undergo dramatic changes in cell wall organization during the early stages of cell culture and can be microinjected without loss of viability (Pechan and Keller, unpublished results). Further investigations will no doubt reveal other cell types that can be used.

B. Protoplasts

The first cell types to be transformed with microinjected DNA were protoplasts from various sources with only partially reformed cell walls (Table 2; Crossway *et al.*, 1986; Reich *et al.*, 1986b). The use of protoplasts overcame the restricions imposed by cell walls and accelerated the development of these experimental procedures. After 16—24 h in culture, alfalfa *(Medicago sativa)* protoplasts have firm cell walls that provided stability during subsequent manipulations but they had not crystallized to the brittle form common for cells in culture (Holbrook *et al.*, 1985) and therefore microinjection was not obstructed. This period also coincided with the reinitiation of DNA synthesis in these protoplasts (Holbrook *et al.*, 1985). Whether this metabolic activity contributed to the integration of microinjected DNA or related to a competent state for transformation (Firoozabady and Galbraith, 1984; Gould and Ashmore, 1982; Meyer *et al.*, 1985) has not been assessed.

C. Cell Culture Conditions

High plating efficiencies and regeneration frequencies are a prerequisite for the recovery of transformants following microinjection. The highest injection rates reported for plant cells are 100/h or about 300/h experiment (Table 1; Reich *et al.*, 1986c). Plating efficiencies of about 10% may be necessary if the transformation frequencies are in the range of 10%. Few cell systems presently meet this requirement. Furthermore, the culture conditions should be compatible with methods that segregate or distinguish the microinjected cells, thereby eliminating the dependence on selectable marker genes for isolating transformants.

Table 1. *The Microinjection of Plant Protoplasts*

Species	Holding method	Microinjection rates cells/h	Microinjected volumes L	References
Nicotiana tabacum	poly-L-lysine	20	$10^{-10}-10^{-12}$	Steinbiss *et al.* (1983, 1985)
Nicotiana tabacum	agarose embedding	50	$10^{-11}-10^{-12}$	Lawrence & Davies (1985)
Euphorbia milli	holding pipette	10	$10^{-12}-10^{-13}$	Morikawa & Yamada (1985)
*Nicotiana tabacum**	holding pipette	—	10^{-12}	Crossway *et al.* (1986)
*Medicago sativa**	poly-L-lysine	100	$\geq 10^{-17}$	Reich *et al.* (1986 a, c)

* The microinjection rates and volumes were estimated for intranuclear microinjection.

Table 2. *Transformation of Protoplasts by Intranuclear Microinjection of Plasmids*

Species	Plasmids injected	No. of experiments	No. of protoplasts injected	No. of calli recovered	No. of transformants	Transformation frequency	References
Medicago sativa	pTiC58[1]	3	523	65	17	26 %	Reich *et al.* (1986 b)
	pMON8015[2]	3	492	90	18	20 %	Reich *et al.* (1986 b)
Nicotiana tabacum	pCGN561[3]	5	113	56	8	14 %	Crossway *et al.* (1986)
Brassica napus	pMON38[4]	1	42	6	4	66 %	Reich *et al.* (unpublished data)

[1] pTiC58 is a wild-type nopaline Ti plasmid.

[2] pMON8015 is a 9 kb plasmid carrying the genes *nos* and chimeric *nptII* but lacking functional T-DNA borders (Fraley *et al.*, 1985).

[3] pCGN561 is a 29 kb plasmid that does not carry genes for expression in plants but possesses T-DNA borders [Crossway *et al.* (1986)].

[4] pMON38 is 6.7 kb plasmid carrying the *nos* gene and T-DNA right border on the Hind III 23 fragment from pTi T37. The plasmid as linearized by digestion with Eco R1 prior to microinjection.

Effective methods for cell culture have been developed which utilize feeder layers separated from the injected cells by filter paper discs (Reich *et al.*, 1986b) or nylon mesh rings (Lawrence and Davies, 1985; Morikawa and Yamada, 1985). Low melting temperature agarose which increases plating efficiency (Holbrook *et al.*, 1985; Shillito *et al.*, 1983) may be used in combination with these techniques (Lawrence and Davies, 1985). Following microinjection, calli have been recovered from tobacco protoplasts with efficiencies of 40 % (Lawrence and Davies, 1985) using this approach. For crop species such as alfalfa and rapeseed, calli have been recovered from microinjected protoplasts with about 15 % efficiency (Table 2). In liquid microdrop cultures, calli have been recovered from injected tobacco protoplasts with 20 % efficiency (Crossway *et al.*, 1986) and embryos have been recovered from rapeseed microspores with about 20 % efficiency (Pechan and Keller, unpublished results). Frequently, the active cells in culture may be identified by their morphological characteristics prior to microinjection. The efficiencies for these subpopulations may be considerably higher than for the whole population of cells.

III. Resolution of Intracellular Compartments

A. Microscopy

Microinjection is the most precise method for delivering macromolecules into specific intracellular compartments (Celis, 1984). Comparisons between intranuclear and cytoplasmic injections of DNA into animal tissue culture cells (Capecchi, 1980) and embryos (Brinster *et al.*, 1985) have shown that intranuclear microinjection is essential for efficient transformation. This conclusion was also confirmed with tobacco protoplasts (Crossway *et al.*, 1986).

Since the nuclei of plant cells are generally difficult to resolve, it is extremely difficult to demonstrate intranuclear microinjection consistently. Micrographs of petunia mesophyll protoplasts and alfalfa protoplasts from suspension culture cells illustrate the poor resolution of the nucleus with bright-field and phase-contrast optics (Fig. 1). A number of approaches have been taken to overcome this problem. Tobacco protoplasts may be selected and oriented with a holding pipette for optimal resolution of the nucleus by bright-field microscopy (Crossway *et al.*, 1986; Fig. 2A). Differential interference contrast optics may enhance the resolution of the nucleus in certain protoplasts (Gordon-Kamm and Bushnell, 1985; Figs. 2C, 2D). Partial staining of nuclei with DNA-specific dyes combined with partial fluorescence microscopy is also very effective (Reich *et al.*, 1986a, b; Fig. 2B). The best approach may depend on the cell types that are being considered for microinjection.

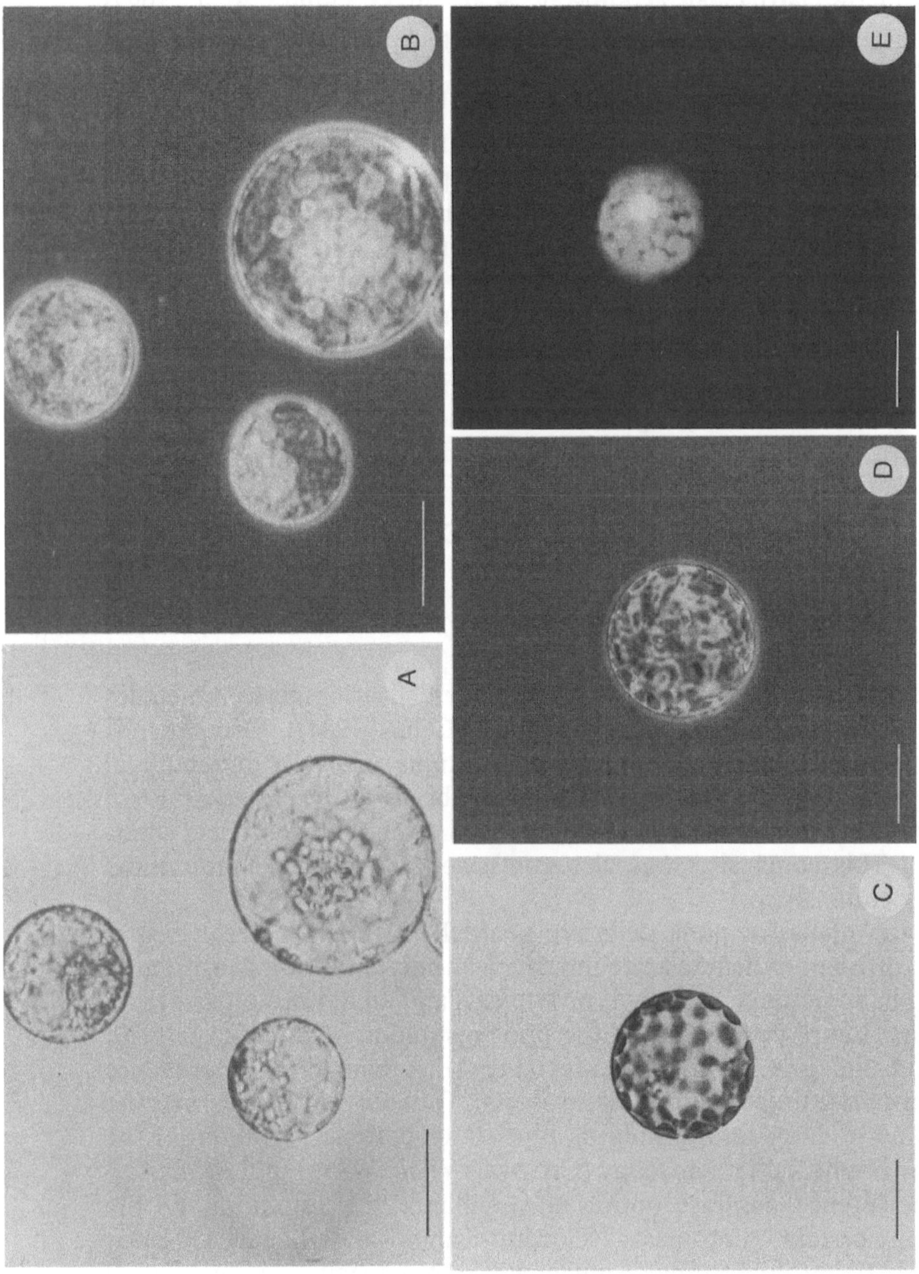

Fig. 1. Micrographs of plant protoplasts. Alfalfa protoplasts from suspension culture cells (A, B) and petunia mesophyll protoplasts (C, D, E) were examined under bright-field (A, C) and phase-contrast (B, D) optics. The location of the nucleus in petunia mesophyll protoplasts was revealed by partial staining with Hoechst 33258 (Reich *et al.*, 1986 a) and fluorescence microscopy using Leitz filter block A (E). Bar = 25 µm

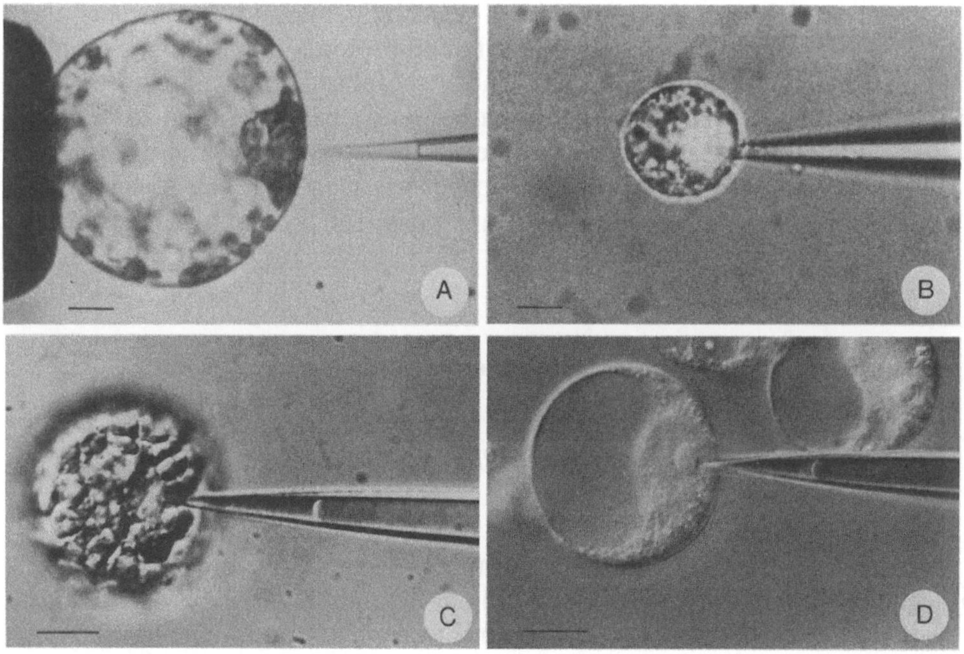

Fig. 2. Micrographs demonstrating intranuclear microinjection of protoplasts. Tobacco mesophyll protoplasts were oriented with a holding pipette for microinjection under bright-field optics (A, courtesy of A. Crossway, Calgene, Inc.). Alfalfa protoplasts were partially stained with Hoechst 33258 to locate the nucleus and microinjection was performed unter combined bright-field and fluorescence optics (B). Barley mesophyll protoplasts (C) and protoplasts from oat (Stout) suspension culture cells (D) were microinjected under differential interference contrast microscopy (courtesy of W. Gordon-Kamm and W. R. L. Bushnell, Agricultural Research Service, U. S. Department of Agriculture). Bar = 10 μm

B. Fluorescent Stains

The use of fluorescent dyes such as Hoechst 33258 and DAPI (4,6-diamidino-2-phenylindole) confers great precision to intranuclear microinjection without adding to the experimental complexity of the procedures (Reich *et al.*, 1986 a, b, c). The flow of DNA solution into the nucleus can be monitored directly by the disturbance in the fluorescence. Nuclear staining is transient and disappears within a day (Reich *et al.*, 1986 a), therefore a variety of fluorescent probes can be used subsequently to examine expression within the microinjected cells. Cell division in alfalfa protoplasts (Reich *et al.*, 1986 a) and embryogenesis in rapeseed microspores (Pechan and Keller, unpublished data) are not inhibited by partial staining, therefore the use of dyes does not seem to be detrimental to plant cells, as it is for animal ova (Hammer *et al.*, 1985). By fluorescence microscopy it is now possible to resolve many intracellular compartments of living plant cells, including the nucleus, cytoplasm and mitochondria (Reich *et al.*, 1986 a).

Fluorescent dyes have also been used to determine the intracellular location of materials injected into unstained protoplasts. The coinjection of Lucifer yellow has been used to identify cytoplasmic injections (Steinbiss and Stabel, 1983). DNA stained with berberine has been used to demonstrate the intranuclear location of injected DNA (Morikawa and Yamada, 1985). Although these approaches do not assist in targeting injected macromolecules, the fluorescent dyes are powerful probes for determining their fate.

IV. Microinjection Methodology

A. Micromanipulation Techniques

Although protoplasts adhere to glass or plastic surfaces, the attachment is not firm enough to anchor them for microinjection. Techniques have been established for attaching somatic animal cells to cover slips with ligands that bind to their surfaces (Graessman *et al.*, 1980b) and for holding mammalian egg cells with blunt holding pipettes (Wagner *et al.*, 1981). Both approaches have been successfully adapted for plant protoplasts (Table 1). The most versatile systems for plant cells promote intranuclear microinjection of large numbers of cells in combination with methods for the efficient recovery and culture of microinjected cells.

Holding pipettes allow precise orientation of the protoplast nucleus relative to the injection pipette (Fig. 2A) and subsequently allow segregation of injected protoplasts into microdrop cultures (Crossway *et al.*, 1986). A variety of pipette tip configurations may be adapted for different cell types. The injection rates reported so far are very low (Table 1) thereby restricting this technology to plant cells with extremely high plating efficiencies.

By attaching protoplasts to cover slips with poly-L-lysine (Steinbiss and Stabel, 1983) microinjection rates of 100/h can be achieved (Table 1). This requires independent methods for distinguishing injected protoplasts from the others at the end of an experiment. One approach has been the coinjection of fluorescent dyes (Steinbiss and Stabel, 1983; Morikawa and Yamada, 1985). Another is the use of reference markings on the cover slip to identify the location of the microinjected protoplasts (Reich *et al.*, 1986c). Figure 3 illustrates a microscopic grid pattern that can be photolithographically applied to glass cover slips. With a video camera the location of each microinjected protoplast can be recorded without interrupting the injections (Fig. 4). At a later time they can be transferred to optimal culture conditions by micromanipulation. Poly-L-lysine is not toxic to protoplasts with partially reformed cell walls. Embedding in agrose (Table 1) has been considered as an alternative for immobilizing fresh protoplasts (Lawrence and Davies, 1985).

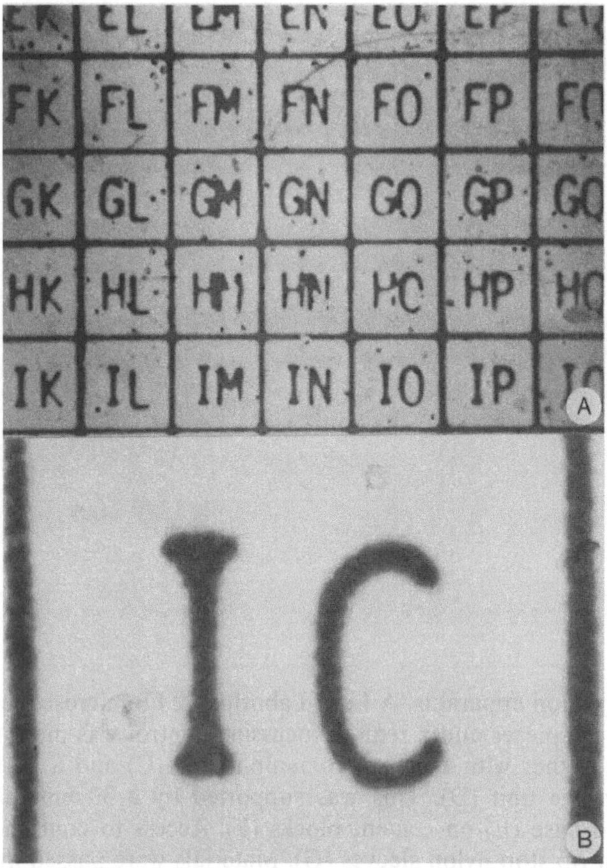

Fig. 3. Micrograph of alfalfa protoplasts attached to coded cover slips. Glass cover slips (24 x 30 mm, No. 1, Canlab) were coated on one surface with a film of Kodak KPR photoresist as described by Lin and Ruddle (1981). These were sandwiched against a photomask containing the image of the grid pattern; exposed with the focussed light beam of a slide projector, and developed with Kodak KPR developer. After drying, a black photoresist dye (Kodak) was applied for 30 s and washed away with distilled water. The cover slips were sterilized by autoclaving. The photomask was prepared by reducing the image of the grid pattern onto Agfa-Gvert, HDU1-P lithographic film using a process camera. From the negative, 25 positives were prepared by contact printing and mounted onto a clear acetate sheet. From this the photomask was obtained by contact printing onto sheet film which was mounted onto a sheet of glass. Each grid on the cover slip was 0.38 mm x 9.38 mm. Protoplasts that were in culture for 24 h (Holbrook *et al.,* 1985) were attached to the other surface of the cover slip with poly-L-lysine (m. w. 30,000—900,000, Sigma Chemical Co.) using the method described by Steinbiss and Stabel (1983). Both the protoplasts and grid pattern were visible under (A) 10X and (B) 40X lens objectives without refocusing. Reproduced from Reich *et al.* (1986 c) with permission

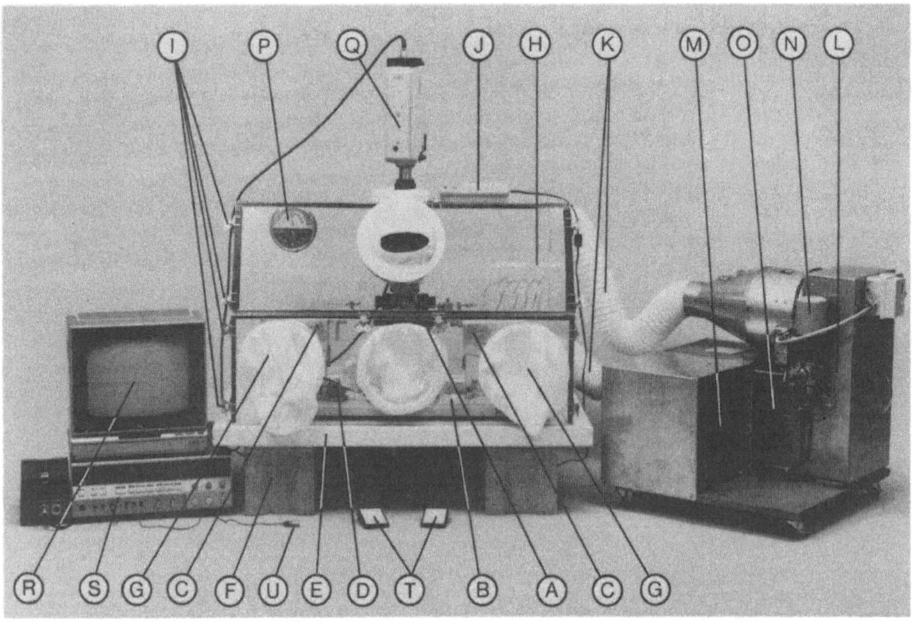

Fig. 4. Microinjection apparatus. A Leitz Laborlux 12 FS microscope (A) equipped with extended eye pieces and a remote focussing control was mounted on a Leitz baseplate (B) together with Leitz micromanipulators (C) and a custom-made microinjection syringe unit (D). This was supported by a 30 mm thick fibreglass-coated plywood base (E) on cement blocks (F). Access to controls was attained through airtight rip stop nylon sleeves (G). Materials were passed into the cabinet by a petrie dish drawer (H) that seals when open or closed. The front of the cabinet is removeable with quick release fasteners (I) and sealed with gaskets. Sterilization of the chamber was initially achieved with a germicidal lamp (J). The chamber is connected by 100 mm diameter flexible hose (K) and separate air inlet and outlet ports to an environmental control unit which circulates air through the cabinet and maintains a positive pressure within. Return air is humidified by a humidifier controlled by a humidistat in the plenum (L). The air is cooled by a 150 W air-cooled condensing unit that runs continuously (M). A modulating hot gas control system (N) and electronic controller cycles the refrigerant between heating and cooling to achieve temperature control within 1 °C inside the cabinet. The air is passed through a (HEPA) high efficiency filter (O) back to the cabinet. A certified temperature and humidity indicator (P) permits monitoring of environmental parameters. The microscope was equipped with a JVC model KV-1950 3-tube colour video camera (Q), 14″ monitor (R), and ½″ VHS industrial recorder (S). It was modified for remote operation of pause and record functions with foot controls (T). Audio commentaries were recorded with a lapel microphone (U). Reproduced from Reich *et al.* (1986 c) with permission

B. Equipment

The equipment used for plant cell microinjection is basically the same as that used for animal cell microinjection (Graessman *et al.*, 1980a; Diacumakos, 1973). For plant cells it is preferable to work in open droplets of media rather than oil chambers for ease of subsequent cell manipulations and culture. A sterile working environment free from vibrations and regulated for temperature and humidity is therefore essential. An environmental cabinet designed for this purpose (Fig. 4) may be the simplest and least expensive means of achieving these conditions (Reich *et al.*, 1986c). An environmental control unit maintains a positive air pressure within the cabinet. The air is filter sterilized and regulated for 25° C ±1° and 80% relative humidity. Protoplasts in 0.5 mL droplets of medium can be microinjected for up to 3 h without contamination or changes in the volume of the droplet. This system supports maximum injection rates for plant protoplasts and provides conditions needed for efficient recovery of viable cells.

Using a variety of syringe systems, volumes as large as 10^{-10} to 10^{-13}L have been introduced into protoplasts without loss of viability (Table 1). Techniques exist for controlling the microinjected volumes to levels as small as 10^{-17}L (Reich *et al.*, 1986a, c) therefore the dose delivered to a protoplast can be finely controlled. Figure 5 shows oil droplets of 10^{-17}L and 10^{-16}L microinjected into an alfalfa protoplast nucleus and cytoplasm, respectively. At this time, data are insufficient to assess whether genetic damage is incurred in regenerated plants as a result of the physical process of microinjection.

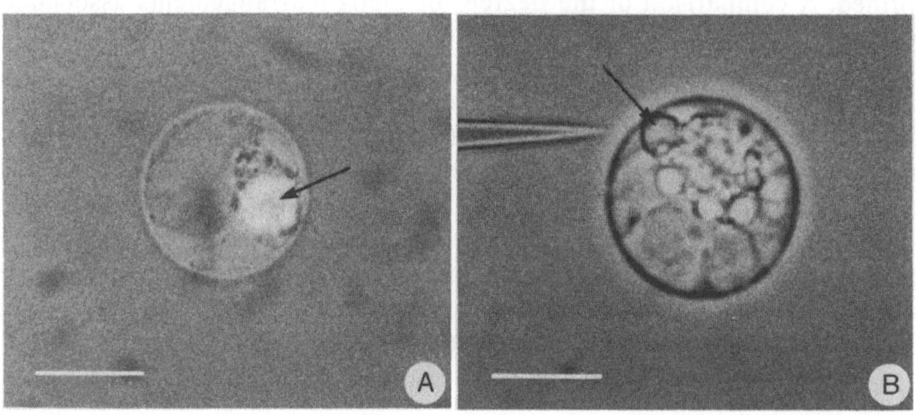

Fig. 5. Micrographs illustrating the microinjection of oil droplets (arrows) into the nucleus of alfalfa protoplasts partially stained with Hoechst 33258 (A) and the cytoplasm partially stained with fluorescein diacetate (B). The protoplasts were examined by fluorescence microscopy using Leitz filter blocks A (A) and I_2 (F) with a low level of bright-field illumination to reveal the unstained portions of the cells. The volumes of the oil droplets are approximately 10^{-17}L (A) and 10^{-16}L (B). Bar = 25 µm. Reproduced from Reich *et al.* (1986a) with permission

V. Genetic Transformation

Plant protoplast transformation by microinjection has now been achieved with three plant species: alfalfa, tobacco and rapeseed. For alfalfa and tobacco, using supercoiled plasmid DNA preparations, the transformation frequencies have ranged from 14 to 26 % (Table 2). Precise delivery of DNA to the nucleus rather than the cytoplasm appears to be essential to achieve this efficiency (Crossway *et al.*, 1986). There is also an indication that the injection of linear DNA as opposed to supercoiled circular DNA is preferred for high frequencies of transformation (Table 2, last line); however, it is not certain at the present time that this extremely high frequency of transformation (66 %) is solely a consequence of the form of DNA that was microinjected. These transformation frequencies are sufficient to make the detection of transformants practical without their selection. Furthermore, it has been observed that single copies of only part of the microinjected DNA may be integrated into the recipient genome (Crossway *et al.*, 1986). Assuming that in such instances only a single and random uninterrupted segment of DNA is inserted, this suggests that by a judicious choice of mapped and selectable markers, one may exercise some control on the selection of fragments that are integrated and recovered in transformants.

Transformation system that utilize cultured plant cells or protoplasts are associated with some variation in the structure of the DNA that is stably integrated (Van Lijsebettens *et al.*, 1986; Krens *et al.*, 1985; Shillito *et al.*, 1986). The mechanism by which this variation arises is seldom understood. Instability could arise from structural features of the DNA that is primarily inserted, from genes contained and expressed from this DNA and also from the properties of the particular cell system being transformed. A comparison of the degrees of DNA rearrangements associated with the various transformation methodologies is therefore difficult. Prelimlinary analysis of the fate of microinjected Ti plasmids indicate that the T-DNA borders were not preferred sites for integration (Reich *et al.*, 1986b). Furthermore, these sequences could be deleted with no appreciable effect on transformation frequency (Table 2, Reich *et al.*, 1986b). Ti plasmid sequences other than T-DNA are integrated and show no evidence for extensive rearrangements (Reich *et al.*, 1986b). This suggests that all or part of any DNA that survives microinjection into nuclei is likely to be stably integrated and that special vectors or carrier DNA are not important for the introduction of genes into plant cells.

VI. Other Applications

In the plant sciences, microinjection as a general technique for depositing molecules into specific cells or cell compartments has not been exploited. In the past, studies were confined to physiological examinations of intercellular transport by tracing the movement of fluorescent dyes following microinjection into specific cells of the plant (Goodwin and Erwee, 1985;

Palevitz and Hepler, 1985). To study the post-synthetic modification and intracellular transport of plant proteins, such as phaseolin, frog oocytes were used as hosts for the microinjection of specific mRNA (Matthews *et al.*, 1981). Generally, microinjection has been more widely applied in the animal sciences (Celis *et al.*, 1986). The microinjection of structural proteins, their antibodies, or specific drugs has been performed extensively to examine the functional organization of cytoskeletal components in living cells (Kreis and Birchmeier, 1982; Mangeat and Burridge, 1985; Wehland *et al.*, 1980, 1981). Microinjection provided a direct approach to study the influence of specific proteins on cell differentiation in culture (Bar-Sagi and Feramisco, 1985). Furthermore, larger structures, such as isolated cell organelles have been transferred into animal cells by microinjection (Berkenstam *et al.*, 1986). The microinjection technology that now exists for plant cells is sufficient for many of these kinds of experiments.

VII. Concluding Remarks

In this review we have described the recently developed techniques of microinjecting plant cells and illustrated that these techniques are now sufficiently well-developed to be exploitable. It is likely that the use of these techniques will be expanded and that they will be used in a fashion that will be complementary to other techniques that now exist or can be developed for the delivery of genes, gene products, chromosomes or organelles into plant cells. The efficiencies of transformation using microinjection are as good or better than any other method that is currently available. Its major advantage is that it allows the targeting of specific molecules or supramolecular structures to a particular cell and cell compartment. It could therefore be particularly useful for the analyses of diverse intracellular or intranuclear events such as extrachromosomal replication of DNA or the processing of DNA that precedes or accompanies integration. The ability to integrate an exogenous gene into the genome of a plant can be considered to be no more than a preliminary first step to gene manipulation. Site-directed integration at a desired location, the replacement of an endogenous gene with an exogenous one and the faithful maintenance and regulation of the exogenous gene are questions that will need to be addressed in the future and for which this technology may prove to be useful.

For many crop species, microinjection may be the most practical approach to genetic transformation at this time. Species such as alfalfa are difficult to infect with *Agrobacterium tumefaciens* (Mariotti *et al.*, 1984) and antibiotic selection systems developed with petunia (Fraley *et al.*, 1983) and tobacco (Herrera-Estrella *et al.*, 1983) are not effective. Microinjection circumvents these biologically imposed barriers. For the transformation of cereals, the microinjection of embryogenic microspores or cells in suspension culture may be necessary given the general inability to regenerate plants from protoplasts. This technology is being developed and preliminary experiments indicate that it will be successful.

Acknowledgements

We are grateful to our colleagues who shared their results of unpublished research and who provided illustrations that are included in this chapter.

VIII. References

Ammirato, P. V., 1983: Embryogenesis. In: Evans, D. A., Sharp, W. R., Ammirato, P. V., Yamada, Y. (eds.), Handbook of plant cell culture (Vol. 1), pp. 82—123. New York: Macmillan Publishing Co.

Bar-Sagi, D., Feramisco, J. R., 1985: Microinjection of the *ras* oncogene protein into PC12 cells induces morphological differentiation. Cell **42**, 841—848.

Berkenstam, A., Ahlberg, J., Glaumann, H., 1986: Lysosomal uptake of isolated cell organelles microinjected into Hela cells. Exp. Cell Res. **163**, 301—308.

Brinster, R. L., Chen, H. Y., Trumbauer, M. E., Yagle, M. K., Palmiter, R. D., 1985: Factors affecting the efficiency of introducing foreign DNA into mice by microinjecting eggs. Proc. Natl. Acad. Sci., U.S.A. **82**, 4438—4442.

Capecchi, M. R., 1980: High efficiency transformation by direct microinjection of DNA into cultured mammalian cells. Cell **22**, 479—488.

Celis, J. E., 1984: Microinjection of somatic cells with micropipettes: comparison with other transfer techniques. Biochem. J. **223**, 281—291.

Celis, J. E., Graessman, A., Loyter, A., 1986: Microinjection and organelle transplantation techniques. London - Orlando: Academic Press.

Crossway, A., Oakes, J. V., Irvine, J. M., Ward, B., Knauf, V. C., Shewmaker, C. K, 1986: Integration of foreign DNA following microinjection of tobacco mesophyll protoplasts. Mol. Gen. Genet. **202**, 179—185.

Deshayes, A., Herrera-Estrella, L., Caboche, M., 1985: Liposome-mediated transformation of tobacco mesophyll protoplasts by an *Escherichia coli* plasmid. EMBO J. **4**, 2731—2737.

Diacumakos, E. G., 1973: Methods for micromanipulation of human somatic cells in culture. Methods Cell Biol. **7**, 287—311.

Dunwell, J. M., 1985: Anther and ovary culture. In: Bright, S. W. J., Jones, M. G. K. (eds.), Cereal tissue and cell culture (Advances in Agricultural Biotechnology), pp. 1—44. Dordrecht - Boston - Lancaster: Martinus Nijhoff - Dr. W. Junk Publishers.

Evans, D. A., Bravo, J. E., 1983: Protoplast isolation and culture. In: Evans, D. A., Sharp, W. R., Ammirato, P. V., Yamada, Y. (eds.), Handbook of plant cell culture, pp. 124—176. New York: Macmillan Publishing Co.

Firoozabady, E., Galbraith, D. W., 1984: Presence of a plant cell wall is not required for transformation of *Nicotiana* by *Agrobacterium tumefaciens.* Plant Cell Tissue Organ Culture **3**, 175—188.

Flavell, R., Mathias, R., 1984: Prospects for transforming monocot crop plants. Nature (London) **307**, 108—109.

Fraley, R. T., Rogers, S. G., Horsch, R. B., Eichholtz, D. A., Flick, J. S., Fink, C. L., Hoffman, N. L., Sanders, P. R., 1985: The SEV system: A new disarmed Ti plasmid vector system for plant transformation. Bio/Technology **3**, 629—635.

Fraley, R. T., Rogers, S. G., Horsch, R. B., Sanders, P. R., Flick, J. S., Adams, S. P., Bittner, M. L., Brand, L. A., Fink, C. L., Fry, J. S., Gallupi, G. R., Goldberg, S. B., Hoffmann, N. L., Woo, S. C., 1983: Expression of bacterial genes in plant cells. Proc. Natl. Acad. Sci., U.S.A. **80**, 4803—4807.

Goodwin, P. B., Erwee, 1984: Intercellular transport studied by microinjection methods. In: Robards, A. W. (ed.), Botanical Microscopy. 1985, pp. 335—358. Oxford - New York - Tokyo: Oxford University Press.

Gordon, J. W., Scangos, G. A., Plotkin, D. J., Barbosa, J. A., Ruddle, F. H., 1980: Genetic transformation of mouse embryos by microinjection of purified DNA. Proc. Natl. Acad. Sci. U.S.A. **77**, 7380—7384.

Gordon-Kamm, W. J., Bushnell, W. R., 1985: Intranuclear microinjection of isolated protoplasts using differential interference contrast microscopy. (Abstracts of the first international congress of plant molecular biology, Savannah 1985), pp. 109. Athens: The University of Georgia Center for Continuing Education.

Gould, A. R., Ashmore, S. E., 1982: Interaction of purified DNA with plant protoplasts of different cell cycle stage: The concept of a competent phase for plant cell transformation. Theor. Appl. Genet. **64**, 7—12.

Graessman, A., Graessman, M., Mueller, C., 1980a: Microinjection of early SV40 DNA fragments and T antigen. Methods Enzymol. **65**, 816—825.

Graessmann, A., Wolf, H., Bornkamm, G. W., 1980b: Expression of Epstein-Barr viral genes in different cell types after microinjection of viral DNA. Proc. Natl. Acad. Sci., U.S.A. **77**, 433—436.

Gurdon, J. B., Lane, C. D., Woodland, H. R., Marbaix, G., 1971: Use of frog eggs and oocytes for the study of messenger RNA and its translation in living cells. Nature (London) **233**, 177—182.

Halliwell, R. S., Gazaway, W. S., 1975: Quantity of microinjected tobacco mosaic virus required for infection of single cultured tobacco cells. Virology **65**, 583—587.

Hammer, R. E., Pursel, V. G., Rexroad, C. E., Wall, R. J., Bolt, D. J., Ebert, K. M., Palmiter, R. D., Brinster, R. L., 1985: Production of transgenic rabbits, sheep and pigs by microinjection. Nature (London) **315**, 680—683.

Heberle-Bors, E., 1985: In vitro haploid formation from pollen: a critical review. Theor. Appl. Genet. **71**, 361—374.

Herrera-Estrella, L., DeBlock, M., Messens, E., Hernalsteens, J.-P., Van Montagu, M., Schell, J., 1983: Chimeric genes as dominant selectable markers in plant cells. EMBO J. **2**, 987—995.

Holbrook, L. A., Reich, T. J., Iyer, V. N., Haffner, M., Miki, B. L., 1985: Induction of efficient cell division in alfalfa protoplasts. Plant Cell Rep. **4**, 229—232.

Jones, M. G. K., 1985: Cereal protoplasts. In: Bright, S. W. J., Jones, M. G. K. (eds.), Cereal tissue and cell culture (Advances in Agricultural biotechnology), pp. 204—230. Dordrecht - Boston - Lancaster: Martinus Nijhoff - Dr. W. Junk Publishers.

Kreis, T. E., Birchmeier, W., 1982: Microinjection of fluorescently labelled proteins into living cells with emphasis on cytoskeletal proteins. Int. Rev. Cytol. **75**, 209—227.

Krens, F. A., Mans, R. M. W., van Slogteren, T. M. S., Hoge, J. H. C., Wullems, G. J., Schilperoort, R. A., 1985: Structure and expression of DNA transferred to tobacco via transformation of protoplasts with Ti plasmids DNA: co-transfer of T-DNA and non T-DNA sequences. Plant Mol. Biol. **5**, 223—234.

Krens, F. A., Molendijk, L., Wullems, G. J., Schilperoort, R. A., 1982: *In vitro* transformation of plant protoplasts with Ti-plasmid DNA. Nature (London) **296**, 72—74.

Kressmann, A. Clarkson, S. G., Telford, J. L., Birnstiel, M. L., 1978: Transcription of *Xenopus* $tDNA_1^{met}$ and sea urchin histone DNA injected into the *Xenopus* oocyte nucleus. Cold Spring Harbor Symp. Quant. Biol. **42**, 1077—1082.

Lawrence, W. A., Davies, D. R., 1985: A method for the microinjection and culture of protoplasts at very low densities. Plant Cell Rep. **4**, 33—35.

Lin, P.-F., Ruddle, F. H., 1981: Photoengraving of coverslips and slides to facilitate monitoring of micromanipulated cells or chromosome spreads. Exp. Cell Res. **134**, 485—488.

Maddock, S. E., 1985: Cell culture, somatic embryogenesis and plant regeneration in wheat, barley, oats, rye and triticale. In: Bright, S. W. J., Jones, M. G. K. (eds.), Cereal tissue and cell culture (Advances in agricultural biotechnology), pp. 131—174. Dordrecht - Boston - Lancaster: Martinus Nijhoff - Dr. W. Junk Publishers.

Mangeat, P. H., Burridge, K., 1985: Microinjection and immunofluorescence microscopy as tools to study cytoskeletal organization in tissue culture cells. Techn. Immunocytochem. **3**, 79—96.

Mariotti, D., Davey, M. R., Draper, J., Freeman, J. P., Cocking, E. C., 1984: Crown gall tumorigenesis in the forage legume *Medicago sativa* L. Plant Cell Physiol. **25**, 473—482.

Matthews, J. A., Brown, J. W. S., Hall, T. C., 1981: Phaseolin mRNA is translated to yield glycosylated polypeptides in *Xenopus* oocytes. Nature (London) **294**, 175—176.

Meyer, P., Walgenbach, E., Bussmann, K., Hombrecher, G., Saedler, H., 1985: Synchronized tobacco protoplasts are efficiently transformed by DNA. Mol. Gen. Genet. **201**, 513—518.

Morikawa, H., Yamada, Y., 1985: Capillary microinjection into protoplasts and intranuclear localization of injected materials. Plant Cell Physiol. **26**, 229—236.

Nims, R. C., Halliwell, R. S., Rosberg, D. W., 1967a: Disease development in cultured cells of *Nicotiana tabacum* L. var. Samsun NN injected with tobacco mosaic virus. Cytologia **32**, 224—235.

Nims, R. C., Halliwell, R. S., Rosberg, D. W., 1967b: Wound healing in cultured tobacco cells following microinjection. Protoplasma **64**, 305—314.

Palevitz, B. A., Hepler, P. K., 1985: Changes in dye coupling of stomatal cells of *Allium* and *Commelina* demonstrated by microinjection of Lucifer yellow. Planta **164**, 473—479.

Paszkowski, J., Shillito, R. D., Saul, M., Mandák, V., Hohn, T., Hohn, B., Potrykus, I., 1984: Direct gene transfer to plants. EMBO J. **3**, 2717—2722.

Reich, T. J., Iyer, V. N., Haffner, M., Holbrook, L. A., Miki, B. L., 1986a: The use of fluorescent dyes in the microinjection of alfalfa protoplasts. Can. J. Bot. **64**, 1259—1267.

Reich, T. J., Iyer, V. N., Miki, B. L., 1986b: Efficient transformation of alfalfa protoplasts by the intranuclear microinjection of Ti plasmids. Bio/Technology **4**, 1001—1004.

Reich, T. J., Iyer, V. N., Scobie, B., Miki, B. L., 1986c: A detailed procedure for the intranuclear microinjection of plant protoplasts. Can. J. Bot. **64**, 1255—1258.

Rubin, G. M., Spradling, A. C., 1982: Genetic transformation of *Drosophila* with transposable element vectors. Science **218**, 348—353.

Rusconi, A., Schaffner, W., 1981: Transformation of frog embryos with a rabbit β-globin gene. Proc. Natl. Acad. Sci., U.S.A. **78**, 5051—5055.

Shillito, R. D., Paszkowski, J., Potrykus, I., 1983: Agarose plating and a bead type culture technique enable and stimulate development of protoplast-derived colonies in a number of plant species. Plant Cell Rep. **2**, 244—247.

Shillito, R. D., Saul, M. W., Paszkowski, J., Müller, M., Potrykus, I., 1986: High efficiency direct gene transfer to plants. Bio/Technology **3**, 1099—1103.

Steinbiss, H.-H., Stabel, P., 1983: Protoplast derived tobacco cells can survive capillary microinjection of the fluorescent dye Lucifer yellow. Protoplasma **116**, 223—227.

Steinbiss, H.-H., Stabel, P., Töpfer, R., Hirtz, R. D., Schell, J., 1984: Transformation of plant cells by microinjection of DNA. In: Experimental manipulation of ovule tissue: their manipulation, tissue culture and physiology (Proceedings of the 1984 Wye International Symposium, Wye College, University of London, 1984), pp. 64—75. London: England.

Van Lijsebettens, M., Inzé, D., Schell, J., Van Montagu, M., 1986: Transformed cell clones as a tool to study T-DNA integration mediated by *Agrobacterium tumefaciens*. J. Mol. Biol. **188**, 129—145.

Wagner, T. E., Hoppe, P. C., Jollick, J. D., Scholl, D. R., Hodinka, R. L., Gault, J. B., 1981: Microinjection of a rabbit β-globin gene into zygotes and its subsequent expression in adult mice and their offspring. Proc. Natl. Acad. Sci., U.S.A. **78**, 6376—6380.

Wehland, J., Osborn, M., Weber, K., 1980: Phalloidin associates with microfilaments after microinjection into tissue culture cells. Eur. J. Cell Biol. **21**, 188—194.

Wehland, J., Weber, K., 1981: Actin rearrangement in living cells revealed by microinjection of a fluorescent phalloidin derivative. Eur. J. Cell Biol. **24**, 176—183.

Zhou, G.-Y., Weng, J., Zeng, Y., Huang, J., Qian, S., Liu, G., 1983: Introduction of exogenous DNA into cotton embryos. Methods Enzymol. **101**, 433—481.

Chapter 11

Transformation of *Chlamydomonas Reinhardtii*

J.-D. Rochaix

Departments of Molecular Biology and Plant Biology, University of Geneva, CH-1211 Geneva 4, Switzerland

With 5 Figures

Contents

I. Introduction

A great deal of progress has been achieved during the past years in the transformation of higher plants. In spite of the impressive advances in this field, a major drawback remains the slow growth of these organisms. In most cases it still takes several weeks or even months to regenerate a plant. While higher plants are uniquely suited for developmental studies, lower plants, especially unicellular algae, offer attractive possibilities for investigating basic features of plants such as photosynthesis and chloroplast-nucleocytoplasmic interactions. These unicellular organisms grow rapidly and some of them can be manipulated with ease both at the biochemical and genetic level. However, transformation of lower plants is still in its infancy and only limited success has been reported. This article will provide a progress report on the transformation of *Chlamydomonas reinhardtii*. A short review on this transformation system was published earlier (Rochaix, 1983).

C. reinhardtii is a green unicellular heterothallic alga containing a single chloroplast. Cells of mating-type (mt) + or − can be grown vegetatively. After transfer to a medium lacking a reduced nitrogen source, the vegetative cells differentiate into gametes (Fig. 1). Gametes of opposite mating-type fuse to form a zygote in which nuclear fusion is followed by chloroplast fusion, thus allowing for chloroplast DNA recombination to occur. After meiosis four daughter cells are produced which can initiate a new cycle. While nuclear genes are transmitted according to mendelian rules, chloroplast genes are generally inherited from the mt^+ parent. In rare cases exceptional zygotes arise which transmit chloroplast genes of both parents to the offspring. These zygotes allow one to study chloroplast DNA recombination.

A large collection of nuclear and chloroplast mutants have been isolated from *C. reinhardtii* (cf. Gillham [1978] for an extensive discussion of *C. reinhardtii* genetics and Harris [1982] for list of available nuclear mutants). Nuclear mutants include ribosomal, flagellar, auxotrophic, drug resistant and photosynthetic mutants of which several have been mapped on the 18 linkage groups. Chloroplast mutants comprise mostly ribosomal mutants (including drug-resistant mutants) and photosynthetic mutants. Several of these chloroplast mutations have been localized on the physical map of the chloroplast genome (Fig. 2, Table I) which consists of 190 kb DNA circles (Rochaix, 1978). At the present time the ribosomal RNA genes (Rochaix and Malnoe, 1978), 9 tRNA genes (Bergmann *et al.*, 1985) and 10 protein genes (cf. Fig. 2) have been mapped on this chloroplast DNA. It is important to note that in contrast to most higher plants, photosynthetic function is dispensable in *C. reinhardtii* when the cells are grown on a reduced carbon source such as acetate. It is therefore relatively easy to isolate and to maintain photosynthetic mutants.

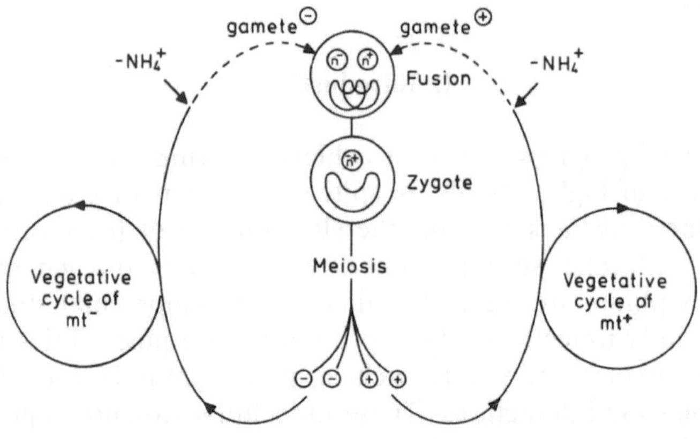

Fig. 1. Life cycle of *Chlamydomonas reinhardtii*. mt, mating type; n, nucleus (from Rochaix, 1981, with permission)

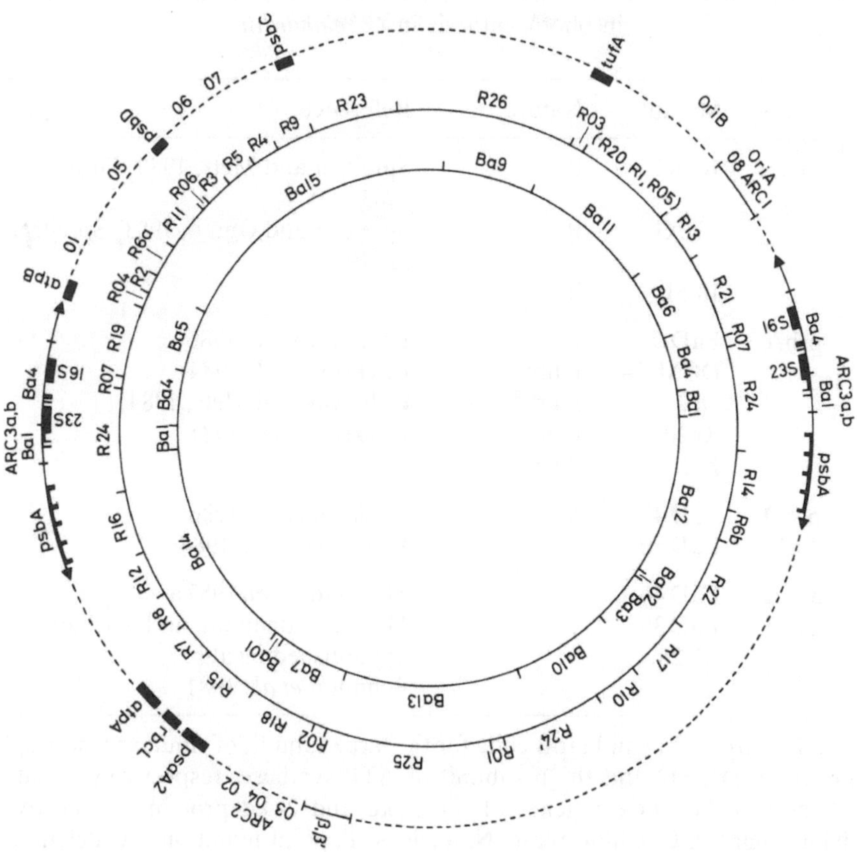

Fig. 2. Restriction map of the chloroplast genome of *Chlamydomonas reinhardtii*. The two inner circles represent the BamHI (Ba) and EcoRI (R) fragments (Rochaix, 1978). The two segments of the inverted repeat containing the ribosomal RNA (Rochaix and Malnoe, 1978) and *psbA* genes coding for the photosystem II D1 protein (Erickson *et al.*, 1984) are drawn on the outside and bounded by arrows. Introns are drawn in thinner lines relative to the coding sequences. Characterized genes in the single copy region are marked: *psaA2* (apoprotein A2 of P700; Schneider *et al.*, unpublished results), *rbcL* (large subunit of ribulose 1,5-bisphosphate carboxylase/oxygenase; Dron *et al.*, 1982); *atp A, B* (α, β subunits of ATP synthase; Woessner *et al.*, 1984), *psbD* (photosystem II D2 protein; Rochaix *et al.*, 1984; Erickson *et al.*, 1986) *psbC* (photosystem II 47 kd protein, Kuchka *et al.*, unpublished results), *tuf* A (elongation factor EF-Tu; Watson and Surzycki, 1982). β, β′ refer to sequences homologous to the genes of the β and β′ subunits of *E. coli* RNA polymerase (Watson and Surzycki, 1983). The eight identified chloroplast ARS sequences are indicated by 01 to 08 (Vallet *et al.*, 1984; Loppes and Denis, 1983; Vallet and Rochaix, 1985). The four chloroplast DNA sequences promoting autonomous replication in *C. reinhardtii* are marked by ARC1, ARC2, ARC3a, b (Rochaix *et al.*, 1984). Two authentic origins of replication are indicated by OriA and OriB (Waddell *et al.*, 1984)

Table I. Characterized mutations affecting chloroplast and nuclear genes involved in photosynthesis in *C. reinhardtii.*

Site	Locus	Mutants	Nature	Reference
C	rbcL	10-6C	P	Spreitzer and Mets, 1980; Dron *et al.*, 1983
		18-7G	P	Spreitzer and Ogren, 1983; Spreitzer *et al.*, 1985
		18-5B	P	
C	psbA	FuD-7	Δ	Bennoun *et al.*, 1986
		DCMU4	P hrb^R	Erickson *et al.*, 1984
		Dr2	P hrb^R	Galloway and Mets, 1984;
		Ar207	P hrb^R	Erickson *et al.*, 1985
		Br2	P hrb^R	
C	psbD	FuD47	∇	Erickson *et al.*, 1986
C	atpB	FuD50	Δ	Woessner *et al.*, 1984
N	33kd	FuD44	∇	Mayfield *et. al.* 1987a
N	24kd	FuD39	?	Mayfield, Bennoun and Rochaix, unpublished results
		BF25	?	Bennoun *et al.*, 1981

The rbcL, psbA, psbD and atpB code for the large subunit of ribulose bisphosphate carboxylase, D1, D2 and the β subunit of ATP synthase, respectively. 33kd and 24kd refer to the nuclear genes of the 33kd and 24kd proteins of the oxygen evolving complex. C, Chloroplast; N, nucleus; P, point mutation; Δ, deletion; ∇, insertion; hrb^R, mutation conferring herbicide resistance.

Because of its easy growth and of its extensively developed genetics, *C. reinhardtii* is an interesting model system for studying basic cellular problems such as flagellar assembly and chloroplast biogenesis.

The development of an efficient transformation system is an important task. First, such a system would allow one to isolate genes by complementing defined mutations with wild-type genomic libraries. Second, it would be a powerful tool for understanding gene function and regulation by inserting modified genes or regulatory elements into cells. Third, it could be used for studying the expression of higher plant genes.

II. Nuclear Transformation

The first step in the transformation of *C. reinhardtii* is to remove the cell wall. This can be achieved by incubating vegetative cells with autolysin, an extract which can be easily obtained from a zygotic supernatant (Claes, 1971). Alternatively the cell wall deficient strain cw15 (Davies and Plaskitt, 1971) can be used. It behaves like a natural protoplast although it is less sensitive to osmoticum.

The second step is the delivery of DNA to the cells. We have used several methods in which the cells were treated with either poly-L-ornithine, polyethylene glycol, chloroquine, Ca-phosphate, dimethylsulfoxide or subjected to electroporation. Most of these methods produce similar transformation yields although we have obtained more reproducible results with poly-L-ornithine (Rochaix and van Dillewijn, 1982).

A. Selection

i) ARG 7 Locus

The ARG7 locus was the first nuclear locus of *C. reinhardtii* to be characterized at the genetic and biochemical level. Several mutations have been isolated at this locus (Gillham, 1965; Loppes *et al.*, 1972; Loppes and Matagne, 1972) that codes for argininosuccinate lyase, the last enzyme of the arginine biosynthetic pathway which converts arginino succinate into arginine and fumarate. This locus corresponds to the yeast ARG4 and *E. coli* argH loci. Because it has not yet been possible to clone the *C. reinhardtii* ARG7 locus, the ARG4 gene of yeast in the plasmid pYe arg4 (Clarke and Carbon, 1978) was used to transform a cw15 arg7 double mutant (Rochaix and van Dillewijn, 1982). Arginine prototrophs were obtained at a frequency ranging between 10^{-6} and 10^{-7}. It could be shown that the yeast DNA was stably integrated in the nuclear genome of *C. reinhardtii.* However the transformation yield was only slightly above the reversion rate of the arg7 mutation.

ii) Resistance to Kanamycin

Since the *C. reinhardtii* cell wall deficient strain is sensitive to kanamycin at concentrations as low as 5 µg/ml in liquid medium, drug resistance should be a valuable selective marker for transformation. We have used several plasmid constructions as transforming DNA in which the gene of neophosphotransferase II (NPTII), an enzyme that inactivates kanamycin, is under the control of the promoter of either of the following genes: those of β tubulin and the small subunit of ribulose 1,5 bisphosphate carboxylase of *C. reinhardtii,* of nopaline synthase from the T-DNA of *Agrobacterium tumefaciens* and of gene VI from cauliflower mosaic virus. Although our transformation attempts have yielded clones resistant to high levels of kanamycin (up to 100 µg/ml in liquid medium) no NPTII gene sequences could be detected in these drug resistant cells.

However two groups have recently reported successful transformation of *C. reinhardtii* by using kanamycin resistance as selection. Hasnain *et al.* (1985) used a vector containing the origin of replication of the yeast 2 µ circle and the NPTII gene under the control of the SV40 early promoter. They obtained kanamycin resistant colonies at a frequency of 10^{-6} of which about two thirds were shown to contain free plasmids undistinguishable from the original plasmid. By transforming *E. coli* with the DNA of the transformants the plasmids could be recovered. Transcripts of the plasmids were detected on dot blots but no attempts to size the RNA products were made. The transformants were stable for more than 200 generations. Similarly Cox *et al.* (1985) were also able to transform *C. reinhardtii* to kanamycin resistance by using a plasmid containing a chloroplast ARS sequence and the NPT II gene with its own bacterial promoter. The plasmid of the transformants could be recovered in *E. coli* and it was shown to have suffered a small deletion. In both of these studies no attempt was made to assay for NPT II activity in the transformants.

It is noteworthy that *C. reinhardtii* nuclear DNA contains multiple regions homologous to the bacterial transposon Tn 5. A 8.1 kb nuclear DNA fragment was cloned and characterized that includes regions that are 70 to 80% homologous to three distinct sequences within the unique sequence of Tn 5 and to one sequence from the inverted repeats of Tn 5 (R. Ausich, personal communication). A stable transcript originating from this nuclear region could be detected. These findings are intriguing but their significance is not clear.

iii) Other Selective Markers

Sequencing of several nuclear genes of *C. reinhardtii* has revealed that the codon usage is surprisingly restricted and biased towards C and G in the wobble position of the codons (Table II). It is especially noteworthy that only one codon is used among the six codons of leucine. Therefore the possibility that heterologous selective marker genes are poorly expressed has to be considered. It is important to isolate genes for which mutations are known to exist and that could be used for selection in the transformations. Only a few nuclear genes of *C. reinhardtii* have been isolated and characterized: those encoding tha α and β tubulin genes (Brunke *et al.,* 1984; Silflow and Rosenbaum 1981; Youngblom *et al.,* 1984) the small subunit of ribulose 1,5 bisphosphate carboxylase (Goldschmidt-Clermont, 1986) and the chlorophyll a/b binding protein (Sheperd *et al.,* 1983). Unfortunately no mutation has yet been reported in these genes. On the other hand the locus coding for arginino succinate lyase where several mutations have been characterized (Loppes *et al.,* 1972) has not yet been cloned.

Recently the three nuclear genes coding for the proteins of 33, 24 and 18 kd of the oxygen evolving complex of photosystem II have been cloned from *C. reinhardtii* (Mayfield *et al.,* 1987 a, b). Three nuclear mutants with lesions on the oxidizing side of photosystem II were isolated by Bennoun *et al.* (1981, unpublished results). Two of them, FUD 39 and BF 25 appear to be specifically affected in the gene of the 24 kd protein while the third, FuD 44, is missing the 33 kd protein and has an alteration in its gene. All three mutants are unable to grow on medium lacking a reduced carbon source (minimal medium). These are the first cloned nuclear genes of *C. reinhardtii* for which mutations exist. Experiments are in progress to use the wild-type alleles of these genes for transforming the mutants using growth on minimal medium as selection. Table I provides a list of nuclear mutations which will be of interest for devising selection schemes for transformation.

Table II. Codon Usage in Nuclear Genes of *Chlamydomonas reinhardtii*

Phe		Ser		Tyr		Cys	
UUU	0	UCU	9	UAU	0	UGU	0
UUC	100	UCC	45	UAC	100	UGC	100
		UCA	1				
Leu		UCG	28	His		Trp	
UUA	0	AGU	0	CAU	10	UGG	100
UUG	0	AGC	17	CAC	90		
CUU	0					Arg	
CUC	0	Pro		Gln		CGU	7
CUA	0	CCU	4	CAA	0	CGC	93
CUG	100	CCC	82	CAG	100	CGA	0
		CCA	0			CGG	0
Ile		CCG	14	Asn		AGA	0
AUU	14			AAU	0	AGG	0
AUC	86	Thr		AAC	100		
AUA	0	ACU	11			Gly	
		ACC	86	Lys		GCU	15
Met		ACA	0	AAA	0	GGC	85
AUG	100	ACG	3	AAG	100	GGA	0
						GGG	0
Val		Ala		Asp			
GUU	16	GCU	19	GAU	14	Stop	
GUC	31	GCC	76	GAC	86	UAA	100
GUA	15	GCA	0			UAG	0
GUG	38	GCG	5	Glu		UGA	0
				GAA	0		
				GAG	100		

The percentage of codons used is based on the nucleotide sequences of the genes of β1 and β2 tubulin (Youngblom *et al.*, 1984) of the two genes of the small subunit of ribulose 1,5 bisphosphate carboxylase (Goldschmidt-Clermont, unpublished results) and the gene of the 24 kd protein from the oxygen evolving complex (Mayfield and Rochaix, unpublished results).

B. ARS Sequences of C. reinhardtii

ARS sequences have been defined as DNA sequences that promote autonomous replication in yeast (Struhl *et al.*, 1979). Plasmids carrying ARS sequences transform yeast at a high frequency and they are unstable in mitosis and meiosis (Stinchcomb *et al.*, 1979; Hsiao and Carbon, 1979). Although it has not yet been proven in yeast that ARS elements are authentic origins of DNA replication, several observations suggest that there may be a close relationship between the two. ARS sequences occur at a frequency in the yeast genome which is similar to that found for origins of DNA replication (Beach *et al.*, 1980; Chan and Tye, 1980; Newlon and Burke, 1980). There appears to be an overlap between the ARS site of yeast 2 μ circles and the origin of replication (Broach *et al.*, 1980; Celniker and Campbell, 1982; Newton *et al.*, 1981; Kojo *et al.*, 1981).

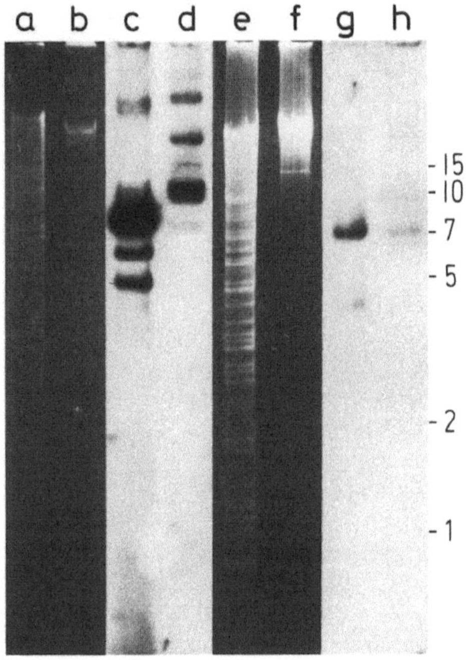

Fig. 3. Autonomously replicating plasmids in *C. reinhardtii.* Agrose gel electrophoretic fractionations of EcoRI digested DNA (lanes a and e) and undigested DNA (lanes b and f), extracted 25 generations (lanes a and b) and 60 generations (lanes e and f) after transformation. Autoradiograms of Southern hybridizations of the DNAs in lanes a, b, e and f with ^{32}P-labeled vector plasmid DNA are shown in lanes c, d, g and h, respectively. Numbers on the right indicate the size of the DNA fragments in kb (from Rochaix *et al.,* 1984, with permission)

Mitochondrial DNA fragments from several organisms have been shown to contain ARS elements (Zakian, 1981; Blanc and Dujon, 1980; Hyman *et al.,* 1982; Tudzynski and Esser, 1983) although this property is not shared by all mitochondrial DNAs (Zakian and Kupfer, 1982). When total DNA from *C. reinhardtii* was tested for ARS activity by inserting restriction fragments into a plasmid which contained the yeast ARG4 gene as a selective marker but was unable to replicate in yeast, most ARS elements identified after transformation of yeast were of chloroplast origin and only two nuclear ARS elements were isolated (Rochaix *et al.,* 1983; Loppes and Denis, 1983; Vallet *et al.,* 1984). At least eight distinct ARS elements have been mapped on the chloroplast genome of *C. reinhardtii* (Fig. 2). Four of these ARS regions comprising between 400 and 2300 bp have been sequenced and shown to have a high AT content, numerous short direct and inverted repeats and at least one element that is related to the yeast ARS core consensus sequence 5′ A/T TTT ATPu TTT A/T (Broach *et al.,* 1982). Recently two authentic origins of replication have been localized on the *C. reinhardtii* chloroplast genome (Waddell *et al.,* 1983). It is interesting to note that one ARS element (08 in Fig. 2) is

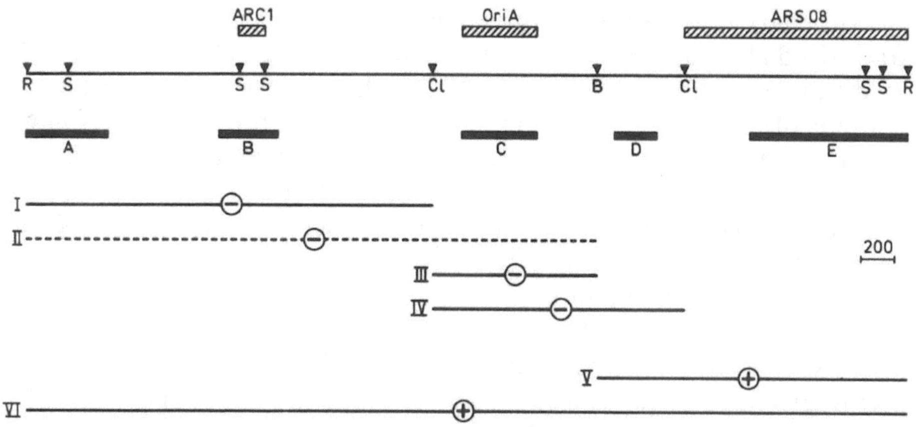

Fig. 4. Chloroplast ARS and ARC sequences are distinct from one origin of DNA replication. Restrictions sites of the chloroplast EcoRI fragment R13 are indicated by R, EcoRI; S, Sau3A; Cl, ClaI; B, BamHI. Fragments able and unable to promote autonomous replication in yeast are marked with + and −, respectively. Homologous regions between *C. reinhardtii* and *Chlamydomonas* strain WXM are indicated by A, B, C, D, E. Region C contains the origin of replication OriA (Wang *et al.*, 1984). Region B contains the ARC1 sequence. The ARS 08 sequence is located on the 1.3 kb ClaI-EcoRI fragment (from Vallet and Rochaix, 1985, with permission)

localized on the same EcoRI fragment R13 as the origin oriA. However fine structure mapping of this element has revealed that the ARS site is distinct from the origin of replication (Fig. 4, Vallet and Rochaix, 1984). Similar results have been obtained with the region containing the origin of chloroplast DNA replication of *Euglena gracilis* which is unable to promote autonomous replication in yeast (Vallet and Rochaix, 1984). It appears therefore that chloroplast ARS sequences are unrelated to authentic chloroplast origins of replication. The frequent occurrence of ARS elements in the chloroplast genome of *C. reinhardtii* may simply be a consequence of the high AT content of this DNA and of the yeast ARS consensus sequence.

C. ARC Sequences

Since ARS elements appear to be distinct from authentic origins of DNA replication, at least in the case of the chloroplast DNA, it may be more promising to look for autonomously replicating plasmids. No free plasmids have been found in this alga except for chloroplast and mitochondrial DNA. We have therefore transformed *C. reinhardtii* with the same plasmid pools that were used for the isolation of ARS elements in yeast. The presence of free plasmids in the transformants was detected by hybridizing the DNA of the transformants with a labelled vector-specific probe (Fig. 4). These ARC plasmids (*a*utonomous *r*eplication in *C*hlamydomonas) could be recovered by transforming *E. coli* with the DNA of the transformants (Rochaix *et al.*, 1984). Surprisingly, among four plasmids that were characterized all hybridized exclusively with chloroplast DNA. The hybridizing regions are shown on the map of the chloroplast genome in Fig. 2. The inserts of these plasmids of 415, 257, 153 and 102 bp, respectively have been sequenced and in at least two cases these sequences were undistinguishable from the corresponding cloned chloroplast DNA sequences indicating that no sequence rearrangements have occurred during passage of these plasmids in *C. reinhardtii.* These ARC regions contain short direct and inverted repeats and two semi-conserved AT rich elements of 19 and 12 bp (Rochaix *et al.*, 1984). Only one of the ARC sequences (ARC 2 in Fig. 2) also displays ARS activity. Interestingly the ARCI element maps on the EcoRI fragment R 13, 1.5 kb away from the authentic origin of DNA replication oriA (Vallet & Rochaix, 1984). Comparison of the R 13 regions of two strains of *C. reinhardtii* has revealed five conserved DNA stretches (Wang *et al.*, 1984). It is noteworthy that both oriA and ARCI coincide with two of these conserved regions (Fig. 4). Whether the ARCI element exerts some cis effect on the origin remains an open question. Although the ARC sequences are of chloroplast origin, it is likely that the plasmids are localized in the nucleus. This point has not been explored because of the difficulty of obtaining a clean nuclear and chloroplast fractionation.

D. *Natural Plasmids*

It was recently reported that the ribosomal DNA of *Physarum polycephalum,* which exists as a 60 kb linear palindromic extrachromosomal molecule transforms yeast with high efficiency (Künzler, 1985) where it appears to replicate autonomously. The structure of this DNA has been characterized extensively (Ferris & Vogt, 1982, Fig. 5). These observations prompted us to use this ribosomal DNA for transformation in *C. reinhardtii.* After incubation of the cells with this DNA, they were plated on non-selective plates. Pools of cells were grown and their DNA was extracted and hybridized with the probe 103 (Fig. 5). Weak signals could be detected in some of the cell pools. We estimate that 0.1—1 % of the cells contained *Physarum* rDNA fragments. Co-Transformation with *Physarum* rDNA and plasmids containing the NPT II gene and selection for kanamycin resistant clones has also allowed us to recover cells that were positive for *Physarum* rDNA. In some cases the *Physarum* rDNA appears to have suffered extensive rearrangements and exists as 7 kb plasmids, in other cases it is considerably larger so that it is difficult to determine whether is is free or integrated in the nuclear genome (Fig. 5). However after prolonged growth the hybridization signals caused by the *Physarum* rDNA become fainter until they disappear completely. Obviously there is no selective pressure to keep this DNA in the cells. These results indicate that the limiting step in the transformation of *C. reinhardtii* is not the delivery of the DNA to the cells, but its stable maintenance. Apparently the *Physarum* rDNA has some properties (DNA modification and/or sequence organization) which convey protection against degradation in the cells. It

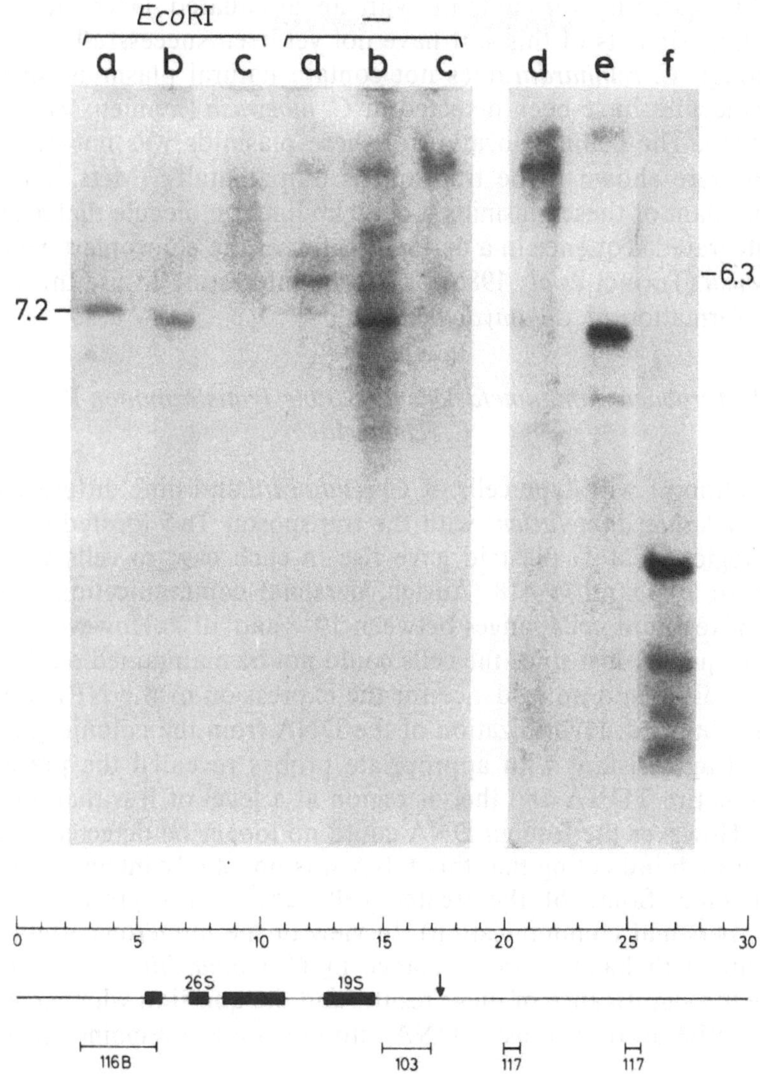

Fig. 5. Transformation of *C. reinhardtii* with *Physarum polycephalum* ribosomal DNA. The restriction map of the region corresponding to half of the *P. polycephalum* ribosomal DNA is shown in the lower part and includes the 19 S and 26 S ribosomal RNA genes and the transcription initiation site (marked by an arrow) established by Ferris and Vogt (1982). The scale is given in kb. Bars at the bottom indicate cloned fragments (116 B, 103, 117) that were used as probes. The two upper panels represent autoradiograms of Southern hybridizations of DNA from transformants electrophoresed on agarose gels and hybridized with the 103 probe. The a lanes contain EcoRI digested and undigested DNA from a single transformant that was obtained on kanamycin plates after cotransformation with a plasmid conferring kanamycin resistance and *P. polycephalum* ribosomal DNA. The DNA in the b lanes was obtained from a plate of cells grown to near confluence that had been treated with *P. polycephalum* ribosomal DNA. The c lanes contain DNA from non-transformed cells. Lanes d, e and f show similar hybridizations with undigested, Hind III and Tag I digested DNA, respectively, from a pool of 24 unselected clones obtained after treatment of these cells with *P. polycephalum* ribosomal DNA. Numbers on the left and the right indicate the size of the DNA fragments in kb

may therefore be possible to increase the transformation yield in *C. reinhardtii* by searching for mutants with an attenuated DNA degradation system. Experiments of this sort have not yet been successfull.

Although *C. reinhardtii* does not contain natural plasmids, small free DNA molecules have been detected in *C. moewusii* (Lemieux *et al.*, 1980; Mets, 1980). The cellular location of these plasmids was not determined but they were shown to be transmitted uniparentally (Mets, 1980). The most abundant of these plasmids is a 5.9 kb linear molecule that also exists as an integrated sequence in a defined region of the chloroplast genome of *C. moewusii* (Turmel *et al.*, 1986). It will be interesting to use this plasmid for transformation of *Chlamydomonas.*

E. Is Agrobacterium tumefaciens a Possible Transformation Vector for C. reinhardtii?

Cocultivation of wild-type cells of *C. reinhardtii* and nine different strains of *Agrobacterium tumefaciens* with the transposon Tn 5 located within the TDNA region of a Ti plasmid gave rise in each case to cells transiently resistant to 10 μg/ml G 418 (Ausich, personal communication). The frequency of resistant cells ranges between 10^{-2} and 10^{-3}. However the resistance was quickly lost since the cells could not be maintained on drug-containing medium and no evidence for the expression of the NPT II gene of Tn 5 was obtained. Hybridization of the DNA from the colonies that were initially drug resistant with appropriate probes revealed the presence of Tn 5, the entire TDNA and the vir region at a level of less than one copy per cell. However the foreign DNA could no longer be detected after prolonged growth indicating that this DNA was not stably integrated into the algal genome. Some of the treated cells could only grow in clusters. (Ausich, personal communication). In view of the numerous artefacts one can obtain with kanamycin resistance in *C. reinhardtii,* it is difficult to evaluate the significance of these results and the question whether *A. tumefaciens* can be used to transfer DNA into green algae remains open.

III. Prospects of Chloroplast Transformation in *C. reinhardtii*

It is well documented that chloroplasts possess a highly selective envelope consisting of an outer and inner membrane. At first sight it would appear that transfer of nucleic acids across this double membrane is a difficult task although sequences homologous to chloroplast DNA have been found in the mitochondrial and nuclear genomes of plants (Stern and Lonsdale, 1982; Timmis and Scott, 1983). These results strongly suggest that DNA transfers must have occurred between organellar and nucleocytoplasmic compartments. Whether these transfers continue to occur remains an open question. A recent report claiming chloroplast transformation in tobacco by *Agrobacterium tumefaciens* with a Ti plasmid containing the gene of chloramphenicol acetyltransferase (De Block *et al.*, 1985) has raised considerable interest and hope.

C. reinhardtii offers considerable advantages for attempts at chloroplast transformation. First, its large unique chloroplast may provide a reasonable target for microinjection of DNA into the organelle. Second, several chloroplast mutations have been characterized at the gene level. The *psbA* locus coding for the photosystem II D1 protein has been under intense investigation. Although the gene is located in the chloroplast inverted repeat and therefore present in two copies per genome (Erickson *et al.*, 1984), several mutants have been obtained that have double deletions covering both *psbA* genes (Bennoun *et al.*, 1986). The D1 protein also appears to be the target for s-triazine and urea type herbicides. In *C. reinhardtii* four distinct sites of the 352 amino acid D1 protein have been shown to be altered in mutants with different levels of resistance and cross-resistance to atrazine, diuron and bromacil (Galloway and Mets, 1984; Erickson *et al.*, 1985; Erickson, Mets and Rochaix, unpublished results). Attempts to transform the deletion mutants with *psbA* genes from the herbicide resistant mutants have not yet been successful. A mutation in another photosystem II gene, *psbD*, coding for the D2 polypeptide has recently been characterized (Erickson *et al.*, 1986). Because of the unstability of the mutation, it does not appear to be ideal for transformation experiments. Three point mutations have been characterized at the *rbcL* locus of *C. reinhardtii* (Spreitzer and Mets, 1980; Dron *et al.*, 1983; Spreitzer *et al.*, 1985). These mutations revert at a low frequency and may be appropriate for transformation. Deletions have been obtained at the *atpb* locus coding for the β subunit of ATP synthase (Myers *et al.*, 1982). However an alternate pathway for ATP synthesis can be switched on in these mutants which allows them to grow on minimal medium (P. Bennoun, personal communication). Table I provides a list of chloroplast mutations which have been characterized at the gene level. It is hoped that some of these mutants will be used successfully for chloroplast transformation.

IV. Conclusions

C. reinhardtii can be transformed with an efficiency around 10^{-6} transformants per treated cell using either arginine prototrophy or kanamycin resistance for selection. Foreign DNA can be stably integrated and autonomously replicating plasmids can be constructed although they do not markedly improve the transformation frequency. The preliminary results obtained with *Physarum* ribosomal DNA are encouraging since they suggest that much higher transformation yields can be reached. Clearly, the limiting step in the *C. reinhardtii* transformation is not the delivery of DNA to the cells, but the stable maintenance of the transforming DNA and/or its expression. All the reported transformations of *C. reinhardtii* have used heterologous genes as selective markers. Because of the unusual codon usage in the nuclear genes of this alga, foreign genes may be poorly expressed. Indeed no direct proof for the expression of the yeast ARG4 gene product or of the neophosphotransferase II has been provided. An urgent task for the future is to clone nuclear genes for which mutations have been isolated. The recent cloning of the nuclear genes of the oxygen evolving complex of *C. reinhardtii* where several mutations are known constitute a first step into this direction. Studies on *C. reinhardtii* have reached a stage where the biochemical, molecular and genetic approaches can be fruitfully coupled. Whether *C. reinhardtii* can qualify as the "green yeast" of plants will depend on how much its transformation system can be improved.

Acknowledgements

I am grateful to R. Ausich and P. Bennoun for communicating unpublished results. I thank J. Erickson and M. Goldschmidt-Clermont for critical reading of the manuscript. Part of this work was supported by grant 3.587.0.84 from the Swiss National Science Foundation.

V. References

Blanc, H., Dujon, B., 1981: Replication regions of the yeast mitochondrial DNA active *in vitro* and in yeast transformants. In: Slonimski *et al.* (eds.), Mitochondrial Genes, Cold Spring Harbor Laboratory Press N. Y., pp. 279—294.

Beach, D., Piper, M., Shall, S., 1980: Isolation of chromosomal origins of replication in yeast. Nature **284,** 185—187.

Bennoun, P., Diner, B. A., Wollman, F. A., Schmidt, G., Chua, N. H., 1981: Thylakoid polypeptides associated with photosystem II in *Chlamydomonas reinhardtii:* comparison of system II mutants and particles. In: Akoyunoglou, G. (ed.), Photosynthesis III. Structure and Molecular Organisation of the Photosynthetic apparatus. Balaban International Science Services, Philadelphia, PA, pp. 839—849.

Bennoun, P., Spierer-Herz, M., Erickson, J., Girard-Bascou, J., Pierre, Y., Delosme, M., Rochaix, J.-D., 1986: Characterization of photosystem II mutants of *Chlamydomonas reinhardtii* lacking the *psbA* gene. Plant Molec. Biol. **6,** 141—160.

Bergmann, P., Schneider, M., Burkard, G., Weil, J. H., Rochaix, J.-D., 1985: Transfer RNA Gene Mapping Studies on Chloroplast DNA from *Chlamydomonas reinhardtii.* Plant Science **39,** 133—140.

Broach, J. R., Li, Y. Y., Feldman, J., Jayaram, M., Abraham, J., Nasmyth, K. A., Hicks, J. B., 1980: Localization and Sequence Analysis of Yeast Origins of DNA replication. Cold Spring Harbor Symp. Quant. Biol. **47,** 1165—1173.

Brunke, K. J., Young, E. E., Buchbinder, B. U., Weeks, D. P., 1982: Coordinate regulation of the four tubulin genes of *Chlamydomonas reinhardtii.* Nucl. Acids Res. **10,** 1295—1310.

Celniker, S. E., Campbell, J. L., 1982: Yeast DNA replication *in Vitro:* Initiation and Elongation Events Mimic *in Vivo* Processes. Cell **31,** 201—203.

Chan, C. S. M., Tye, B. R. K., 1980: Autonomously replicating sequences in Saccharomyces cerevisiae. Proc. Natl. Acad. Sci., U.S.A. **77,** 6329—6333.

Claes, H., 1971: Autolysin of the cell wall from the gametes of *Chlamydomonas reinhardtii.* Arch. für Mikrobiologie **78,** 180—188.

Clarke, L., Carbon, J., 1978: Functional Expression of Cloned Yeast DNA in *Escherichia coli:* Specific Complementation of Argininosuccinate Lyase (argH) Mutations. J. Mol. Biol. **120,** 517—532.

Cox, J. C., Bingham, S. E., Bishop, R. J., 1985: Transformation of *Chlamydomonas reinhardtii* with plasmids conferring antibiotic (Kanamycin) resistance. Abstracts 1st Intl. Congr. Plant Molec. Biol. p. 115.

Davies, D. R., Plaskitt, A., 1971: Genetical and Structural Analysis of Cell Wall formation in *Chlamydomonas reinhardtii.* Genet. Res. **17,** 33—43.

De Block, M., Schell, J., Van Montagu, M., 1985: Chloroplast transformation by *Agrobacterium tumefaciens.* EMBO J. **4,** 1367—1372.

Dron, M., Rahire, M., Rochaix, J.-D., Mets, L., 1983: First DNA Sequence of a Chloroplast Mutation. A Missense Alteration in the Ribulosebisphosphate Carboxylase Large Subunit Gene. Plasmid **9,** 32—324.

Erickson, J. M., Rahire, M., Rochaix, J.-D., 1984: *Chlamydomonas reinhardtii* gene for the 32000 mol wt protein of photosystem II contains four large introns and is located entirely within the chloroplast inverted repeat. EMBO J. **3,** 2753—2762.

Erickson, J. M., Rahire, M., Rochaix, J.-D., Mets, L., 1985: Herbicide Resistance and Cross-Resistance: Changes at three distinct Sites in the Herbicide-Binding Protein. Science **228,** 204—207.

Erickson, J. M., Rahire, M., Malnoe, P., Girard-Bascou, J., Pierre, Y., Bennoun, P., Rochaix, J.-D., 1986: Lack of the D2 protein in a *Chlamydomonas reinhardtii* psbD mutant affects photosystem II stability and D1 expression. EMBO J., submitted.

Ferris, P. J., Vogt, V. M., 1982: Structure of the Central Spacer Region of Extrachromosomal Ribosomal DNA in *Physarum polycephalum.* J. Mol. Biol. **159,** 359—381.

Galloway, R., Mets, L., 1984: Atrazine, bromacil and diuron resistance in *Chlamydomonas:* A single non-mendelian genetic locus controls the structure of the thylakoid binding site. Plant Physiol. **74,** 469—474.

Gillham, N. W., 1965: Induction of chromosomal and non chromosomal mutations in *Chlamydomonas reinhardtii* with N-methyl-N'-Nitro-N-Nitrosoguanidine. Genetics **52,** 529—537.

Gillham, N. W., 1978: Organelle Heredity, Raven, New York.

Goldschmidt-Clermont, M., 1986: The two genes for the small subunit of RuBP Carboxylase/oxygenase are closely linked in *Chlamydomonas reinhardtii.* Plant Molec. Biol. **6,** 13—21.

Harris, E. H., 1982: Nuclear gene Loci of *Chlamydomonas reinhardtii.* In: O'Brien, S. J. (ed.), Genetic Maps, Vol. **2,** pp. 168—174, Laboratory of Viral Carcinogenesis, National Cancer Institute, NIH, Frederick, MD 21701.

Hasnain, S. E., Manavathu, E. K., Leung, W. L., 1985: DNA mediated Transformation of *Chlamydomonas reinhardtii* cells: Use of Aminoglycoside 3' — phosphotransferase as a selectable marker. Mol. Cell Biol. **5,** 3647—3650.

Hsiao, C., Carbon, J. M., 1979: High-frequency transformation of yeast by plasmids containing the cloned yeast ARG4 gene. Proc. Natl. Acad. Sci., U.S.A. **76,** 3829—3833.

Hyman, B. C., Cramer, J. H., Rownd, R. H., 1982: Properties of a *Saccharomyces cerevisiae* mtDNA segment conferring high-frequency yeast transformation. Proc. Natl. Acad. Sci., U.S.A. **79,** 1578—1582.

Koyo, H., Greenberg, B. D., Sugino, A., 1981: Yeast 2 μ plasmid DNA replication *in vitro:* Origin and direction. Proc. Natl. Acad. Sci., U.S.A. **78,** 7261—7265.

Künzler, P., 1985: The linear extrachromosomal DNA of *Physarum polycephalum* replicates and is maintained under non selective conditions in two different lower eukaryotes. Nucl. Acids Res. **13,** 1855—1869.

Lemieux, C., Turmel, M., Lee, R. W., 1980: Characterization of chloroplast DNA in *Chlamydomonas eugametes* and *C. moewusii* and its inheritance in hybrid progeny. Curr. Genet. **2,** 139—147.

Loppes, R., Matagne, R., Strijkert, P. J., 1972: Complementation at the ARG7 locus in *Chlamydomonas reinhardtii.* Heredity **28,** 239—251.

Loppes, R., Matagne, R., 1972: Allelic complementation between ARG7 mutants in *Chlamydomonas reinhardtii.* Genetica [The Hague] **43,** 422—430.

Loppes, R., Denis, C., 1983: Chloroplast and Nuclear DNA Fragments from *Chlamydomonas* Promoting High Frequency Transformation of Yeast. Curr. Genet. **7,** 473—480.

Mayfield, S. P., Bennoun, P., Rochaix, J.-D., 1987a: Expression of the nuclear encoded OEE1 protein is required for oxygen evolution and stability of photosystem II particles in *Chlamydomonas reinhardtii.* EMBO J. **6,** 313—318.

Mayfield, S. P., Rahire, M., Frank, G., Zuber, H., Rochaix, J.-D., 1987b: Expression of the nuclear gene encoding oxygen-evolving enhancer protein 2 is required for high levels of photosynthetic oxygen evolution in *Chlamydomonas reinhardtii.* Proc. Natl. Acad. Sci., U.S.A. **84,** 749—753.

Mets, L., 1980: Uniparental inheritance of chloroplast DNA sequences in interspecific hybrids of *Chlamydomonas.* Curr. Genet. **2,** 131—138.

Myers, A. M., Grant, D. M, Rabert, D. K., Harris, E. H., Boynton, J. E., Gillham, N. W., 1982: Mutants of *Chlamydomonas reinhardtii* with physical alterations in their chloroplast DNA. Plasmid **7,** 133—151.

Newlon, C. S., Burke, W., 1980: Replication of small chromosomal DNAs in yeast. In: Alberts, B., Fox, C. C. (eds.), ICN-UCLA Symposia on Molecular Cellular Biology **19B;** pp. 339—409, Academic Press, N. Y.

Newlon, C. S., Devenish, R. J., Suci, P. A., Roffis, C. J., 1981: Replication origins used *in vivo* in yeast. In: Rays, D. S., Fox, C. C. (eds.), ICN-UCLA Symposia on Molecular and Cellular Biology **22,** pp. 501—506, Academic Press, N. Y.

Rochaix, J.-D., 1978: Restriction Endonuclease Map of the Chloroplast DNA of *Chlamydomonas reinhardtii.* J. Mol. Biol. **126,** 597—617.

Rochaix, J.-D., 1981: Organization, function and expression of the chloroplast DNA of *Chlamydomonas reinhardtii.* Experientia **37,** 323—332.

Rochaix, J.-D., 1983: Transformation in the green Alga *Chlamydomonas reinhardtii.* In: Genetic Engineering Vol. 6. Setlow, J. K. and Hollaender, A. (eds.), Plenum Press, New York - London.

Rochaix, J.-D., Malnoe, P. M., 1978: Anatomy of the chloroplast ribosomal DNA of *Chlamydomonas reinhardtii.* Cell **15,** 661—670.

Rochaix, J.-D., van Dillewijn, J., 1982: Transformation of the green alga *Chlamydomonas reinhardtii* with yeast DNA. Nature **296,** 70—72.

Rochaix, J.-D., Dron, M., Schneider, M., Vallet, J. M., Erickson, J., 1983: *Chlamydomonas reinhardtii:* A model system for studying the biosynthesis of the photosynthetic apparatus. Downey *et al.* (eds.). In: Advances in Gene Technology: Molecular Genetics of Plants and Animals, Miami Winter Symposia, vol. 20, pp. 81—100, Academic Press.

Rochaix, J.-D., van Dillewijn, J., Rahire, M., 1984: Construction and Characterization of Autonomously Replicating Plasmids in the Green Unicellular Alga *Chlamydomonas reinhardtii.* Cell **36,** 925—931.

Sheperd, H. S., Ledoigt, G., Howell, S. H., 1983: Regulation of light-harvesting chlorophyll-binding protein (LHCP) mRNA accumulation during the cell cycle in *Chlamydomonas reinhardtii*. Cell **32,** 99—107.

Silflow, C. D., Rosenbaum, J. L., 1981f: Multiple α and β tubulin genes in *Chlamydomonas* and regulation of tubulin mRNS levels after deflagellation. Cell **24,** 81—88.

Spreitzer, R. J., Mets, L., 1980: Non-Mendelian mutation affecting ribulose 1,5-bisphosphate carboxylase structure and activity. Nature **285,** 114—115.

Spreitzer, R. J., Ogren, W. L., 1983: Rapid recovery of chloroplast mutations affecting ribulosebisphosphate carboxylase/oxygenase in *Chlamydomonas reinhardtii*. Proc. Natl. Acad. Sci., U.S.A. **80,** 6293—6297.

Spreitzer, R. J., Goldschmidt-Clermont, M., Rahire, M., Rochaix, J.-D., 1985: Nonsense mutations in the *Chlamydomonas* chloroplast gene that codes for the large subunit of ribulose bisphosphate carboxylase/oxygenase. Proc. Natl. Acad. Sci., U.S.A. **82,** 5460—5464.

Stern, D. B., Lonsdale, D. M., 1982: Mitochondrial and chloroplast genomes of maize have a 12 kb DNA sequence in common. Nature **299,** 698—702.

Stinchcomb, D. T., Struhl, K., Davis, R. W., 1979: Isolation and characterization of a yeast chromosomal replicator. Nature **282,** 39—43.

Struhl, K., Stinchcomb, D. T., Scherer, S., Davis, R. W., 1979: High frequency transformation of yeast: autonomous replication of hybrid DNA molecules. Proc. Natl. Acad. Sci., U.S.A. **76,** 1035—1079.

Timmis, J. N., Scott, N. S., 1983: Sequence homology between spinach nuclear and chloroplast genomes. Nature **305,** 65—67.

Tudzynski, P., Esser, K., 1983: Nuclear Association in Yeast of a Hybrid Vector Containing Mitochondrial DNA. Curr. Genet. **7,** 165—166.

Turmel, M., Bellemare, G., Lee, R. W., Lemieux, C., 1986: A linear DNA molecule of 5.9 kb is highly homologous to the chloroplast DNA in the green alga *Chlamydomonas moewusii*. Plant Molec. Biol. **6,** 313—319.

Vallet, J. M., Rahire, M., Rochaix, J.-D., 1984: Localization and sequence analysis of chloroplast DNA sequences of *Chlamydomonas reinhardtii* that promote autonomous replication in yeast. EMBO J. **3,** 415—421.

Vallet, J. M., Rochaix, J.-D., 1985: Chloroplast origins of DNA replications are distinct from chloroplast ARS sequences in two green algae. Curr. Genet. **9,** 321—324.

Waddell, J., Wang, X.-M., Wu, M., 1984: Electron microscopic localization of the chloroplast DNA replicative origins in *Chlamydomonas reinhardtii*. Nucl. Acids Res. **12,** 3842—3856.

Wang, X.-M., Chang, C. H., Waddell, J., Wu, M., 1984: Cloning and delimiting one chloroplast DNA replicative origin of *Chlamydomonas*.

Watson, J. C., Surzycki, S. J., 1982: Extensive sequence homolgy in the DNA coding for elongation factor Tu from *Escherichia coli* and the *Chlamydomonas reinhardtii* chloroplast. Proc. Natl. Acad. Sci., U.S.A. **79,** 2264—2267.

Watson, J., Surzycki, S. J., 1983: Both the chloroplast and Nuclear genomes of *Chlamydomonas reinhardtii* share Homology with *Escherichia coli* Genes for Transcription and Translational components. Current Genet. **7,** 201—210.

Woessner, J. P., Masson, A., Harris, E. H., Bennoun, P., Gillham, N. W., Boynton, J. E., 1984: Molecular and genetic analysis of the chloroplast ATPase of *Chlamydomonas*. Plant Molec. Biol. **3,** 177—190.

Youngblom, J., Schloss, J. A., Silflow, C. D., 1984: The two β tubulin genes of *Chlamydomonas reinhardtii* Code for identical Proteins. Mol. Cell Biol. **4,** 2682—2696.

Zakian, V., 1981: Origin of replication from Xenopus laevis mitochondrial DNA promotes high-frequency transformation of yeast. Proc. Natl. Acad. Sci., U.S.A. **78,** 3128—3132.
Zakian, V., Kupfer, D. M., 1982: Replication and Segregation of an Unstable Plasmid in Yeast. Plasmid **8,** 15—28.

Chapter 12

Induction of Expression in and Stable Transformation of an Algal Cell by Nuclear Microinjection with Naked DNA

H.-G. Schweiger and G. Neuhaus

Max-Planck-Institut für Zellbiologie, D-6802 Ladenburg bei Heidelberg, Federal Republic of Germany

With 11 Figures

Contents

I. Introduction

Transformation of animal and prokaryotic cells is well studied (Garapin and Colbère-Garapin, 1983; Harris, 1983) but comparatively little is known on plant cell transformation (Schell *et al.*, 1984; Ohta, 1986) and particularly about that of algae (Rochaix and van Dillewijn, 1982). In plant cells vector bound DNA is preferentially used for transformation. The Ti-plasmid of *Agrobacterium tumefaciens* (for review see Schell *et al.*, 1984) and the Cauliflower mosaic virus DNA (Brisson *et al.*, 1984; Fromm *et al.*, 1986) have proved to be useful in this respect. There have been only a few studies concerned with transforming effects of naked DNA (Paszkowski *et al.*, 1984; Hain *et al.*, 1985; Potrykus *et al.*, 1985; Lörz *et al.*, 1985; Ohta, 1986).

Efforts to transform plant cells focus on controlled conditions and high efficiency. In order to fulfill these requirements the transforming DNA has to come into close contact with the genome, preferably it should be introduced directly into the cell nucleus. Such an approach presumably increases the probability of integration and protects the exogenous DNA from degradation. Controlled conditions means that among other things one would like to know what particular cell is being transformed.

In the following, experiments are summarized which were performed on individual cells of the unicellular marine green alga Acetabularia. They were aimed at studying the expression of foreign genes and gene constructions and the transformation of these cells. These experiments started from naked DNA, thereby excluding any limitations imposed by the narrow host range which may play a major role in transformation with vector bound DNA (Zambryski *et al.*, 1984).

II. Acetabularia

The unicellular and uninuclear marine green alga Acetabularia is a member of the family Dasycladaceae (for references see Berger *et al.*, 1974). Since all Dasycladaceae are highly calcified in nature the family of Dasycladaceae can be traced back for more than 500 million years (Kamptner, 1958). This has proved useful in studies on evolutionary problems (for review see Schweiger *et al.*, 1986).

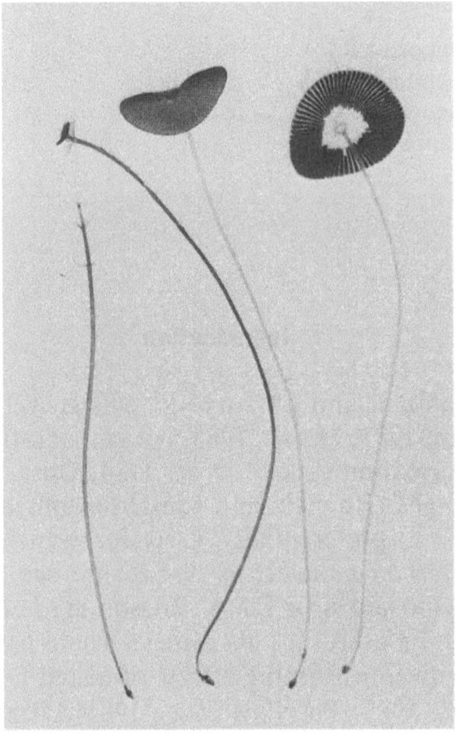

Fig. 1. Four cells of A. mediterranea. The four cells represent different stages of development: prior to cap formation (note the whirls); early cap formation; maximum diameter cap; cap with cysts as indicated by the radial structure

The Acetabularia cell is unique in its size and can grow to a length of up to 20 cm depending on the species. It consists of a stalk carrying whorls of hairs. The stalk bears a rhizoid at the basal end accomodating the single primary nucleus. At the end of its vegetative phase the Acetabularia cell forms a cap-like structure at the apical end of the stalk (Fig. 1). In the early generative phase the primary nucleus starts to divide, giving rise to a large number of secondary nuclei, which migrate through the stalk into the cap, proceeding to divide, forming cysts and eventually gametes. The gametes are released and they fuse and form zygotes (for review see Schweiger and Berger, 1979). Acetabularia cells can easily be grown in an artificial seawater-like medium (Schweiger *et al.*, 1977).

III. Techniques

A highly advantageous feature of the Acetabularia cell is, that due to its size and to the fact that it is uninuclear, it can be used for a number of basic surgical experiments. The nucleus can easily be removed by amputating the rhizoid (Hämmerling, 1932; Schweiger and Berger, 1979). Nucleate and anucleate apical cell fragments of Acetabularia are capable of surviving and of performing species-specific morphogenesis for several weeks (Hämmerling, 1932). Grafting of cell fragments of Acetabularia has been the basis for a number of classical experiments in cell biology (for references see Schweiger and Berger, 1979). The contents of the amputated rhizoid can also be squeezed out and the nucleus can be identified in and

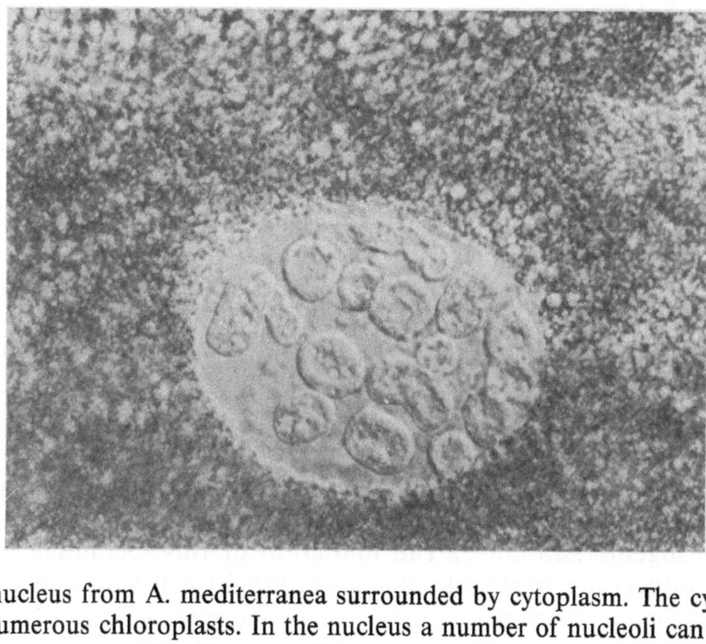

Fig. 2. A nucleus from A. mediterranea surrounded by cytoplasm. The cytoplasm contains numerous chloroplasts. In the nucleus a number of nucleoli can be seen. The diameter of the nucleus: 70 μm

isolated from the cytoplasm at low magnification (Fig. 2) (Hämmerling, 1955; Richter, 1959; Sandakchiev *et al.*, 1973; Schweiger and Berger, 1979).

The isolated nucleus is extremely useful for microinjection experiments. Due to its size, reaching a diameter of 80 μm, the isolated nucleus of the species Acetabularia mediterranea can easily be fixed on a holding capillary by gentle suction and microinjected by means of an injection capillary mounted on a micromanipulator (Fig. 3) (Cairns *et al.*, 1978a; Neuhaus *et al.*, 1984). Microinjected nuclei can be recombined with cytoplasm in two ways (Cairns *et al.*, 1978a). Firstly the nucleus can be implanted into an anucleate apical cell fragment (Fig. 4 a). A cell nucleus pretreated and implanted in this way will survive and exert its activities in the cytoplasm. Such a nucleus is normally capable of undergoing divisions and forming secondary nuclei which subsequently result in gamete formation.

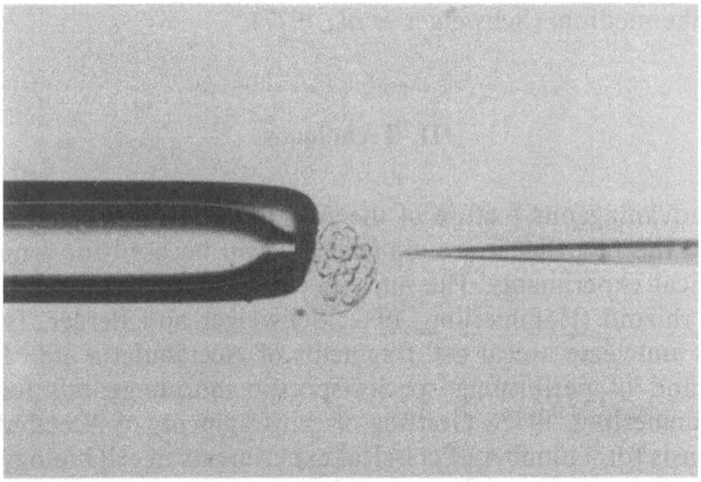

Fig. 3. Microinjection of an Acetabularia nucleus. On the left side: the holding capillary carrying a nucleus. On the right side: injection capillary

The second possibility is to fuse an isolated nucleus with a droplet of cytoplasm (Fig. 4b) (Cairns *et al.*, 1978a; Primke *et al.*, 1978; Neuhaus *et al.*, 1984). Cytoplasts are obtained in the following way: A stalk is cut into small, 1 to 3 mm long cell fragments. Within minutes after cutting, droplets of cytoplasm will emerge from the open ends of the stalk fragments and form cytoplasts. Such cytoplasts can be fused either with other cytoplasts (Gibor, 1965) or with isolated nuclei, by bringing the two partners into contact (Primke *et al.*, 1978). This type of fusion does not need the addition of any fusogenic substance. The limitation of this technique is that the fusion products are incapable of regenerating a cell wall and performing nuclear division. Therefore studies on morphogenesis and on progeny can not be performed with such fusion products.

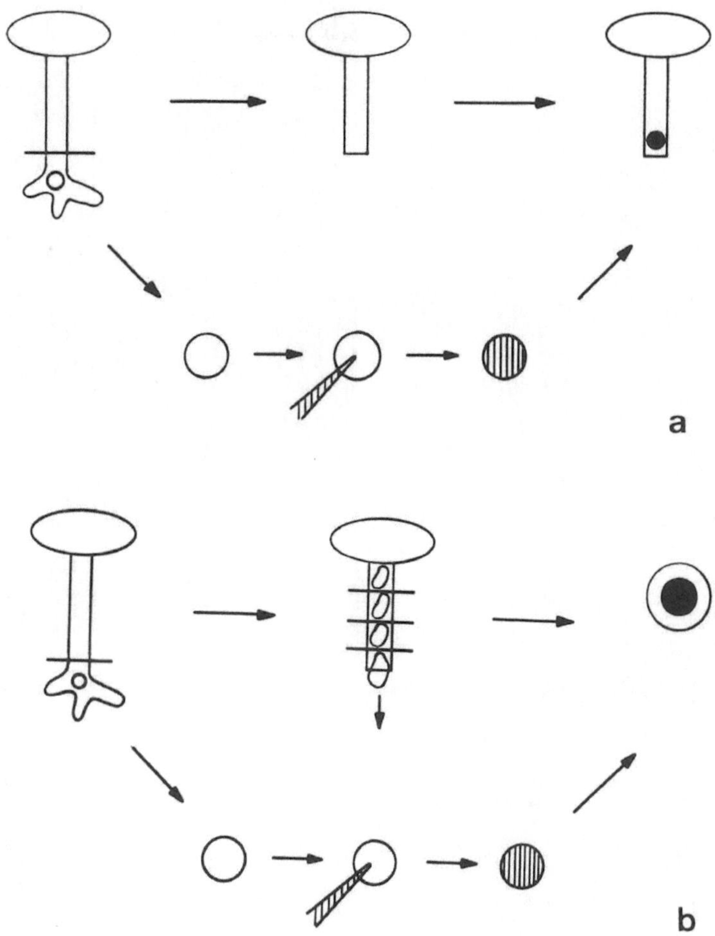

Fig. 4. Two ways for recombining an isolated nucleus with cytoplasm. The nucleus is isolated from the amputated rhizoid of an Acetabularia cell and microinjected. Cytoplasm is either represented by an anucleated cell fragment (a) or by cytoplasts prepared by cutting a stalk (b)

IV. Expression of Genomic RNA

In experiments devoted to the expression of genomic RNA of plant and animal viruses, TMV-RNA and mengovirus-RNA were studied (Cairns *et al.*, 1978 a, b). RNA was microinjected into an isolated nucleus, and after combining the nucleus with cytoplasm by using either one of the two methods mentioned above the existence of expression products was demonstrated by indirect immunofluorescence (Fig. 5). These results showed that the Acetabularia cell is not only capable of surviving this type of manipulation but also of translating the microinjected RNA and that this RNA is protected from the attack of ribonucleases (Schweiger, 1966).

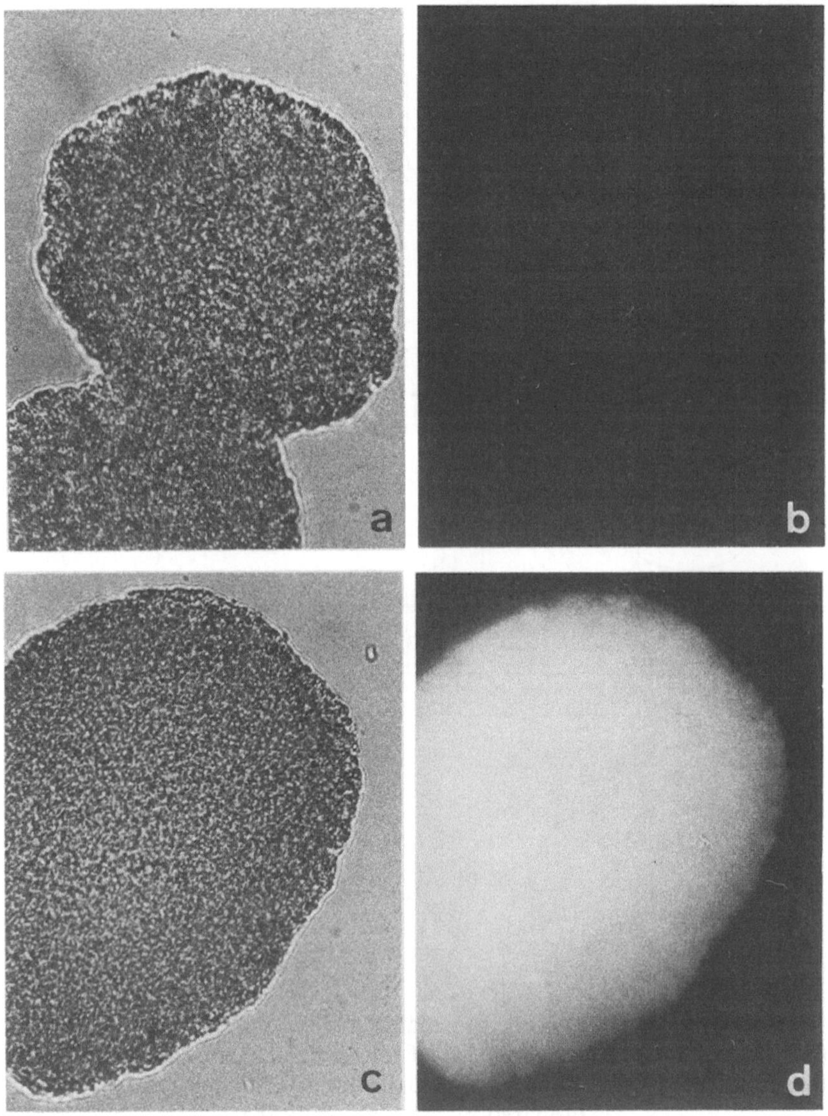

Fig. 5. Expression of TMV RNA in an Acetabularia cell. The expression was proved by indirect immunofluorescence in cytoplasmic droplets. a) Buffer injected control: light microscopy. b) The same as in a: fluorescence microscopy. c) Three days after injection of TMV RNA: light microscopy. d) The same as in c: fluorescence microscopy

These experiments also indicated that, after implantation of the nucleus into an anucleate cell fragment, the microinjected RNA is released intact from the nucleus to the site of translation in the cytoplasm. Another conclusion was that in Acetabularia the translation apparatus is capable of recognizing and translating foreign viral plant as well as animal RNA and that there is no species-dependent compatibility problem. These experiments also suggested that the translation product of the microinjected RNA was not subjected to immediate degradation at least for the time needed for the test. These experiments can only show that antigenic epitopes are expressed. The question of whether the total genome is expressed or only parts of it and whether this expression is subjected to regulation still remains unanswered.

V. Expression of Genomic DNA

In addition to the two virus RNA species the expression of two DNAs in Acetabularia has been investigated. One of them was the adenovirus II DNA representing an animal type genome (Cairns *et al.,* 1978c). Expression products were studied by means of indirect immunofluorescence in a similar way as used for the RNA genomes. Expression products of the adenovirus II DNA were found to be present in the cytoplasm. In addition to the conclusions drawn from experiments on the expression of RNA these experiments demonstrated that microinjected DNA can be transcribed, and that the transcription apparatus of the Acetabularia cell is capable of recognizing and transcribing foreign DNA. This finding again underlines the lack of specificity of the gene expression apparatus in Acetabularia.

In addition to the expression of the adenovirus II DNA, expression of SV40 DNA was investigated (Neuhaus *et al.,* 1984). In this case the specific early expression product of the virus genome, the large T-antigen, was studied by means of indirect immunofluorescence. From studies on animal cells it is known (Tegtmeyer *et al.,* 1975; Das *et al.,* 1985) that the large T-antigen of SV40 must migrate from the site of translation in the cytoplasm into the nucleus, where it is accumulated. Studies on the expression of the SV40 gene for the large T-antigen demonstrated that the nuclear envelope of Acetabularia is capable of allowing the passage of the large T-antigen, which then is accumulated in the nucleus (Fig. 6).

No accumulation of T-antigen in the nucleus is observed (Kalderon *et al.,* 1984) when using a mutant with a single amino acid substitution in the large T-antigen for transformation of animal cells. Acetabularia cells injected with this mutated T-antigen-gene showed the same specificity in that the expression product was also not accumulated in the cell nucleus (unpublished results). This result indicates that the specificity in the amino acid sequence of the translation product as well as the recognition apparatus of the Acetabularia nucleus are at least in some ways similar to those of animal cells. The experiments using SV40 DNA support the view that the gene for the large T-antigen is expressed correctly.

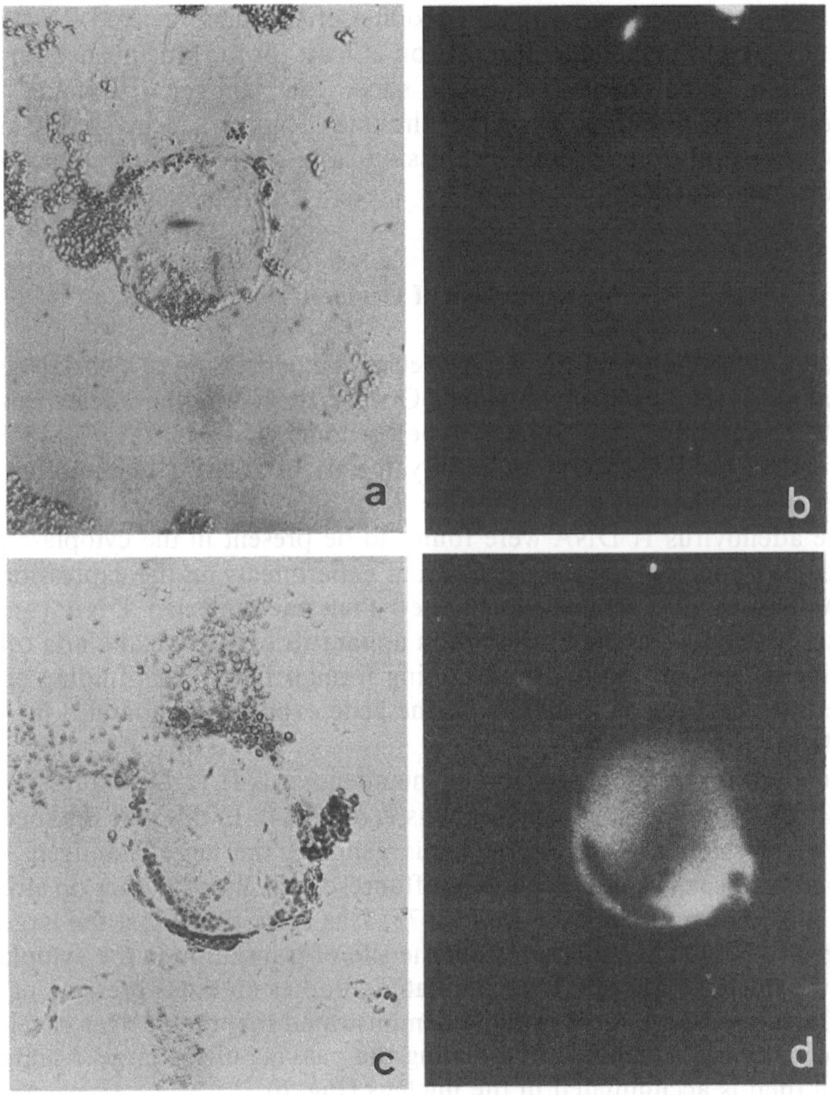

Fig. 6. Expression of the gene for the SV40 large T-antigen in an Acetabularia cell. The expression was proved by indirect immunofluorescence in nuclei from cells of A. mediterranea. a) Buffer injected control: light microscopy. b) The same as in a: fluorescence microscopy. c) Twenty-four h after injection of SV40 DNA: light microscopy. d) The same as in c: fluorescence microscopy

VI. Expression of Genes and Gene Constructions

In addition to whole genomes a number of gene constructions have been studied. Expression of the gene for the small subunit of the ribulose1,5-bisphosphate carboxylase from wheat could be detected by using highly specific antibodies distinguishing between the wheat and the Acetabularia protein (Neuhaus *et al.*, 1983, Broglie *et al.*, 1983). Genes for the zeins, storage proteins of maize, were also expressed in Acetabularia (Langridge *et al.*, 1985). Again the expression of these proteins was detected by means of indirect immunofluorescence (Fig. 7). In this context one particular feature of zeins should be mentioned. Zeins are highly hydrophobic proteins which are insoluble in aqueous solutions. In Acetabularia the expression products of zein genes as evidenced by indirect immunofluorescence exhibited a microscopic appearance resembling the

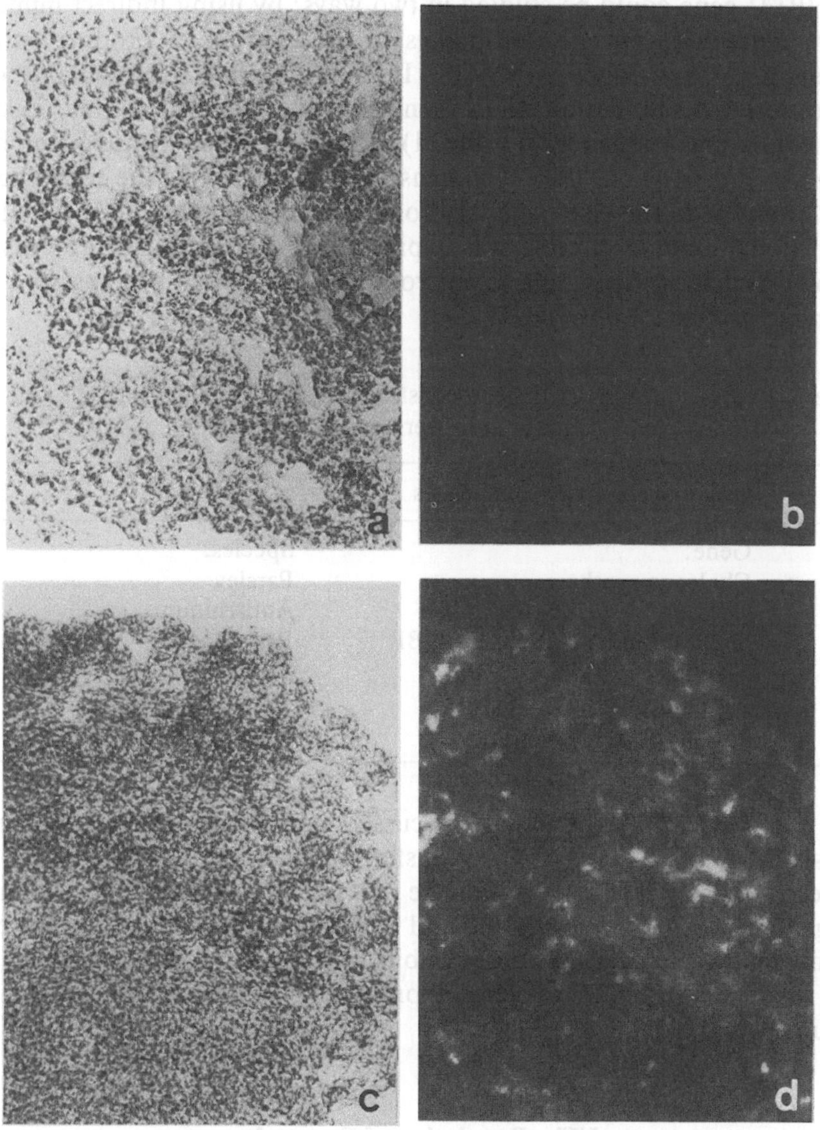

Fig. 7. Expression of a gene for zein in an Acetabularia cell. The expression was proved by indirect immunofluorescence in cytoplasmic droplets. a) Buffer injected control: light microscopy. b) The same as in a: fluorescence microscopy. c) Ten days after injection of a zein gene: light microscopy. d) The same as in c: fluorescence microscopy

structures found in maize endosperm cells indicating that the hydrophobic character of the protein is conserved, and that assembly of protein bodies may occur as in maize endosperm.

Another gene expressed in Acetabularia cells after microinjection was that for neomycin phosphotransferase II (NPT II) from the E. coli transposon 5 (Jimenez and Davies, 1980). This codes for an enzyme, which by transferring a phosphate group from ATP to antibiotics such as neomycin, kanamycin or G 418 inactivates the antibiotic (Haas and Dowding, 1975). Since Acetabularia cells are sensitive to the antibiotic G418, expression of the NPT II gene could be studied in two ways: by using indirect immunofluorescence with specific antibodies to the expression product and by screening for resistance to G 418. In microinjected Acetabularia cells expression of a number of NPT II gene constructions with various eukaryotic regulatory sequences (Table 1) could be demonstrated by both methods (Weeks *et al.*, 1985; Neuhaus *et al.*, 1986). Since the resistance to G 418 was due to the enzyme NPT II one can conclude that in Acetabularia a functionally intact enzyme was expressed. This result also supports the hypothesis that, in Acetabularia, microinjected genes can be expressed into complete proteins.

Table 1. Expression of NPT II gene constructions by Acetabularia cells. All constructions were expressed.

Origin of regulatory sequences	
Gene:	Species:
Chalcone synthase	Parsley
	Antirrhinum
Small subunit of ribulose-1,5-BP	Pea
T-antigen	SV 40
Nopaline synthase	Ti-plasmid
β-tubuline	Chlamydomonas

The microinjection experiments discussed here also allow some conclusions on the specificity of the transfer of macromolecules through the nuclear envelope. The results indicate that the nuclear envelope separating the genome, with its transcription and replication apparatus, from the cytoplasm and the site of translation exhibits little if any species specificity and that its regulatory role in gene expression can only be a limited one (Neuhaus and Schweiger, 1986).

VII. Regulation of Expression

The time expression products need to appear in Acetabularia depends on the nucleic acid or genes being studied. Expression products of RNA occur within a few days. A different situation is observed with a number of gene constructions. For example, genes coding either for the zeins or for the small subunit of the ribulose-1,5-bisphosphate carboxylase from wheat take ten days or longer before the corresponding expression product can be detected. However, the expression of the large T-antigen of SV40 appears as early as a couple of hours after microinjection (Neuhaus *et al.,* 1984). One may ask whether this difference is due to the fact that the large T-antigen-gene is under the control of an enhancer sequence, as can be demonstrated in animal cells (for review see Khoury and Gruss, 1983). Experiments were performed in Acetabularia with constructions in which the so-called enhancer sequences were either present or absent. In these experiments it was demonstrated that the Acetabularia cell is capable of recognizing the SV40 enhancer sequences (Neuhaus *et al.,* 1984). Interestingly the expression of the gene for the large T-antigen in Acetabularia is slow if it is combined with the promoter and the terminating sequences for β-tubulin genes from Chlamydomonas (Weeks *et al.,* 1985). These findings lead to the conclusion that the fast expression of the large T-antigen in Acetabularia is due to the enhancer sequences of SV40.

VIII. Transformation

The expression of microinjected genes (see chapter on techniques) in the Acetabularia cell leads to the question of whether the expression is transient or stable. Stable expression can only be studied if microinjected cells are able to reproduce. Since microinjected Acetabularia cells do not exhibit significant morphological or functional differences to control cells, this requirement is fulfilled.

Manipulated Acetabularia cells perform normal morphogenesis characteristic for the vegetative phase, eventually developing cysts. These cysts release gametes; they fuse and form zygotes which are capable of germinating.

Studies on cells of the F_1 generation obtained in this way revealed that all progeny cells originating from SV40 microinjected cells expressed the large T-antigen and that the large T-antigen was accumulated in the nucleus (Neuhaus *et al.,* 1986). The same was true for the F_2 and F_3 generations. The conclusion that the SV40 genome was replicated in some way in these cells is based on the fact that in every generation more that 10^6 gametes are formed by the cysts of one cell resulting in more than 10^{18} gametes within three generations (Schulze, 1939). The number of molecules microinjected into one nucleus, however, was only 10^6—10^7.

Integration of the SV40 DNA into the Acetabularia genome leads to this stable expression of the T-antigen as has been shown by Southern hybridization experiments. Detailed analysis revealed that the large T-antigen coding sequence was present in the high molecular DNA if undigested DNA was used (Fig. 8). This result excluded the possibility of autonomous replication of the SV40. After restriction enzyme treatment with BamHI and HpaII a hybridization signal was obtained in a band which corresponded in size to that obtained by similar treatment of SV40

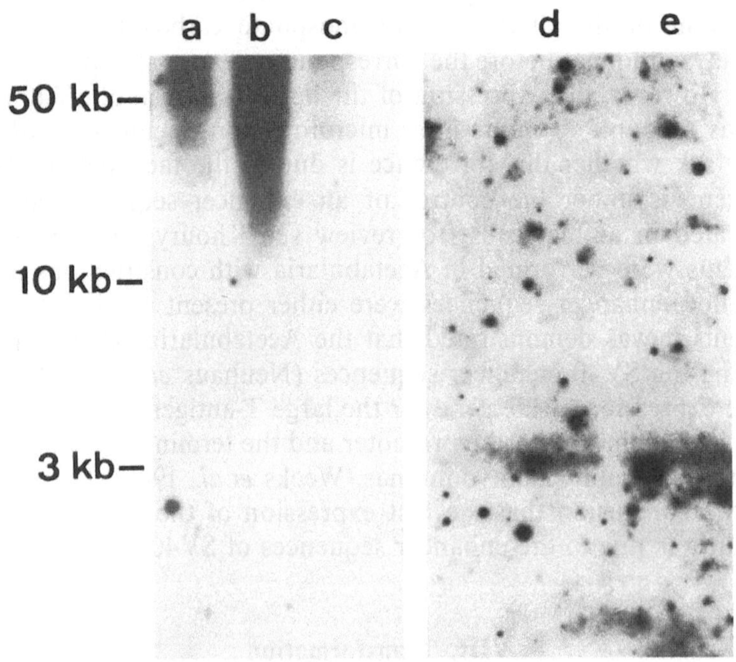

Fig. 8. Southern analysis of the DNA from SV40 transformed cells. Undigested (a, b, c) or Bam HI and Hpa II restricted DNA (d, e) was subjected to gel electrophoresis and hybridized to SV40 DNA. a) and b): DNA from 2 different transformed Acetabularia cell lines. c) Control: pBR 322 transformed line. d) Bam HI/Hpa II insert from the plasmid pSVL6 (Neuhaus *et al.,* 1984). e) Bam HI/Hpa II digested DNA from transformed A. mediterranea DNA

DNA (Fig. 8), indicating that the large T-antigen gene was correctly integrated into the nuclear genome of Acetabularia. Digestion of DNA from transformed F_2 generation cells with Bcl I cutting the integrated gene once resulted in three bands. Two of these bands with molecular weights different from the integrated gene (10.2 and 3.8 kbp) apparently represent the junction fragments (Neuhaus *et al.,* 1986).

Moreover, the finding that one of the bands corresponded to a molecular weight larger than the SV40 genome also underlined that the SV40 DNA was incorporated into the genome. The fact that another band hybridizing to the large T-antigen coding sequence had a size corresponding to the SV40 genome might indicate that more than one SV40 genome is integrated into the Acetabularia genome and that this integration is tandemly organized (Neuhaus *et al.,* 1986).

Moreover it could be shown by nuclear transplantation experiments that integration had to occur in the nuclear genome rather than in the chloroplast or mitochondrial genome. If a nucleus from a transformed cell was implanted into a non-transformed cytoplasm the large T-antigen was expressed. If, however, a nontransformed nucleus was implanted into the cytoplasm from a transformed cell this did not occur (Fig. 9).

The stable expression of a gene in F_3 generation cells was also observed

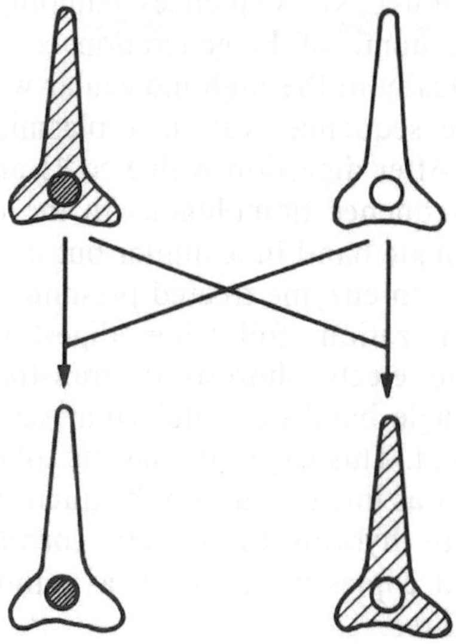

Fig. 9. Nuclear exchange experiment between transformed (shaded) and nontransformed (empty) cells. Nuclei were isolated from transformed F_2 generation and from nontransformed cells and implanted into anucleate fragments obtained from non-transformed and transformed cells respectively. Shaded nucleus indicates expression of the large T-antigen gene

by injecting the plasmid pSV2neo (Southern and Berg, 1982), containing the NPT II gene, into Acetabularia cells (Neuhaus *et al.*, 1986). In this case expression was demonstrated not only by indirect immunofluorescence but also by resistance of these cells to the antibiotic G 418 (Fig. 10). By means

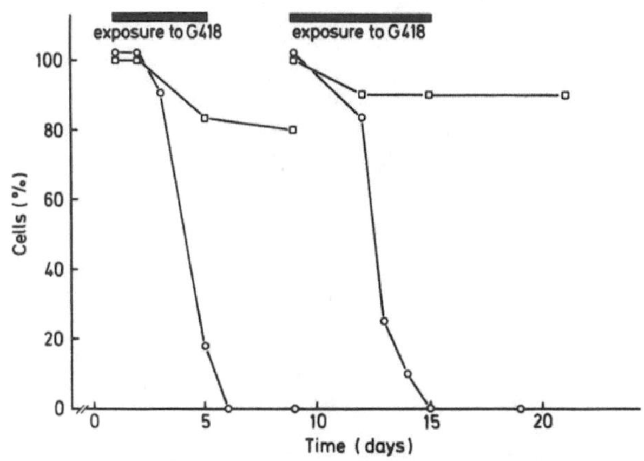

Fig. 10. Induction of stable resistance to G 418 by microinjection of the NPT II gene. F_3 generation cells of NPT II gene injected A. mediterranea were exposed to 2×10^{-3} M G 418 (bar). Almost 80 % of the transformed cells survived a first round and 90 % of the survivors a second round of G 418 treatment while 100 % of two sets of untreated control cells did not survive. The ordinate represents the percentage of cells surviving

of Southern hybridization DNA sequences homologous to the pSV2neo were detected in the genome of F_3 generation cells. Unrestricted DNA migrated electrophoretically in the high molecular weight range excluding the possibility that the sequence was in a plasmid or in another low molecular component. After digestion with Eco RI and Pvu I, excising the coding sequence, the sequence homologous to the coding sequence was found to migrate in a single band in a similar but not identical position as compared to the restriction enzyme treated plasmid.

In Southern hybridization following digestion with appropriate restriction enzymes and electrophoresis of transformed genomic DNA from Acetabularia a single band of a molecular size similar to the introduced gene was observed. This suggests that the total coding sequence of the transforming DNA was integrated into the nuclear genome. The results obtained with Southern hybridization were corroborated by dot blot analysis. The number of copies per genome, as estimated by this method, was between one and three.

IX. Genetics

An interesting result was obtained by crossing experiments starting from two cysts. From a single cyst only either (+) or (−) mating type gametes are released, preventing gametes from a single cyst from fusing. The two cysts were either both transformed or nontransformed or one was transformed and the other one not (Fig. 11). In all crossings in which both cysts

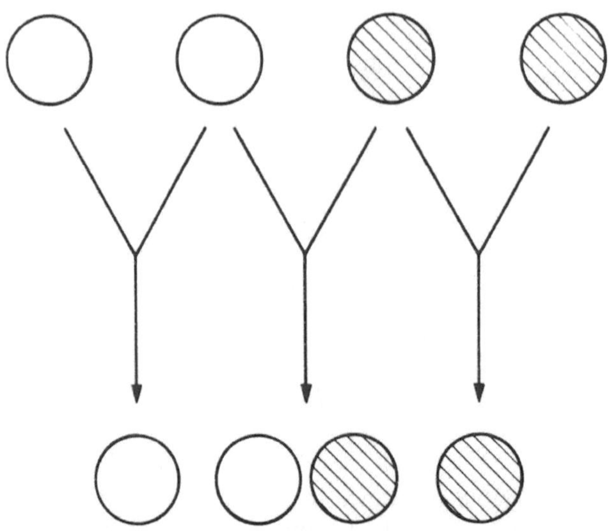

Fig. 11. Scheme of genetic experiments. Cysts from nontransformed (empty) and transformed (dashed) F_3 generation Acetabularia cells were fused. The progeny was either nontransformed, transformed or partially transformed

were either transformed or nontransformed all cells of the progeny were either transformed or nontransformed, respectively. From a transformed and a nontransformed cyst 20—50% transformed cells originated (Neuhaus *et al.,* 1986).

X. Discussion

The experiments described here demonstrate that Acetabularia cells can be transformed by microinjection with a high yield of transformed cells. The yield in the first step, i. e., expression of a microinjected gene or genome is almost 100 % of the nuclei surviving microinjection and implantation. In the second step, starting from the microinjected cell and leading to transformed progeny, the yield is also almost 100 %. The extremely high yield and the fact that the DNA is integrated is striking since DNA without vector has been injected (Neuhaus *et al.,* 1986). Presently it is not known where in the nuclear genome the microinjected genes are integrated.

Since the technique described here has a high efficiency, one may even think of using Acetabularia cells for the expression of distinct genes at a preparative level. A major limiting factor, however, might be that the generation time of Acetabularia is about 100—120 days.

Another important point worthwhile discussing is the following: The fact that transformation of a plant cell can be obtained under controlled conditions using naked DNA and resulting in high yield, raises the question whether similar results can be obtained for higher plant cells by using this technique. This would require culture of individual selected protoplasts, fragmentation of these cells into nuclei and cytoplasm, microinjection of the isolated nucleus, fusion of the microinjected nucleus and cytoplasm and regeneration of the fusion product to calli and eventually whole plants (Melnikov *et al.,* 1985; Spangenberg *et al.,* 1985; Crossway *et al.,* 1986; Spangenberg *et al.,* 1986). Substantial progress has been made in this direction. Regeneration of fusion products of cell fragments after microinjection has been obtained at an acceptable rate. Even with the crop plant rape the yield is significantly higher than 10 %.

In this case the experiments performed on Acetabularia may gain general importance allowing the development of novel approaches and techniques which eventually may be applied to higher plants.

Acknowledgements

We express our gratitude to Drs. F. Kreuzaler and P. Schreier and Ms. H. Kaulen, Köln, to Drs. H. J. Bohnert, Tucson, Arizona, P. Gruss, Heidelberg, and D. Weeks, Palo Alto, California, for supplying us with different NPT II gene constructions.

We are grateful to Dr. G. Neuhaus-Url for helpful discussions, to Mrs. B. Gernert for critically reading the manuscript, to Mrs. R. Fischer for photography and Ms. C. Schardt for typing the manuscript.

XI. References

Berger, S., Sandakhchiev, L., Schweiger, H. G., 1974: Fine structural and biochemical markers of Dasycladaceae. J. Microsc. **19,** 89—104.

Brisson, N., Paszkowski, J., Penswick, J. R., Gronenborn, B., Potrykus, I., Hohn, T., 1984: Expression of a bacterial gene in plants using a viral vector. Nature **310,** 511—514.

Broglie, R., Coruzzi, G., Lamppa, G., Keith, B., Chua, N.-H., 1983: Structural analysis of nuclear genes coding for the precursor to the small subunit of wheat ribulose-1,5-bisphosphate carboxylase. Biotechnol. **1,** 55—61.

Cairns, E., Gschwander, H. H., Primke, M., Yamakawa, M., Traub, P., Schweiger, H. G., 1978a: Translation of animal virus RNA in the cytoplasm of a plant cell. Proc. Natl. Acad. Sci., U.S.A. **75,** 5557—5559.

Cairns, E., Sarkar, S., Schweiger, H. G., 1978b: Translation of tobacco-mosaic-virus RNA in *Acetabularia* cell cytoplasm. Cell Biol. Int. Rep. **2,** 573—578.

Cairns, E., Doerfler, W., Schweiger, H. G., 1978c: Expression of a DNA animal virus genome in a plant cell. FEBS Lett. **96,** 295—297.

Crossway, A., Oakes, J. V., Irvine, J. M., Ward, B., Knauf, V. C., Shewmaker, C. K., 1986: Integration of foreign DNA following microinjection of tobacco mesophyll protoplasts. Mol. Gen. Genet. **202,** 179—185.

Das, G. C., Niyogi, S. K., Salzman, N. P., 1985: SV40 promoters and their regulation. Progr. Nucl. Acid Res. **32,** 217—236.

Fromm, M. E., Taylor, L. P., Walbot, V., 1986: Stable transformation of maize after gene transfer by electroporation. Nature **319,** 791—793.

Garapin, A. C., Colbère-Garapin, F., 1983: DNA transfer into animal cells. In: Lurquin, P. F., Kleinhofs, A. (eds.), Genetic engineering in eukaryotes, pp. 37—42. Plenum Press New York.

Gibor, A., 1965: Surviving cytoplasts *in vitro.* Proc. Natl. Acad. Sci., U.S.A. **54,** 1527—1531.

Haas, M. I., Dowding, I. E., 1975: Aminoglycoside-modifying enzymes. Meth. Enzymol. **43,** 611—628.

Hain, R., Stabel, P., Czernilofsky, A. P., Steinbiss, H. H., Herrera-Estrella, L., Schell, J., 1985: Uptake, integration, expression and genetic transmission of a selectable chimeric gene by plant protoplasts. Mol. Gen. Genet. **199,** 161—168.

Hämmerling, J., 1932: Entwicklung and Formbildungsvermögen von *Acetabularia mediterranea.* II. Das Formbildungsvermögen kernhaltiger und kernloser Teilstücke. Biol. Zentralblatt **52,** 42—61.

Hämmerling, H., 1955: Neuere Versuche über Polarität und Differenzierung bei *Acetabularia.* Biol. Zentralbl. **74,** 545—554.

Harris, T. J. R., 1983: Expression of eukaryotic genes in *E. coli.* In: Williamson, R. (ed.), Genetic Engineering, 127—185. New York: Academic Press.

Jimenez, A., Davies, I., 1980: Expression of a transposable antibiotic resistance element in Saccharomyces. Nature **287,** 869—871.

Kalderon, D., Richardson, W. D., Markham, A., Smith, A. E., 1984: Sequence requirements for nuclear location of simian virus 40 large T-antigen. Nature **311,** 33—35.

Kamptner, E., 1958: Über das System und die Stammesgeschichte der Dasycladaceen *(Siphoneae verticillatae).* Ann. Naturhist. Museum Wien **62,** 95—122.

Khoury, G., Gruss, P., 1983: Enhancer elements. Cell **33,** 313—314.

Langridge, P., Brown, J. W. S., Pintor-Toro, J. A., Feix, G., Neuhaus, G., Neuhaus-Url, G., Schweiger, H. G., 1985: Expression of zein genes in *Acetabularia mediterranea.* Eur. J. Cell Biol. **39,** 257—264.

Lörz, H., Baker, B., Schell, J., 1985: Gene transfer to cereal cells mediated by protoplast transformation. Mol. Gen. Genet. **199,** 178—182.

Melnikov, P. V., Pasternak, T. P., Gleba, Y. Y., Sytnik, K. M., 1985: DNA microinjection into higher plant cells. Dokl. Akad. Nauk SSR Ser. B Geol., Khim. Biol. Nauki **10,** 69—71.

Neuhaus, G., Neuhaus-Url, G., Schweiger, H. G., Chua, N.-H., Broglie, R., Coruzzi, G., Chu, N., 1983: Expression of higher plants in the green alga *Acetabularia mediterranea.* Eur. J. Cell Biol. Suppl. **2,** 30.

Neuhaus, G., Neuhaus-Url, G., Gruss, P., Schweiger, H. G., 1984: Enhancer-controlled expression of the simian virus 40 T-antigen in the green alga *Acetabularia.* EMBO J. **3,** 2169—2172.

Neuhaus, G., Neuhaus-Url, G., de Groot, E. J., Schweiger, H. G., 1986: High yield and stable transformation of the unicellular green alga Acetabularia by microinjection of SV40 DNA and pSV2neo. EMBO J. **5,** 1437—1444.

Neuhaus, G., Schweiger, H. G., 1986: Two-way traffic between nucleus and cytoplasm: cell surgery studies on *Acetabularia.* In: Peters, R., Trendelenburg, M. (eds.), Nucleocytoplasmic transport. Berlin: Springer.

Ohta, Y., 1986: High efficiency genetic transformation of maize by a mixture of pollen and exogenous DNA. Proc. Natl. Acad. Sci., U.S.A. **83,** 715—719.

Paszkowski, J., Shillito, R. D., Saul, M., Mandak, V., Hohn, T., Hohn, B., Potrykus, I., 1984: Direct gene transfer to plants. EMBO J. **3,** 2717—2722.

Potrykus, I, Paszkowski, J., Saul, M. W., Petruska, J., Shillito, R. D., 1985: Molecular and general genetics of a hybrid foreign gene introduced into tobacco by direct gene transfer. Mol. Gen. Genet. **199,** 169—177.

Primke, M., Berger, S., Schweiger, H. G., 1978: Protoplasts from *Acetabularia:* Isolation and fusion. Cytobiologie **16,** 375—380.

Richter, G. 1959: Die Auslösung kerninduzierter Regeneration bei gealterten kernlosen Zellteilen von Acetabularia und ihre Auswirkungen auf die Synthese von Ribonucleinsäure und Cytoplasmaproteinen. Planta **52,** 554—564.

Rochaix, J.-D., Van Dillewijn, J., 1982: Transformation of the green alga *Chlamydomonas reinhardtii* with yeast DNA. Nature **296,** 70—72.

Sandakhchiev, L., Niemann, R., Schweiger, H. G., 1973: Kinetics of changes of malic dehydrogenase isoenzyme pattern in different regions of *Acetabularia* hybrids. Protoplasma **76,** 403—415.

Schell, J., Herrera-Estrella, L., Zambryski, P., de Block, M., Joos, H., Willmitzer, K., Eckes, P., Rosahl, S., van Montagu, M., 1984: Genetic engineering of plants. In: Schell, J. S., Starlinger, P. (eds.), The impact of gene transfer techniques in eucaryotic cell biology, pp. 73—90. Berlin: Springer.

Schulze, K. L., 1939: Cytologische Untersuchungen an *Acetabularia mediterranea* und *Acetabularia wettsteinii.* Arch. Protistenk. **92,** 179—225.

Schweiger, H. G., 1966: Ribonuklease-Aktivität in *Acetabularia.* Planta **68,** 247—255.

Schweiger, H. G., Dehm, P., Berger, S., 1977: Culture conditions for *Acetabularia.* In: Woodcock, C. L. F. (ed.), Progress in *Acetabularia* Research, pp. 319—330. New York: Academic Press.

Schweiger, H. G., Berger, S. 1979: Nucleocytoplasmic interrelationships in *Acetabularia* and some other *Dasycladaceae.* Int. Rev. Cytol. Suppl. **9,** 11—44.

Schweiger, H. G., de Groot, E. J., Leible, M. B., Tymms, M. J., 1986: Conservative and variable features of the chloroplast genome of *Acetabularia.* In: Akoyunoglou, G., Senger, H. (eds.), Regulation of chloroplast differentiation. Plant Biology Series, New York: Alan R. Liss Inc. Publ.

Southern, P. I., Berg, P., 1982: Transformation of mammalian cells to antibiotic resistance with a bacterial gene under the control of SV40 early region promoter. J. Mol. Appl. Genet. **1,** 327—341.

Spangenberg, G., Koop, H.-U., Schweiger, H. G., 1985: Different types of protoplasts from *Brassica napus* L.: analysis of conditioning effects at the single-cell level. Eur. J. Cell Biol. **39,** 41—45.

Spangenberg, G., Neuhaus, G., Schweiger, H. G., 1986: Expression of foreign genes in a higher plant cell after electrofusion-mediated cell reconstitution of a microinjected karyoplast and a cytoplast. Eur. J. Cell Biol. **42,** 236—238.

Tegtmeyer, P., Schwartz, M., Collins, J. K., Rundell, K., 1975: Regulation of tumor antigen synthesis by simian virus 40 gene. A. J. Virol. **16,** 168—178.

Weeks, D., Brunke, K., Beerman, N., Anthony, J., Neuhaus, G., Neuhaus-Url, G., Schweiger, H. G., 1985: Promoter regions of four coordinately regulated tubulin genes of Chlamydomonas and their use in construction of fused genes which are expressed in *Acetabularia.* In: Freeling, M. (ed.), Plant Genetics, pp. 477—490. New York: Alan R. Liss.

Zambryski, P., Herrera-Estrella, L., de Block, M., van Montagu, M., Schell, J., 1984: The use of the Ti-plasmid of *A. tumefaciens* to study the transfer of foreign DNA in plant cells: new vectors and methods. In: Hollaender, A., Setlow, J. (eds.), Genetic engineering, Vol. **6**A, pp. 253—278. New York: Plenum Press.

Chapter 13

Transient Expression of DNA in Plant Cells

Michael Fromm and Virginia Walbot

Department of Biological Sciences, Stanford University, Stanford, CA 94305, U.S.A.

With 1 Figure

Contents

Techniques for transferring DNA into cells often introduce much more DNA into the cell nucleus than the amount that becomes stably incorporated into the host's chromosomes. If non-replicating, this extrachromosomal DNA is lost over a one to two week period as a result of dilution by cell division and susceptibility to intracellular degradation. During its transient existence in the cell, the extrachromosomal DNA is known to be transcriptionally active because typically 1 to 10% of the cells express the introduced DNA but only 0.01 to 0.1% of the cells stably integrate and maintain the introduced gene. This "transient expression" is extremely useful in studying gene structure and function relationships for a number of reasons. The most important is the speed with which results on gene expression can be obtained. The transfected cells are typically ready for analysis of gene expression 24 to 48 hours after DNA transfer. A second important aspect is the lack of host chromosomal flanking sequences that may influence gene expression. Quite often there is variation in gene expression in stable transformants depending on the site of insertion; such variation can be greater than the variation between the gene structures being compared (Odell *et al.*, 1985; Velten *et al.*, 1984). Transient DNA is free of host flanking sequences allowing a much simpler comparison of normal and altered gene structures.

I. Overview of Transient Assay Applications

Because of the paucity of reports using transient assays in plant cells, we briefly describe examples of transient assays in animal cells to indicate the versatilty of the technique. Transient expression assays have been used to measure changes in protein levels, often by enzymatic activity, or RNA abundance and structure, or alterations in DNA structure depending on the purpose of the experiment. The structure and function of promoters has been analyzed by quantitating the steady state level of mRNA, or the corresponding level of the mRNA encoded protein, produced from the normal and mutant constructs, as well as determining the 5′ end of the mRNA (Fromm and Berg, 1982; McKnight *et al.*, 1981). Regulatory protein interactions with promoters *in vivo* have been determined by transiently introducing the promoter of interest into cells containing the regulatory protein (Velcich and Ziff, 1985; Foster *et al.*, 1985; Lewis *et al.*, 1985). The strength of different promoters and enhancers has been examined by attaching them to the same transcription unit and determining the resulting levels of expression (Mosthaf *et al.*, 1985). the signals for RNA splicing (Wieringa *et al.*, 1984) and the 3′ cleavage and poly(A) addition to mRNA (Danner and Leder, 1985) have been analyzed by mutation of the DNA template and determination of the resulting RNA structure. The effect of RNA structure and sequence on mRNA translation has been measured by altering the DNA template coding for the mRNA (Johansen *et al.*, 1984; Kozak, 1986). Experiments demonstrating that anti-sense RNA inhibits expression of mRNA have also been carried out transiently (Izant and Weintraub, 1985).

Transient assays have been used for purposes besides analyzing gene expression *per se*. The sequences defining the SV40 origin of replication have been determined by transient DNA replication assays (Fromm and Berg, 1982; Myers and Tjian, 1980). *In vivo* alterations in the introduced DNA have been used to monitor DNA recombination (Subramani and Berg, 1983), transposable element excision (Rio *et al.*, 1986) and DNA mutagensis and repair (Lebkowski *et al.*, 1985). In the latter two cases, the altered DNA forms were examined by subsequent rescue into bacteria. Bacterial genetics and selection systems can then be used to find the altered DNA templates. Although plant cell transient assays are relatively recent, there is every indication that experiments similar to those performed in animal cells will be possible.

II. Transient Assays in Plant Cells

Published examples of transient assays in plant cells have used a number of DNA transfer techniques. Fraley *et al.* (1984) found that nopaline synthase gene expression could be detected as early as 12 hours after co-cultivation of *Agrobacterium* and petunia protoplasts. Whether the state of the introduced DNA is extrachromosomal or integrated is uncertain but clearly rapid analysis of gene expression is possible using co-cultivation. For unspliced genes, however, some concern about expression from residual contaminating bacteria is warranted and appropriate controls need to be done.

Electroporation was used to compare the strengths of the nopaline synthase promoter and the cauliflower mosaic virus (CaMV) 35 S promoter in carrot and maize protoplasts. The promoter strengths were comparable in carrot cells but the CaMV 35 S promoter was 30-fold stronger than the nopaline synthase promoter in the maize cells (Fromm *et al.*, 1985). The maize heat shock hsp 70 promoter has been found to be heat inducible in maize protoplasts using transient assays (J. Callis, M. Fromm and V. Walbot, unpublished data) indicating this is a rapid and convenient method for studying heat shock gene expression. Ecker and Davis (1986) used electroporation-mediated transient assays to demonstrate that an anti-sense CAT expression plasmid inhibited the expression of a CAT expression plasmid *in vivo,* indicating that anti-sense RNA inhibition of gene expression can occur in plants as it does in animal cells (Izant and Weintraub, 1985).

Fusion of *E. coli* spheroplasts carrying a plasmid containing a complete CaMV genome resulted in detection of transient expression of the CaMV capsid protein by immunofluorescence (Tanaka *et al.*, 1984). Similarly, Walden and Howell (1983) and Lebeurier *et al.* (1982) performed transient DNA recombination assays on CaMV genomes using subsequent viral propagation in plant cells as a detection system.

Several direct DNA transfer techniques for obtaining stably transformed plant cells have been described (Krens *et al.*, 1982; Deshayes *et al.*, 1985; Hain *et al.*, 1985; Potrykus *et al.*, 1985; Shillito *et al.*, 1985). Most of these techniques probably produce transient expression as well but such results have not been reported to date. Electroporation-mediated gene transfer appears to be one of the more general and efficient direct DNA transfer techniques for obtaining transient expression and stable transformants in plant cells (Fromm *et al.*, 1985, 1986 a; Shillito *et al.*, 1985). We have described our electroporation protocol in detail elsewhere (Fromm *et al.*, 1985, 1986 b) and just mention some of its basic features here.

III. Transient Expression after Electroporation-Mediated Gene Transfer

Elektroporation refers to the process of reversibly permeabilizing cell membranes by briefly applying a high intensity electric field (for review, see Zimmermann and Vienken, 1982). The resulting membrane polarization leads to membrane compression that when severe enough will form localized pores in the cell membrane. These pores in the membrane exist long enough for macromolecules to diffuse through the pores before the membrane reseals. Electroporation permeabilized membranes have proven useful for gene transfer into both plant (Fromm *et al.*, 1985, 1986 a; Shillito *et al.*, 1985) and animal cells (Neumann *et al.*, 1982; Potter *et al.*, 1984; Smithies *et al.*, 1985). The technique may be applicable to all cell types because

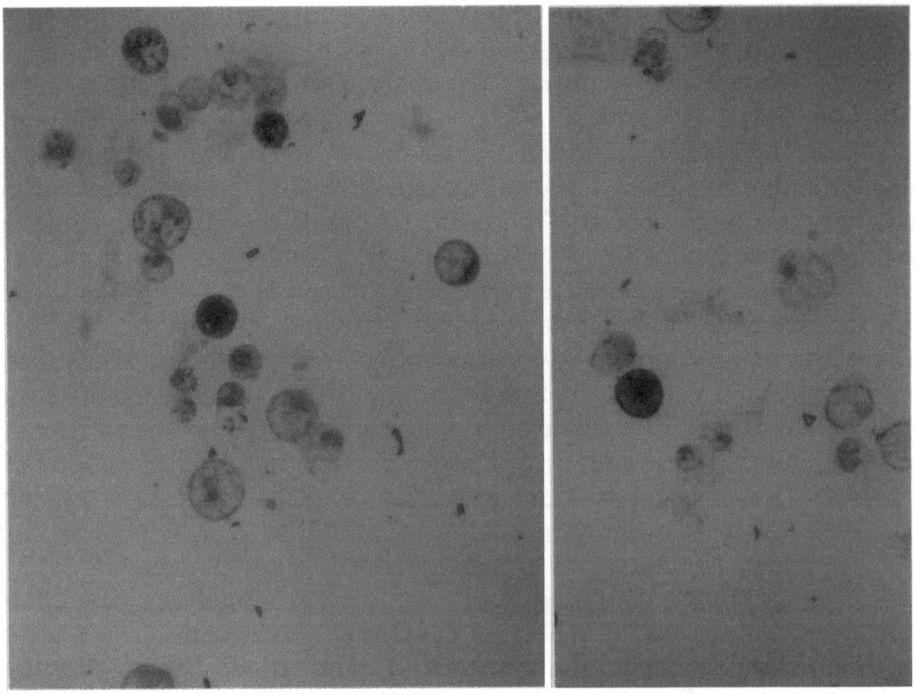

Fig. 1. Transient expression of a beta-galactosidase expression vector in tobacco protoplasts. Tobacco protoplasts were electroporated with or without the beta-galactosidase vector CaMVLACZ present. 48 hours after electroporation a solution containing 0.2% 5-bromo-4-chloro-3-indolyl beta-D-galactopyranoside and 100 mM Tris, pH 7.5 was added; cells expressing beta-galactosidase cleave the substrate and turn blue while non-expressing protoplasts remain colorless (Helmer *et al.*, 1984). *a*, protoplasts electroporated without CaMVLACZ DNA present. *b*, protoplasts electroporated with 140 μg/ml of CaMVLACZ DNA present.

Methods.

CaMVLACZ. The structure of the beta-galactosidase expression vector used is identical to that described for CaMVNEO (Fromm *et al.*, 1986a) except that the neomycin phosphotransferase II coding region has been replaced with the *E. coli* beta-galactosidase coding region. Both constructs use the cauliflower mosaic virus (CaMV) 35S promoter and the nopaline synthase (NOS) polyadenylation site. The coding region consists of the a 3.3 kb fragment from the *E. coli lacZ* gene that codes for the entire beta-galactosidase coding region except for the first 8 amino acids (Casadaban *et al.*, 1983). This amino terminus was fused to the first 16 amino acids and the 35 bp 5′ untranslated leader region of the NOS gene (Depicker *et al.*, 1982). The structure of CaMVLACZ is CaMV promoter/NOS 5′ leader and first 16 amino acids/lacZ coding region/NOS polyadenylation site.

Electroporation. Tobacco protoplasts were isolated from a rapidly growing suspension culture using the same protocol described for isolating corn protoplasts (Fromm *et al.*, 1986b). Protoplasts in 1 ml of 2 mM phosphate, pH 7.1, 128 mM NaCl, 4 mM $CaCl_2$, and 0.2 M mannitol were electroporated (Fromm *et al.*, 1985, 1986b) using an electric pulse with a 7.5 msec. e^{-1} decay constant and 675 volt/cm electric field intensity, either with or without CaMVLACZ DNA present. After 10 min at 0 °C and 10 min at room temperature, 8 ml of protoplast media and the all biomembranes can be electrically permeabilized (Zimmermann and Vienken, 1982).

Operationally electroporation-mediated gene transfer is performed by applying an electric pulse to a solution containing protoplasts and DNA. After a few minutes recovery time during which the cell membranes reseal, the transfected protoplasts are diluted with medium and incubated to allow time for gene expression prior to analysis. Transient expression was used as a rapid and sensitive assay for gene transfer into plant cell nuclei to develop efficient conditions for electroporation-mediated gene transfer into plant protoplasts (Fromm *et al.*, 1985). The important electroporation parameters were found to be DNA concentration, the intensity and duration of the electric pulse, and isolating transcriptionally active protoplasts (Fromm *et al.*, 1985, 1986b). Because the electric pulse has two parameters (duration and intensity), a large number of eletric pulse field intensity and duration combinations are possible. We have found electric pulses with 500 to 1000 V/cm field intensities and with e^{-1} decay constants of 5 to 15 milliseconds to be the most effective for gene transfer. Electroporation efficiently transferred genes into *Daucus carota, Nicotiana tabacum* and *Zea mays* cells (Fromm *et al.*, 1985) demonstrating the technique functions with both monocot and dicot cells. R. Hauptmann (personal communication) has observed electroporation-mediated transient expression in protoplasts of eleven different monocot species indicating the technique is generally useful in plants.

In our experience the level of gene expression observed from cells of an electroporation-mediated transient assays is equivalent to that from an equal number of stably transformed cells containing one to a few copies of the same construct. Assuming chromosomal and extrachromosomal copies express at equivalent levels, this suggests that the transiently expressing cell population averages a few DNA copies per cell. This could result from either a small fraction of cells expressing a large number of introduced

---

protoplasts were added to a 10 cm plastic petri dish and incubated at 26 °C in the dark. The protoplast incubation media consists of Murashige and Skoog salts (Gibco), 30 g/l sucrose, 3 mg/ml (2,4-dichlorophenoxy)acetic acid, 100 mg/l inositol, 1.3 mg/l niacin, 0.25 mg/l thiamine, 0.25 mg/l pyridoxine, 0.25 mg/l calcium pantothenate, 0.3 mg/l kinetin, 0.3 M mannitol and 20 % tobacco conditioned media with a final pH of 5.8. After 48 hours to allow time for gene expression, the protoplasts were pelleted at 200 × g and resuspended in the electroporation buffer described above containing 0.2 % 5-bromo-4-chloro-3-indolyl beta-D-galactopyranoside and 100 mM Tris, pH 7.5 to detect beta-galactosidase expression. Blue color appears in three to four hours at room temperature in protoplasts expressing beta-galactosidase. Control protoplasts do not exhibit any beta-galactosidase activity in this assay. Approximately 10 % of the initial protoplasts received and expressed the CaMVLACZ DNA as measured by this assay. About 50 % of the initial protoplasts are destroyed during isolation and electroporation and the assay only detects high levels of beta-galactosidase. Therefore the actual number of surviving protoplasts that contain and express the CaMVLACZ DNA is probably much higher than 10 %

DNA molecules, or a large fraction of cells expressing a few DNA templates. The number of protoplasts receiving and expressing DNA has been estimated by transferring a beta-galactosidase expression vector (CaMVLACZ) into tobacco protoplasts by electroporation. Protoplasts expressing beta-galctosidase become blue after cleavage of the colorless substrate 5-bromo-4-chloro-3-indolyl beta-D-galactopyranoside to yield an insoluble blue reaction product, while non-expressing protoplasts remain colorless. Using this chromogenic assay on tobacco protoplasts electroporated with CaMVLACZ DNA, 10 % of the protoplasts were observed to turn blue (Fig. 1). This probably represents a lower estimate of transfer because high levels of beta-galactosidase expression are needed for a visible color. From this experiment we infer that transient expression in electroporated protoplasts results from a high fraction (> 10 %) of the cell population containing a small number of introduced DNA molecules.

Note that the protoplasts expressing beta-galactosidase appear intact and viable (round with distinct cytoplasmic strands) indicating that the cell membranes reseal and the cell is not permanently damaged. We have observed protoplasts expressing beta-galactosidase that had divided by 48 hours after electroporation; this indicates that viable cells capable of cell division are obtained after electroporation-mediated gene transfer. This observation was confirmed by obtaining stably transformed tobacco (Shillito *et al.*, 1985) and maize (Fromm *et al.*, 1986 a) cells after electroporation.

IV. Discussion

In this report we have highlighted the utility and versatility of transient expression for studying plant genes. This technology has become available only recently and many applications will probably be reported soon. Transient gene expression usually correlates with the average gene expression in stable transformants. This allows rapid characterization of a gene before undertaking more intensive studies of stable transformants. We found that transient gene expression in maize protoplasts was a successful indicator of which vectors would function in maize callus, leading to the isolation of stably transformed maize calli resistant to kanamycin (Fromm *et al.*, 1986 a).

The possibility of performing transient assays on protoplasts isolated from different plant tissues is an aspect of transient assays whose potential has yet to be determined. This might be a rapid method for observing regulated gene expression in the tissue type of interest before obtaining transformed plants. For example, the expression of the alpha-amylase gene increases after giberellic acid treatment but is suppressed by abscisic acid in barley aleurone protoplasts (Jacobsen and Beach, 1985). These responses are the same as the intact aleurone indicating it should be possible to study hormone-regulated expression of the alpha-amylase gene transiently in aleurone protoplasts. However, some types of protoplasts may be regulated differently than the cells they were derived from, so this approach will probably not be useful for all genes and cell types.

Acknowledgements

We thank Pioneer Hi-bred International and the Nationals Institutes of Health (GM 32422) for support.

V. References

Casadaban, M., Martinez-Arias, A., Shapira, S. K., Chou, J., 1983: Beta-galactosidase gene fusions for analyzing gene expression in *Escherichia coli* and yeast. Meth. Enz. **100,** 293—308.

Danner, D., Leder, P., 1985: role of an RNA cleavage/poly(A) addition site in the production of membrane-bound and secreted iM mRNA. Proc. Natl. Acad. Sci., U.S.A. **82,** 8658—8662.

Depicker, A., Stachel, S., Dhaese, P., Zambryski, P., Goodman, H. M., 1982: Nopaline synthase: Transcript mapping and DNA sequence. J. Molec. Appl. Genet. **1,** 561—573.

Deshayes, A., Herrera-Estrella, L., Caboche, M., 1985: Liposome-mediated transformation of tobacco mesophyll protoplasts by an *Escherichia coli* plasmid. EMBO J. **4,** 2731—2737.

Ecker, J. and Davis, R., 1986: Inhibition of gene expression in plant cells by expression of antisense RNA. Proc. Natl. Acad. Sci., U.S.A. **83,** 5372—5376.

Foster, J., Stafford, J., Queen, C., 1985: An immunoglobin promoter displays cell-type specificity independently of the enhancer. Nature **315,** 423—425.

Fraley, R. T., Horsch, R. B., Matzke, A., Chilton, M. D., Sanders, P. R., 1984: *In vitro* transformation of petunia cells by an improved method of co-cultivation with *A. tumefaciens.* Plant Mol. Biol. **3,** 371—378.

Fromm, M., Berg, P., 1982: Deletion mapping of DNA regions required for SV40 early region promoter function *in vivo.* J. Molec. Appl. Gent. **1,** 457—481.

Fromm, M., Taylor, L. P., Walbot, V., 1985: Expression of genes transferred into monocot and dicot plant cells by electroporation. Proc. Natl. Acad. Sci., U.S.A. **82,** 5824—5828.

Fromm, M., Taylor, L. P., Walbot, V., 1986a: Stable transformation of maize after gene transfer by electroporation. Nature **319,** 791—793.

Fromm, M., Taylor, L. P., Walbot, V., 1986b: Electroporation of DNA and RNA into Plant Protoplasts. Meth. Enzym., in press.

Hain, R., Stabel, P., Czernilofsky, A. P., Steinbiss, H. H., Herrera-Estrella, L., Schell, J., 1985: Uptake, integration, expression and genetic transmission of a selectable chimaeric gene by plant protoplasts. J. Mol. Gen. Genet. **199,** 161—168.

Helmer, G. Casadaban, M., Bevan, M., Kayes, L., Chilton, M. D., 1984: A new chimaeric gene as a marker for plant transformation. Bio/Technology **2,** 520—527.

Izant, J. G., Weintraub, H., 1985: Constitutive and conditional suppression of exogenous and endogenous genes by anti-sense RNA. Science **229,** 345—352.

Jacobsen, J. V., Beach, L. R., 1985: control of transcription of alpha-amylase and rRNA genes in barley aleurone protoplasts by gibberellin and abscisic acid. Nature **316,** 275—277.

Johansen, H., Schumperli, D., Rosenberg, M., 1984: Affecting gene expression by altering the length and sequence of the 5′ leader. Proc. Natl. Acad. Sci., U.S.A. **81,** 7698—7702.

Kozak, M., 1986: Point mutations define a sequence flanking the AUG initiator codon that modulated translation by eukaryotic ribosomes. Cell **44,** 283—292.

Krens, F. H., Molendijk, L., Wullems, G. J., Schilperoort, R. A., 1982: *In vitro* transformation of plant protoplasts with Ti-plasmid DNA. Nature **296,** 72—74.

Lebeurier, G., Hirth, L., Hohn, B., Hohn, T., 1982: *In vivo* recombination of cauliflower mosaic virus DNA. Proc. Natl. Acad. Sci., U.S.A. **79,** 2932—2936.

Lebkowski, J. S., Clancy, S. Miller, J. H., Calos, M. P., 1985: The *lac I* shuttle: rapid analysis of the mutagenic specificity of ultraviolet light in human cells. Proc. Natl. Acad. Sci., U.S.A. **82,** 8606—8610.

Lewis, E. D., Manley, J. L., 1985: Repression of simian virus 40 early transcription by viral DNA replication in human 293 cells. Nature **317,** 172—175.

Myers, R., Tjian, R., 1980: Construction and analysis of simian virus 40 origins defective in tumor antigen binding and DNA replication. Proc. Natl. Acad. Sci., U.S.A. **77,** 6491—6495.

McKnight, S. L., Gavis, E. R., Kingsbury, R., Axel, R., 1981: Analysis of transcriptional regulatory signals of the HSV thymidine kinase gene: identification of an upstream control region. Cell **25,** 385—398.

Mosthaf, L., Pawlita, M., Gruss, P., 1985: A viral enhancer specifically active in human haematopoietic cells. Nature **315,** 597—600.

Neumann, E., Schaefer-Ridder, M., Wang, Y., Hofschneider, P. H., 1982: Gene transfer into mouse lyoma cells by electroporation in high electric fields. EMBO J. **1,** 841—845.

Odell, J. T., Nagy, F., Chua, N.-H., 1985: Identification of DNA sequences required for activity of the cauliflower mosaic virus 35 S promoter. Nature **313,** 810—812.

Potrykus, I., Saul, M. W., Petruska, J., Paskowski, J., Shillito, R. D., 1985: Direct gene transfer to cells of a granimaceous monocot. Mol. Gen. Genet. **199,** 183—188.

Potter, H., Weir, L., Leder, P., 1984: Enhancer-dependent expression of human kappa immunolglobulin genes introduced into mouse pre-B lymphocytes by electroporation. Proc. Natl. Acad. Sci., U.S.A. **81,** 7161—7165.

Rio, D. C., Laski, F. A., Rubin, G. M., 1986: Identification and immunochemical analysis of biologically active *Drosophilia P* element transposase. Cell **44,** 21—32.

Shillito, R. D., Saul, M. W., Paskowski, S. J., Muller, M., Potrykus, I., 1985: High efficiency direct gene transfer to plants. Bio/Technology **3,** 1099—1103.

Smithies, O. Gregg, R. G., Boggs, S. S., Koralewski, M. A., Kucherlapati, R. S., 1985: Insertion fo DNA sequence into the human chromosomal beta-globin locus by homologous recombination. Nature **317,** 230—234.

Subramani, S., Berg, P., 1983: Homologous and non-homologous recombination in monkey cells. Mol. Cell. Biol. **3,** 1040—1052.

Tanaka, N., Ikesami, M., Hohn, T., Matsui, C., Watanabe, I., 1984: *E. coli* spheroplast-mediated transfer of cloned cauliflower mosaic virus DNA into plant protoplasts. Mol. Gen. Genet. **195,** 378—380.

Velcich, A., Ziff, E., 1985: Adenovirus E1a proteins repress transcription from the SV40 early promoter. Cell **40,** 705—716.

Velten, J., Velten, L., Hain, R., Schell, J., 1984: Isolation of a dual plant promoter fragment from the Ti plasmid of *Agrobacterium tumefaciens.* EMBO J. **3,** 2723—2730.

Walden, R. M., Howell, S. H., 1983: Uncut recombinant plasmids bearing nested cauliflower mosaic virus genomes infect plants by intragenomic recombination. Plant Mol. Biol. **2,** 27—31.

Wieringa, B., Hofer, E., Weissmann, C., 1984: A minimal intron length but no specific internal sequence is required for splicing the large rabbit beta-globin intron. Cell **37,** 915—925.

Zimmermann, U., Vienken, J., 1982: Electric field induced cell to cell fusion. J. Membrane Biol. **67,** 165—182.

Chapter 14

Plastid Transformation: A Progress Report

M. J. Cornelissen[2-3], M. De Block[2-3], M. Van Montagu[2], J. Leemans[3], P. H. Schreier[1], and J. Schell[1-2]

[1] Max-Planck-Institut für Züchtungsforschung, D-5000 Köln 30, Federal Republic of Germany

[2] Laboratorium voor Genetika, Rijksuniversiteit Gent, B-9000 Gent, Belgium

[3] Plant Genetic Systems, B-9000 Gent, Belgium

With 2 Figures

Contents

I. Introduction

Reproducible methods, allowing the introduction of defined DNA fragments in the plant nuclear genome, are presently available (see previous chapters in this volume) and are contributing significantly to the experimental analysis of the gene structure-function relationships governing regulated gene-expression in plants.

Plants, however, are of particular interest genetically since they have three genomes: the nuclear, the mitochondrial and the plastid. It would be of major scientific interest to be able to introduce defined DNA fragments in the mitochondrial and plastid genomes in such a way that the introduced DNA fragments would become a stable component of the mitochondrial or plastid genomes and therefore amenable to functional analysis *in vivo.* If successful such studies would lead to a broader understanding of plant development and reproduction. In this short chapter we want to review mostly unpublished work, the aim of which was to investigate whether and under which conditions the plastid genome can be made to accept foreign DNA and express its functions and whether such transformed plastids can be stably maintained in plant tissue cultures and regenerated plants.

To put this work in perspective one should be aware of the recent progress made with attempts to target nuclear gene products for transport into chloroplasts (see chapter in this volume by Les J. Szabo and Anthony R. Cashmore). A very elegant example of the potential of this approach is given by the work of A. Cheung *et al.* (1987). The chloroplast gene *psb*A from an atrazine resistant biotype of *Amaranthus hybridus* and coding for an atrazine insensitive photosynthetic membrane protein, was modified by fusing its protein coding region to transcriptional and transit peptide sequences of the nuclear SSU gene and introduced in the nuclear genome of tobacco using *Agrobacterium* Ti-plasmid vectors. The protein made from this chimeric gene was found to be transported into chloroplasts. Transgenic plants harbouring this nuclear *psb*A gene were found to be tolerant to atrazine thus demonstrating that the *psb*A gene product was functional in photosynthesis. These experiments therefore showed that it is possible to convert a chloroplast gene into a nuclear gene the product of which is functionally targetted for the plastid. However, the success of this approach does not diminish the scientific interest in developing methods for transformation of the plastid genome itself.

Even if DNA could readily enter the plastid and be integrated in its genome, the question remains whether it is possible or likely that a single recombinant plastid replicon can become predominant in a given cell-line or plant. Indeed a single plant cell may contain 1000 or more plastid genome copies divided over several plastids.

Furthermore, a recombinant plastid genome may be at a selective replicative disadvantage relative to the wild type genome. Observations described below could indeed be interpreted if one makes the assumption that recombinant plastid genomes could have a selective disadvantage due to either size increase and/or disruption of plastid functions resulting from integration of a foreign DNA sequence. A recombinant replicon may therefore only be successful provided the insertion codes for an essential function and provided the selective advantage given by this function overcomes any selective disadvantage resulting from the insertion itself. It is moreover conceivable that for some reason regeneration of plants from cells in tissue culture, selects against cells containing mostly transformed plastid genomes.

Three different observations argued in favour of the notion that it must be possible to find conditions that satisfy the selective demands. Indeed a single mutant replicon can become the predominant type as seen in mutants of *Nicotiana tabacum* (Cseplo and Maliga, 1984) which are resistant to drugs such as streptomycin or lincomycin and which contain a uniform population of mutant plastids. The fact that recombinant plastids were obtained which combine mutated genes originally present in independent plastid genomes (Medgyesy *et al.*, 1985) also argues in favour of the notion that a recombinant plastid genome can become the dominant plastid genome in a cell-line. The most direct argument was provided by observations (De Block *et al.*, 1985) made with some tobacco plants resulting from a co-cultivation transformation experiment involving Agrobacteria carrying a Ti plasmid vector containing a chimeric gene consisting of the promoter region of the nopaline synthase gene — (this promoter region was found to be able to drive transcription not only in plant nuclei but also in procaryotic cells (Herrera-Estrella *et al.*, 1983) — and therefore presumably also in chloroplasts) fused to the coding sequence of the chloramphenicol acetyltransferase gene of Tn9 from *E. coli.*

Some plants derived from transformed calli selected on 10 μg/ml chloramphenicol, were unable to transmit the Cm^R phenotype through pollen whereas a typically maternal pattern of inheritance was observed. DNA/DNA hybridizations showed that the chimeric Pnos-cat gene was present in the chloroplasts of these resistant plants and that chloramphenicol acetyltransferase activity was associated with the chloroplast fraction.

However, subcultures derived from top-shoots of these Cm resistant plants and grown in the absence of chloramphenicol, lost the transformed plastids indicating that not all plastid genomes were transformed in the Cm resistant plants and that the transformed plastid genomes in these plants might well be at a selective disadvantage relative to the wild-type plastid genome.

At the time we wrote (De Block *et al.*, 1985): "It can be expected that new selectable marker genes specifically designed to be expressed in chloroplasts from chloroplast specific promoter sequences, will provide a stronger selection. It is hoped that the use of such chloroplast-specific genes will yield more stable chloroplast transformants." In the following we will summarize the various attempts made along these lines and show that the proposed goal is not easily achieved!

II. Construction of Vectors for the Transformation of Plastids

Plastid transformation and subsequent identification of the transformants requires a selectable marker gene. Genes that code for resistance to streptomycin, kanamycin, chloramphenicol or lincomycin, which are drugs that interfere with plastid protein synthesis, could be the basis for construction of selectable marker genes. The NPTII gene of Tn5 which confers resistance to kanamycin is a good candidate. Many plant cells are sensitive to kanamycin and the NPTII gene product has proven to be a potent inactivator. The effect of this drug on mitochondria is unknown. The fact, however, that animal cells are not sensitive to kanamycin would argue that possibly mitochondrial protein synthesis is not inhibited by kanamycin in eucaryotic cells. The necessity to select for the recombinant plastid genome can best be satisfied if the selectable marker gene can be expressed only in plastids. The marker gene has therefore to be fitted with a plastid specific promoter sequence. The promoters of the tobacco plastid *psb*A gene and the *rsp*4 gene should meet this requirement. It is important to bear in mind that plastid promoter sequences have thus far been defined by *in vitro* transcription initiation-experiments (Gruissem, 1984), only. The sequence of a functional *in vivo* plastid promoter may differ from the one defined *in vitro*. Chloroplast transformation is an essential method to settle this question. The chimeric genes tested as plastid selectable marker genes are presented in Table 1.

Table 1

	Promoter of Plastid Gene		Fused to Marker Gene
1	*rbc*L	*Zea mays*	NPT II
2	*rbc*L	*Nicotiana tabacum*	NPT II
3	*pbs*A	*Zea mays*	NPT II
4	*pbs*A	*Amaranthus hybridus*	CAT
5	*pbs*A	*Nicotiana tabacum*	NPT II
6	*rps*4	*Nicotiana tabacum*	NPT II

Constructs 1, 2, 3, 4 and 6 were prepared in collaboration with Dr. A. Cheung and Dr. L. Bogorad (Harvard University).

De Block *et al.* (1985) provided evidence that *Agrobacterium* mediated T-DNA transfer can introduce DNA in plastids. In view of our assumption that large inserts in the plastid genome might negatively affect the maintenance of a recombinant plastid genome, a Ti-plasmid vector system specifying the transfer of small DNA fragments was chosen for some of the experiments (pGV2260::pGV1500 derivatives; Deblaere *et al.*, 1985).

In other instances the pGV3850 vector system (Zambryski *et al.*, 1984) was used. It is important to note that the major difference between pGV3850 and the pGV2260::pGV1500 vectors is the size of the expected T-DNA fragment. Because of the position of the border sequences specifying the right and the left end of the expected T-DNA segment, pGV3850 vectors are presumed to transfer the selectable marker gene flanked on either side by a pBR322 sequence. Thus the total size of the expected T-DNA fragment would be at least 9 kb. In contrast the pGV2260::pGV1500 vectors would be expected to transfer a T-DNA of +/− 2 kb without any repeated sequences. In all cases vectors were assumed to be plastid specific because they contained a chimeric selectable marker gene supposed to be driven specifically by plastid transcription initiation signals (see Table 1).

Both "co-cultivation" (Zambryski *et al.*, 1984) and "leaf-disc infection" (Horsch *et al.*, 1985) were used as transformation methods. The main difference between these two methods is the state of the plant cell when infected with *Agrobacterium*. In the first method the plant cell is present as a protoplast and, upon growth, it will give rise to undifferentiated tissue. In the second method the transformed plant cell is part of wounded leaf tissue, that will readily develop into a meristematic tissue. The co-cultivation method has the advantage that the delay between transformation and onset of cell differentiation may provide the correct conditions for the generation of a stable plastid transformant. the recombinant replicon probably needs a number of plastid divisions in order to establish itself as the predominant type. When differentiation takes place before the recombinant replicon has established itself as the dominant population chimaeric plants with a mixed plastid population will result. Kanamycin resistant calli were obtained after co-cultivation experiments using either pGV3850 or pGV2260::pGV1500 derived vectors, at frequencies of about 10^{-4} relative to the number of treated protoplasts. The results with leaf-disc infections were inconclusive.

Most of the putative transformant calli obtained by co-cultivation experiments died after 2—4 months of continued growth in the presence of 100 µg/ml kanamycin. From surviving calli (after co-cultivation with pGV2260::pGV1500 vectors), shoots were regenerated on media containing 100 µg/ml kanamycin. DNA extracted from chloroplasts from one of these plants did not, however, show any hybridization to NPT II specific probes. Total DNA from these plants on the other hand did contain hybridizing bands and a weak NPT II activity was detected in leaf extracts.

In view of these inconclusive results a new vector was later developed with the aim to circumvent the possibility that integration of the vector would disrupt an essential function of the plastid genome. The idea was to offer a site for homologous recombination between the plastid replicon and the (putative) circular T-DNA intermediate (Koukolikova *et al.*, 1985; Stachel *et al.*, 1986). Sequences were chosen so as to contain either an intact 5′ or 3′ part of a plastid transcription unit. In this way a recombinant replicon should maintain all its biological functions (see Fig. 1).

Table 2 summarizes the results obtained with these vectors in co-cultivation experiments. Two elements dramatically influenced the frequency with which kanamycin resistant calli were observed: 1) the type of promoter: only a nuclear promoter (P_{NOS}) or the *psb*A plastid promoter gave significant numbers or surviving calli; 2) the presence to one end of the selectable marker gene of a specific plastid sequence that might allow homologous recombination (see Fig. 1). The construct in which the NPT II gene is flanked to the 5′ side by the psbA promoter and to the 3′ end by the *rbc*L plastid gene gave putative transformation frequencies as high as those obtained with similar vectors containing a nuclear NPT II gene. These results are illustrated in Fig. 2.

These remarkable observations strongly suggested that the strategy aiming at insertion of a short DNA sequence without disruption of an essential function, had indeed resulted in a high frequency of plastid transformation. In order to test whether this was indeed the case, plants were regenerated from these calli under continuous kanamycin selection. Chloroplasts were isolated and purified from three independent kanamycin resistant plants. Total DNA and DNA from purified chloroplasts was hybridized after restriction enzyme digestion and blotting of the fragments onto filters to NPT II specific probes. Although clear hybrid-

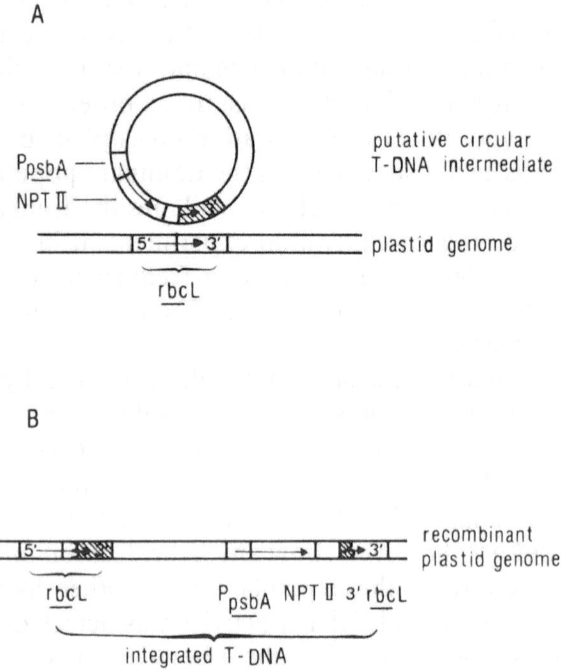

Fig. 1. Schematic representation of vectors designed to integrate in the plastid genome without inactivating any plastid genes. The approach postulates a circular T-DNA intermediate

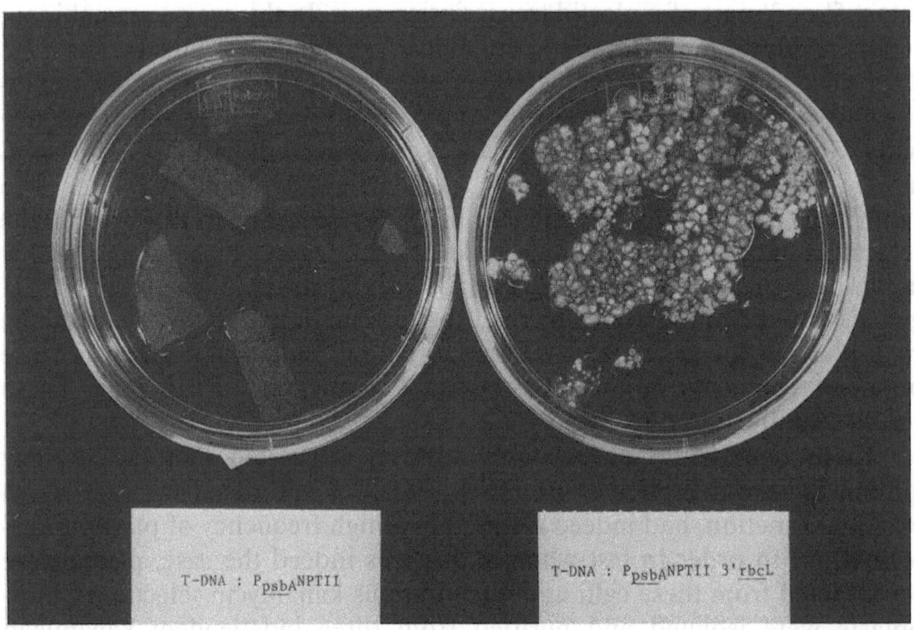

Fig. 2. Picture of calli of tobacco taken after 8 weeks of growth after co-cultivation on media containing 50 µg/ml kanamycin. 10^5 protoplasts were used in the co-cultivation experiments. Callus regeneration in the absence of selection was +/− 50 %

Table 2

Transformation of tobacco SR1 protoplasts with the constructs:		(I)	(II)	(III)
A pGV2260	Negative control	0	0	0
B pGV2260:pGSC1500/1	P*rps*4-NPTII coding seq.	0	0	0
C pGV2260:pGSC1500/1 LS	P*rps*4-NPTII coding seq. +3′ part LS gene*	7	7	1
D pGV2260:pGC1500/1 32K	P*rps*4-NPTII coding seq. +5′ part 32K gene*	2	2	1
E pGV2260:pGSC1500/2 B	Positive control P*nos* NPTII 3′ ocs	80	9	9
F pGV2260:pGSC1500	Negative control (no marker gene)	0	0	0
G pGV2260:pGSC1500/4	P*psb*A-NPTII	36	25	10
H pGV2260:pGSC1500/4 LS	P*psb*A-NPTII +3′ part LS gene*	720	493	80%
I pGV2260:pGSC1500/2 B	Positive control P*nos*-NPTII 3′ ocs	200	14	14

(I) Calli obtained on 50 μg/ml kanamycin. Each experiment was started with 5×10^5 protoplasts.
(II) Calli transferred to selective plates. Calli larger than 1 mm diameter were placed on plates containing 100 μg/ml kanamycin.
(III) Calli surviving after 6 months on 100 μg/ml kanamycin.
* The homology regions are about 1 kb in length.
Note: A, B, C, D, E, and F, G, H, I, are different protoplast preparations. E and I are controls for nuclear transformation.
pGSC1500: series was derived from pGV1500 (R, Deblaere, 1985). The second column summarizes the selectable marker gene located within the T-DNA borders of pGSC1500 intermediate vector.

ization was found in all three cases to total DNA, no hybridization could be observed to chloroplast DNA.

These observations are very confusing: on the one hand they provide interesting evidence suggesting that calli, obtained from protoplasts co-cultivated with Agrobacteria carrying a Ti-plasmid vector with an NPTII gene designed to express specifically in plastids (driven by the *psb*A promoter) and to integrate into plastid DNA without gene disruption (assuming a circular T-DNA intermediate; see Fig. 1), can indeed survive selection by kanamycin as a result of the uptake and expression in their plastids of a plastid specific NPTII gene. Because we are thus far unable to prepare purified plastids from such calli, it was not possible to directly demonstrate the NPTII insertion in the plastids of these calli. Plants were therefore regenerated from the calli under continuous kanamycin selection. In these plants however, NPTII hybridizing DNA sequences were detected in total DNA preparations but not in chloroplast DNA!

If we assume that these results mean that the transformed calli resulted from a nuclear insertion of the NPTII gene and expression through a weak nuclear promoter activity of the *psb*A upstream sequences or possibly through "read through" from adjacent nuclear DNA sequences, we cannot explain the dramatic increase in transformation frequency between experiments G and H reported in Table 2.

If on the other hand we assume that plastid transformation is responsible for the observed high transformation frequency in experiment H (Table 2), we must postulate that the transformed plastid genomes are at a considerable selective disadvantage and that cells containing such transformed chloroplasts cannot regenerate into plants. Possibly only some cells within the calli in which NPTII-DNA from plastids would be released in the cytoplasm and would have been taken up into the nucleus, could form meristems capable of shoot formation in the presence of kanamycin.

III. General Conclusions

The results of the experiments reviewed above are confusing and do not allow clearcut conclusions with regard to plastid transfomation. The occurance of plastid transformation is suggested by circumstantial evidence based on high transformation frequencies obtained with vectors supposed to be specific for plastid transformation. Direct proof for the existence of plants containing primarily if not exclusively transformed chloroplasts, has thus far only been obtained in one example. In the first one reported in De Block *et al.* (1985), the presence and the expression in chloroplasts of a chloramphenicol acetyl transferase gene driven by a promoter from the nopaline synthase gene was demonstrated. However, continued growth of this plant in the absence of selection resulted in the total loss of the transformed chloroplasts and of the transforming DNA. All attempts to obtain further plants with a homogeneously transformed plastid population have thus far failed, although circumstantial evidence suggests that calli containing transformed plastids could readily be obtained using some of the vectors described in this paper. However, these putative transformed plastids appeared to be at a selective disadvantage and would seem not to be able to develop into meristems and shoots. When shoots were obtained under continuous kanamycin selection from kanamycin resistant calli resulting from high frequency transformation with plastid specific selective marker genes (Fig. 1, Table 2) it turned out that these kanamycin resistant plants did not contain the NPTII gene in their chloroplast DNA, but probably in their nuclear DNA.

In consequence we are entertaining the unlikely possibility that under these selective conditions, shoots actually regenerated from cells in which plastid-borne NPTII genes were somehow transferred to the nucleus. Although the original plastid-transformant (De Block *et al.*, 1985) was obtained by selection for chloramphenicol acetyl transferase activity (CAT), the results briefly described in this paper were in fact obtained by selection for kanamycin resistance. Constructs linking the same plastid promoters (e. g. from *psb*A) to the *E. coli cat* gene were however made and used, but chloramphenicol selection turned out to be a very tricky and unreproducible system and was therefore not pursued extensively.

The following experimental strategies are presently being pursued in order to test our explanations for the previous observations and with the hope to be able to establish a reproducible method leading to stable plastid transformation.

Dr. A. Cheung and Prof. L. Bogorad (The Biological Laboratories, Harvard University) have privately communicated to us results that strongly suggest plastid transformation. Using essentially the same type of plastid selectable marker genes as those reported here, these authors obtained kanamycin resistant plants after leaf disc infections. DNA from the chloroplasts of one particular plant was used to transform *E. coli* cells. Since the P*psb*A-NPT II gene is functional in *E. coli,* transformants were selected for in the presence of kanamycin. A circular DNA plasmid of 8 kbp was thus isolated. Furthermore, in this and in other plants, NPT II sequences linked to large plastid DNA sequences must have been present since such sequences were recovered after enzyme digestion, ligation and transformation in *E. coli.* If cloning artefacts can be completely excluded, these observations would prove the occurence of plastid transformation. Dr. A. Cheung and Dr. Bogorad have however also observed that foreign DNA in chloroplasts of kanamycin resistant plants is present in far less than one copy per plastid genome. Important plastid DNA rearrangements were also observed.

We shall use the strategy of rescueing potentially transformed chloroplast DNA by cloning in *e. coli* to test our hypothesis that the transformed calli observed in experiment H (Table 2), indeed harbour NPT II sequences linked to plastid DNA. To test whether transformation of plastid genomes results in a selective disadvantage for these plastid DNAs, we are using original plastid genes, flanked by extensive plastid DNA sequences which would give a selective advantage to plant cells by conferring resistance to an antibiotic or a herbicide. In this case we hope to select for a gene conversion between the mutant plastid gene and its wildtype counterpart resident in the plastid genome. This event, induced by homologous recombination, would leave the genome size and its structure unaltered. There are at least two possible candidates for such experiments: 1) a gene conferring streptomycin resistance and 2) a mutant *psb*A gene conferring resistance to atrazine. It has been shown that recombination between different chloroplast molecules can occur (Medgyesy *et al.,* 1985) and it therefore seems reasonable to expect that a similar recombination may occur after introduction of homologous DNA by Ti plasmid vectors.

Another possibility which we have not yet tested would be that our strategy to use relatively strong plastid gene promoters to drive the expression of the selectable marker genes, was a poor choice. Indeed the original transformant described in De Block *et al.,* was obtained with a gene driven by a non-specific nuclear promotor (from the T-DNA nopaline synthase gene). We might ultimately be forced to use such unspecific and therefore presumably weak promoters to drive expression of genes transferred into plastids.

This chapter was written by J. Schell using data and discussions provided by the authors. J. Schell therefore assumes responsibility for the somewhat controversial nature of this "Progress report".

IV. References

Cheung, A., Schell, J., Van Montagu, M., Bogorad, L., 1986: Relocation of a gene for herbicide tolerance: Converting a chloroplast gene into a nuclear gene. Science, submitted 1987.

Cseplo, A., Maliga, P., 1984: Large scale isolation of maternally inherited lincomycin resistance mutations, in diploid *Nicotiana plumbaginifolia* protoplast cultures. Mol. Gen. Genet. **196,** 407—412.

Deblaere, R., Bytebier, B., De Greve, H., Deboeck, F., Schell, J., van Montagu, M., Leemans, J., 1985: Efficient octopine Ti plasmid-derived vectors for *Agrobacterium*-mediated gene transfer to plants. Nucl. Acids Res. **13,** 4777—4788.

De Block, M., Schell, J., Van Montagu, M., 1985: Chloroplast transformation by *Agrobacterium tumefaciens.* EMBO J. **4,** 1367—1372.

Gruissem, W., 1984: A chloroplast transcription system from higher plants. Plant Molec. Biol., Reporter **2,** 15—23.

Herrera-Estrella, L., De Block, M., Messens, E., Hernalsteens, J.-P., Van Montagu, M., Schell, J., 1983: Chimeric genes as dominant selectable markers in plant cells. EMBO J. **2,** 987—995.

Horsch, R. B., Fry, F. E., Hoffmann, N. L., Eichholtz, D., Rogers, S. G., Fraley, R. T., 1985: A simple and general method for transferring genes into plants. Science **227,** 1229—1231.

Koukolikova-Nicola, Z., Shillito, R. D., Hohn, B., Wang, K., Van Montagu, M., Zambryski P., 1985: Involvement of circular intermediates in the transfer of T-DNA from *Agrobacterium tumefaciens* to plant cells. Nature **313,** 191—196.

Medgyesy, P., Fejes, E., Maliga, P., 1985: Interspecific chloroplast recombination in a *Nicotiana* somatic hybrid. Proc. Natl. Acad. Sci. U.S.A. **82,** 6960—6964.

Stachel, S. E., Timmermann, B., Zambryski, P., 1986: Generation of single-stranded T-DNA molecules during the initial stages of T-DNA transfer from *Agrobacterium tumefaciens* to plant cells. Nature **322,** 706—712.

Zambryski, P., Herrera-Estrella, L., De Block, M., Van Montagu, M., Schell, J., 1984: The use of the Ti plastid of Agrobacterium to study the transfer and expression of foreign DNA in plant cells: New vectors and methods. In: Hollaender, A., Setlow, J. (eds.), Genetic Engineering, Principles and Methods, Vol. 6, pp. 253—278, New York: Plenum Press.

Chapter 15

Targeting Nuclear Gene Products into Chloroplasts

Les J. Szabo and Anthony R. Cashmore

Laboratory of Cell Biology, Rockefeller University, New York, N. Y., U.S.A.

With 3 Figures

Contents

I. Introduction

Plant cells contain several distinct subcellular compartments or organelles that perform specialized functions within the cell. Two of these organelles, chloroplasts and mitochondria, exist in a semi-autonomous state within the cell containing their own genetic system and protein synthesis machinery. However, the majority of chloroplast and mitochondrial proteins are nuclear encoded, synthesized on free cytosolic ribosomes and then imported into their respective organelles. Many of these proteins are synthesized as higher molecular weight precursors that are processed to their mature forms either during or shortly after import. The proper targeting and import of these polypeptides depends on a transit peptide located at the amino-terminus of the precursor (for recent reviews see Ellis, 1981; Cashmore *et al.*, 1985; Schmidt and Mishkind, 1986; Hay *et al.*, 1984 and Hurt and van Loon, 1986).

In the late 1970's *in vitro* studies employing cell free translation products of poly (A) mRNA from *Chlamydomonas* and higher plants pro-

Table I. Identification of Chloroplast Precursor Polypeptides.

Location Protein	Source	*in vitro* MW*	*in vivo* MW*	Transit peptide*	*in Vitro* Import+	Ref.
Outer Envelope						
22 kd	Spinach	32	22	10	N	17
Inner Envelope						
Phosphate translocator	Spinach	40	29	11	N	17
36 kd	Spinach	38	36	2	N	17
Stromal						
Ribosome L-18	Chlamydomonas	18.5	15.5	3	N	51
Ribosome S-16	Pea	20	15	5	Y	18
Ribosome L-6	Pea	31	27	4	Y	18
Ribosome L-12	Pea	25	20.5	4.5	Y	18
Ribosome L-13	Pea	23	19	4	Y	18
Ribosome L-18	Pea	23	17.5	5.5	Y	18
Ribosome L-25	Pea	17	13.5	3.5	Y	18
Acyl carrier protein I	Spinach	20	14.3	5.7	N	41
Acyl carrier protein II	Spinach	21	15	6	N	41
Ferrodoxin	Tobacco	27	22	5	Y	64
Ferritin	Bean	31.5	26.5	5	Y	64
Pyruvate phosphate dikinase	Maize	110	94	16	Y	1
Ribulose 1,5-bisphosphate carboxylase	Chlamydomonas	20	16.5	3.5	N	13
Ribulose 1,5-bisphosphate carboxylase	Pea	20	14	6	N	4
Ribulose 1,5-bisphosphate carboxylase	Pea	20	14	6	Y	9, 24
Ribulose 1,5-biphosphate carboxylase	Spinach	18	14	4	Y	9
Thylakoid						
Chlorophyll a/b binding protein	Pea	32	28	4	Y	50
Ferredoxin-NAPD oxidoreductase	Pea	43	35	8	Y	22
Water splitting protein 34 kd	Spinach	40	34	6	Y	66
23 kd	Spinach	33	23	10	Y	66
16 kd	Spinach	26	16	10	Y	66
Lumen						
Plastocyanin	Pea	25	10	15	Y	22
Plastocyanin	Barley	18	10	8	Y	45
Plastocyanin	Wheat	17	10	7	Y	45

* Molecular weights are in kilodaltons.
+ Denotes whether *in vitro* import of precursor polypeptide was demonstrated.
N = No
Y = Yes.

References see p. 338

vided the first direct evidence that chloroplast proteins are imported post-translationally. Precursors of the small subunit of ribulose-1,5-bisphosphate carboxylase (preRBCS) and the protein components of the light harvesting chlorophyll a/b binding complex (preCAB) were shown to be imported, processed and assembled correctly by isolated chloroplasts (Dobberstein *et al.*, 1977; Cashmore *et al.*, 1978; Highfield and Ellis, 1978; Chua and Schmidt, 1978; Schmidt *et al.*, 1981). Subsequent to these studies many additional chloroplast proteins were also shown to be synthesized as higher molecular weight precursors and transported post-translationally into chloroplasts (Table I). In addition, several nuclear genes encoding chloroplast proteins have been cloned and sequenced (see below), thus enabling a more detailed study of the structure of transit peptides and the import process.

Recently, several reports have appeared in which chloroplast transit sequences have been fused to genes encoding non-chloroplast proteins (Van den Broeck *et al.*, 1985; Schreier *et al.*, 1985; Lubben and Keegstra, 1986). These studies clearly demonstrate that the transit peptide is sufficent to target and mediate transport of foreign gene products into chloroplasts. However, very little is known about the details of the import process and the specific requirements and limitations of this system. Experimentally the import process can be divided into three steps: *i*, binding of the precursor molecule to the outer chloroplast envelope membrane; *ii*, translocation of the polypeptide across the envelope membrane(s); and *iii*, processing of the precursor polypeptide to the mature form. In this article we will review what is known about each of these steps and then we will discuss recent experiments which examine the role of the transit peptide in the import of polypeptides into chloroplasts.

II. Binding of Precursors to the Outer Membrane of the Chloroplast

One of the first events in the import process is the binding of precursor polypeptides to the outer chloroplast membrane. Grossman *et al.*, (1979) observed that at 4 °C intact chloroplasts would bind but not import precursor polypeptides and that upon elevation of the temperature a fraction of these bound polypeptides were imported. Pfisterer *et al.* (1982) incubated isolated envelope membranes with poly (A) mRNA translation products and observed binding of many different polypeptides; included in this group were preRBCS and preCAB. Competition experiments demonstrated that unlabeled translation products would compete with the binding of preRBCS and preCAB but wheat germ extracts showed no effect. Specificity of binding was demonstrated by the inability of thylakoid membranes to selectively bind precursor polypeptides. In similar experiments, Ellis (1983) observed that thylakoid and envelope membrane fractions exhibited equal affinity for binding preRBCS and preCAB polypeptides, but demonstrated that these polypeptides showed no specificity for binding to membranes isolated from red blood cells.

Recently, a more specific binding assay has been developed (Cline *et al.*, 1985). Translation products of hybrid-selected mRNA for preRBCS and preCAB were shown to bind but not be imported by intact chloroplasts in the presence of the ionophore nigericin. This inhibition was reversed by the addition of ATP, resulting in the import of 50—70 % of the bound preRBCS and 30—50 % of the bound preCAB. Typically, 3—4 times more preRBCS molecules were bound per chloroplast than preCAB. Preincubation of chloroplasts with protease greatly reduced the binding and import of these polypeptides, suggesting that proteinaceous receptors are involved (Cline *et al.*, 1985; Chua and Schmidt, 1978). At present there is no direct evidence for the involvement of specific receptors in the binding process of precursor polypeptides to the chloroplast surface.

III. Translocation of Polypeptides Across the Envelope Membranes

Very little is known about the actual mechanism by which proteins are translocated across the two envelope membranes of chloroplasts. It has been demonstrated that this event requires energy in the form of ATP (Grossman *et al.*, 1980). Import of translation products by isolated pea chloroplasts is stimulated by light or the addition of ATP when they are incubated in the dark. This light dependent import process is inhibited by uncouplers of photosynthetic phosphorylation but not inhibitors of electron transport in photosystem II. This inhibition is readily reversed by the addition of exogenous ATP, demonstrating that protein import into chloroplasts requires ATP and not a chemical or electrical membrane potential. This contrasts with the evidence that import of proteins by mitochondria requires an electrochemical potential across the inner membrane (Schleyer *et al.*, 1982; Kolansky *et al.*, 1982; Gasser *et al.*, 1982).

The specific mechanism by which ATP is required for protein import into chloroplasts has not yet been defined. Several chloroplast protein kinases have been characterized (Lin *et al.*, 1982; Soll and Buchanan, 1983), one of which is localized in the outer envelope membrane. Incubation of isolated envelope membranes with this enzyme preparation resulted in the phosphorylation of the mature form of RBCS (Soll and Buchanan, 1983). The exact nature of this membrane bound form of RBCS is not known and may just represent contamination of the membrane fraction. However, it is interesting that RBCS acts as a substrate for this enzyme, raising the possibility that phosphorylation of RBCS (or preRBCS) may be involved in the import mechanism. Currently *in vitro* techniques are available to test if a phosphorylated intermediate is involved and if so, what form.

IV. Processing of Precursors to the Mature Polypeptide

All cytoplasmically synthesized chloroplast proteins characterized to date are produced as precursor polypeptides which are proteolytically cleaved to their mature forms either during or immediately after translocation across the membrane(s). The exact role of processing in the import process is not known, although there is evidence that processing is not essential for import. Experiments employing *Chlamydomonas* mRNA translation products incubated with isolated pea chloroplasts resulted in the accumulation of an 18,000 dalton intermediate of RBCS within the chloroplasts (Mishkind *et al.*, 1985). This intermediate form could be further processed to the mature 16,500 dalton form by incubation with cell extracts from *Chlamydomonas.* The 18,000 dalton intermediate polypeptide was unable to assemble with pea large subunits to form holoenzyme and was degraded. Microsequencing of the amino-terminus of the 18,000 dalton polypeptide showed that cleavage occurred between two alanine residues located in a conserved region in the middle of the transit peptide (Figure 1).

A similar observation was made when translation products from pea mRNA were incubated with iodoacetate prior to import, resulting in the accumulation of a prominent 18,000 dalton intermediate of RBCS polypeptide (Robinson and Ellis, 1984b). Apparently, the carboxymethylation of the cysteine residue at the carboxyl end of the transit peptide blocked the proper processing of preRBCS. These experiments suggest that processing of the precursor to the mature form of RBCS is not required for import. It has been presumed by several authors that this alternate processing site reflects a genuine intermediate processing step. Whereas this indeed may be so, an alternative explanation that proteolytic cleavage occurs only at the alternate site in the absence of efficient processing at the juncture of the mature polypeptide and the transit peptide, has not been rigorously excluded. Evidence against this alternate model would be the demonstration that a homogeneous enzyme processes preRBCS in a two-step manner.

A more definitive demonstration of the separation of import from processing comes from experiments that examined the import of *Lemna* preCAB into chloroplasts isolated from greening barley leaves (Chitnis *et al.*, 1986). Import experiments using chloroplasts isolated from etiolated barley leaves exposed to light for 0—5 hours, contained mostly the precursor form of CAB in the thylakoid fraction. Nondenaturing gel electrophoresis showed that the imported preCAB polypeptides were assembled into the light harvesting complexes containing chlorophyll but whether they were properly assembled and functional is not known. These experiments clearly demonstrate that import of preCAB is independent of processing.

A protease capable of specifically processing preRBCS to its mature form has been partially purified from pea (Robinson and Ellis, 1984a). This protease is soluble, metal-ion dependent and has an apparent molecular weight of 180,000 daltons as determined by gel filtration. The

```
                                                                                 Ref
Ribulose 1,5-bisphosphate carboxylase

CHLAMYDOMONAS
          MAVSAKSSVSAA-VAR--PARSS-VRPMAALK--------PAV-KA------APVVAPAEAND  M  49
            I                                                  A

LEMNA
 SSU 5A   MASSMMASTAA--VARAGPAQSSMVAPFNGLRSSVA--FPATRKANNDLSTLPSNGGRVSC  M  61
 SSU 5B   ...........--...V....TN.............--.......................  .  61
 SSU 26   ...........--........S..............--........KN.........K...  .  61
 Lg SSU   ......V....--...VR...TN..GA...C.....--..................S......  .  58

WHEAT
 W9       MAPAVMASSA-TT-------------VAPFQGLKST-AGL-PISCRSGSIGLSSVSNGGRIRC  M   3
 W4.3     ..........-..-------------..............--.V.R..-RGS.G..........  .   3

PEA
 SS3.6    MAS-MISSSAVTTVSRASRGQSAAVAPFGGLKSSMTG-FP-VKK-NTDITSITSNGGRVKC  M   5
 SS8.0    ...-...............V................-..-...-................  .  60
 pPS-2.4  ...-................................-..-...-................  .  12

SOYBEAN
 SRS1     MASSMISSPAVTTVNRAGAG---MVAPFTGLKS-MAG-FPTRKR-NNDITSIASNGGRVQC  M   2

TOMATO
 RBCS-1   MASSIVSSAAAAT--RSNVAQASMVAPFTGLKSAAS-FPVTKKNNNVDITSLASNGGRVRC  M  44
 RBCS-2A  ....AI....V..--....T.............S.T-...SR.-Q.L....I.......S.  .  44
 RBCS-3A  ....AM....V..--.G.G..............T..-...SR.-Q.L....I.......S.  .  44

TOBACCO
 NtSS23   MASSVLSSAAVAT--RSNVAQANMVAPFTGLKSAAS-FPVSRK-QNLDITSIASNGGRVQC  M  39
                                                     /

PETUNIA
 SSU 8    MASSVISSAAVAT--RTNVAQASMVAPFNGLKSAVS-FPVSSK-QNLDITSIASNGGRVQC  M  62
 SSU 11A  .....M.......--S..A.........T.....A.-....R.-.................  .  62

FERREDOXIN
 SILENE
          MASTLSTLSVSASLLPKQQPMVASSLPTNMGQALFGLKAGSRGRVTAM  A            55

Chlorophyll a/b binding polypeptides

PEA
 AB66     MAASSSSSMALSSPTLAGKQLKLNPSS----QEL-GAARFTMR                    60
 AB80     ...........................----...-........                     6

TOMATO
 1A       MAAAA---MALSSPSFAGQAVKLSPSA----SENSGNGRITMR                    43
 1B       ....T---...................----..I.........                    43
 1C       ....T---...................----..I.........                    43
 3A       ...ST---.....ST...KT...A..S----..IT........                    43
 3B       ...ST---.....ST...K.......S----..I.........                    43
 3C       ..SST---.....ST...K.......S----..IT....V...                    43

ARABIDOPSIS
 AB165    MAAST---MALSSPAFAGKAVNLSPAA----SEVLGSGRVNMR                    36
 AB180    .....---...................----............                    36
 AB140    .....---.............K.....----............                    36

PETUNIA
 91R      MAAAT---MALSSPSFAGKAVKFSPSS----SEITGNGKATMR                    14
 13       .....---.....S.......NV-...----....R...V...                    14
 25       .....---..I..S.......NV-...----.Q..........                    14
 22L      .....---.....ST....V..L....----............                    14
 22R      ...T.---.....S........L.S..----........V...                    14

LEMNA
 30       MAAS----MALSSPSLVGKAVKLAPAA---SEVFGEGRVSTMR                    32
 19       ....------.IQ.SAFA.QTALKQRDELVRLVGVSD..FSTMR                   29

WHEAT
 AB1.6    MAATT---MSLSSSSFAGKAVKNLPSS-----ALIGDARVNMR                    35

Plastocyanin

SILENE
     MATVTSSAAVAIPSFAGLKASSTTRAATVKVAVATPRMSIKASLKDVGVVVAATAAAGILAGNAMA A  56
```

Fig. 1. Transit sequences of proteins transported into chloroplasts. Amino acids are designated by the one letter code. A dot (.) represents an identical amino acid with that in the top sequence. A dash (—) represents a deletion introduced to maximize homology. References see p. 339

exact composition of this enzyme awaits purification to homogeneity, in that the above complex contains at least 10 polypeptides. The soluble nature of this enzyme and the lack of activity found in washed envelope membranes, indicates that this enzyme is located in the stroma or is loosely associated with the envelope membrane. In addition to processing pea preRBCS, the precursors of plastocyanin from wheat and barley are also cleaved to their mature forms by this protease fraction. If this enzyme preparation contains only a single protease, then this enzyme is neither precursor specific nor species specific. It is intriguing that plastocyanin, a protein located in the inner thylakoid membrane space (lumen) maybe processed by the same stromal protease that cleaves the precursor of RBCS. These results suggest that either the mature form of plastocyanin is capable of transporting itself across the thylakoid membrane or that a similar protease is located outside of the stroma.

Precursor modification experiments have resulted in either partial or complete loss of the ability of preRBCS to be processed by this chloroplast enzyme preparation. Incubation of preRBCS with iodoactetate prior to incubation with this enzyme resulted in the formation of an 18,000 dalton intermediate identical to that observed with isolated chloroplasts (Robinson and Ellis, 1984b). Experiments in which canavanine was substituted for arginine in the translation mix resulted in the complete loss of processing for preRBCS (Robinson and Ellis, 1985). This inhibition of cleavage was pH dependent, exhibiting an increasing level of precursor maturation as the pH of the incubation mixture decreased below the pK_a of canavanine. Additional experiments in which amino acid analogues of proline or leucine were used also demonstrated inhibition of the processing of preRBCS. These experiments suggest that changes in the conformation of the precursor polypeptide strongly influence its ability to be a substrate for this proteolytic enzyme.

V. The Transit Peptide Itself Can Mediate Import of Foreign Polypeptides

In order to definitively demonstrate that the transit peptide can mediate import of foreign polypeptides, a hybrid construct was made by fusing the RBCS transit sequence to the bacterial neomycin phosphotransferase II (NPT) gene (Van den Broeck *et al.*, 1985). This chimeric gene was then introduced into tobacco cells and was shown to be expressed in a light regulated manner from the RBCS promoter. Isolated chloroplasts from transformed tissue contained neomycin phosphotransferase activity. The imported polypeptide was protease insensitive and comigrated with bacterial NPT suggesting that it was processed correctly. These results clearly demonstrate that a foreign gene product can be targeted to the chloroplast and that only the information contained within the transit peptide is required. In a second experiment, the transit peptide plus the first 23 amino acids of the mature RBCS polypeptide was fused to neomycin phosphotransferase (Schreier *et al.*, 1985). This region (the first 23 amino acids) of the mature RBCS polypeptide is highly conserved, indicating that it may have a functional role in the import process. This hybrid polypeptide was also expressed in transformed tobacco tissue and shown to be imported into chloroplasts.

Due to the nature of these *in vivo* experiments it is very difficult to quantitate the efficiency of import for chimeric polypeptides. Therefore, an *in vitro* import system has been developed and is now being used in many laboratories. In addition to quantitation, the use of an *in vitro* system makes it possible to analyze the import characteristics of many more precursor polypeptides than would be possible *in vivo*. Employing such a system, we have demonstrated that although the transit peptide of RBCS alone is sufficient to mediate import of neomycin phosphotransferase II, it does so inefficiently (Szabo and Cashmore, unpublished results; Wasman, C. *et al.*, 1986). In comparison to preRBCS, this hybrid polypeptide is imported at a level of 10—20 % of wild-type. Similar results were obtained in import experiments in which the soybean RBCS transit peptide and the first 13 amino acids of the mature RBCS protein were fused to a soybean 17,500 dalton heat shock protein (Lubben and Keegstra, 1986). This hybrid polypeptide was imported at a rate approximately 40 % of the soybean preRBCS polypeptide.

In a second set of NPT fusion experiments the efficiency of two constructs were compared *in vitro*, one containing just the transit peptide and the second containing the transit peptide and the first 23 amino acids of the mature RBCS polypeptide. The inclusion of the first 23 amino acids of the mature polypeptide dramatically improved the import efficiency to approximately wild type levels (Wasman, C. *et al.*, 1986), suggesting that sequences within the mature RBCS polypeptide contribute to the import process either directly or indirectly by influencing the conformation of the transit peptide.

In contrast to the *in vitro* results, the two NPT fusion polypeptides display markedly different results *in vivo* (Kuntz *et al.*, 1986). Transgenic tobacco plants containing both of the hybrid constructs were obtained by crosses and the behavior of each of the NPT fusion polypeptides was analyzed. The majority of NPT activity resulting from the fusion polypeptide containing only the RBCS transit peptide was localized within the chloroplast stroma. However, NPT activity resulting from the second hybrid polypeptide, containing the transit peptide and the first 23 amino acids of the mature RBCS protein, was localized predominantly outside of the chloroplast. This cytoplasmic NPT activity displayed an apparent molecular weight similar to the expected size if proper processing had occurred. No precursor polypeptide was detected for either of these constructs. The authors suggest that *in planta* the efficiency of import is mediated not only by the rate at which a polypeptide is translocated across the chloroplast membranes, but also by the stability of the precursor polypeptide in the cytoplasm. These results underscore the difficulty of extrapolating results obtained from *in vitro* experiments to *in vivo*.

VI. Structural Analysis of Chloroplast Transit Peptides

The exact mechanism by which chloroplast transit peptides mediate the import and processing of precursor polypeptides is not known. Although these transit peptides vary substantially in terms of length and amino acid composition, they share similar properties. The precursor of RBCS from the algae *Chlamydomonas* is imported by isolated chloroplasts from higher plants (Mishkind *et al.*, 1985; Kloppstech *et al.*, 1985). Similarly, precursors of RBCS from monocots can be imported and processed by chloroplasts from dicots (Coruzzi *et al.*, 1983; Mishkind *et al.*, 1985) indicating that the import mechanism has been conserved functionally across the diverse evolutionary range found within the plant kingdom.

In the last few years sequence data has become available for a limited number of different nuclear encoded chloroplast genes and in two cases from several different species. A list of the currently published amino acid sequence data for these transit peptides is given in Figure 1. Comparison of this data has revealed some general characteristics which include: *i*, an abundance of arginine and lysine residues, resulting in a net positive charge; *ii*, a conspicuous lack of negatively charged side chains; *iii*, relative conservation of the position of proline residues; *iv*, the occurrence of a methionine-alanine-serine sequence at the amino terminus; and *v*, relative abundance of alanine, serine and threonine residues, comprising 30—40 % of the total amino acids.

By comparing amino acid transit sequence data for RBCS from evolutionarily divergent species, three conserved regions have been defined (Mishkind *et al.*, 1985; Schmidt and Mishkind, 1986). Domain I consists of the first 12—20 amino-terminal amino acids, which are rich in threonine and serine residues. Domain II consists of 11 to 18 residues surrounding a glycine-leucine-lysine block, is positively charged and contains one or two prolines. Because this region is the most conserved, it was proposed that this region constitutes the binding site for envelope receptors. In addition, this region contains the cleavage site that results in the formation of the 18,000 dalton intermediate of preRBCS. Domain III is also rich in serine and threonine residues, has a net positive charge and terminates in a cysteine-methionine at the transit peptide-mature polypeptide juncture. The only exception for domain III is the *Chlamydomonas* sequence which exhibits very little similarity to the higher plant sequences. This correlates well with the *in vitro* import data suggesting that this region is involved in the final processing of the precursor to the mature form and is of relative minor importance for chloroplast import.

Extending this analysis to include the transit sequence data for CAB and ferredoxin, Karlin-Neumann and Tobin (1986) proposed a shared framework model. This model contains three homology blocks (I, MASASS; II, PFGR; III, GGRV) and two variable interblock regions. They proposed that homology blocks I and/or II are essential for binding and import into chloroplasts and homology block III functions primarily in the final processing event. Any sequences that are involved in specifying intraorganellar sorting would be located in the interblock regions. Sequence data for the precursor of plastocyanin (Smeekens *et al.*, 1985b) contains similar regions to those described in this model, with homology block III being the most different. The most striking difference between the preplastocyanin and the other transit peptides is the large hydrophobic region at the carboxy terminus, flanked by charged residues on either side. This sequence resembles a stop transfer sequence found in mitochondrial presequences. This is the only example of a putative membrane spanning region in a chloroplast transit sequence, suggesting that this region may play an important role in intraorganellar sorting to the lumen.

VII. Experimental Analysis of Transit Peptides

At present there is a small but rapidly growing body of experimental data addressing the functional significance of the conserved regions of transit peptides. Removal of the amino terminal portion or the entire transit peptide of preRBCS *in vitro* resulted in the complete loss of import, demonstrating that the amino terminus, at least, contains a region required for import (Mishkind *et al.*, 1985). Amino acid substitution experiments have demonstrated that modifying proline, leucine or arginine residues drastically reduces the ability of preRBCS to be imported and stably incorporated into holoenzyme (Robinson and Ellis, 1985). Unfortunately, these experiments were unable to discriminate between those that influence the transit peptide from those that influence the mature protein. However, these experiments indicated that structural conformation of the polypeptide is important for efficient import. Analysis of modified transit peptides containing specific amino acid substitutions and deletions will provide a direct method for determining the functional importance of the domains identified in the models proposed above.

As was outlined above, the import process can be divided into 3 distinct steps; binding, translocation, and processing. Each of these steps is thought to be mediated by the transit peptide. Are there distinct functional domains that define each of these steps and if so, do these functional domains correspond to the conserved regions defined above? One approach to test this question directly is to construct a series of deletion mutants within the transit peptide and determine their import characteristics. If there are distinct functional domains, it should then be possible to construct deletion mutants that are only defective in one of the three import steps. For example, one might expect to find a mutant that would bind but not be translocated, or bind and translocate but not be processed.

With this goal in mind, we have constructed a series of amino-terminal and carboxy-terminal deletions within the pea RBCS transit peptide. The exact nature of some of these deletions is shown in Figure 2. Labeled preRBCS polypeptides and deletion derivatives were produced *in vitro* by an SP6 transcription/wheat germ translation system (Szabo, unpublished results). The *in vitro* import characteristics of each of these derivatives is

```
              5        15        25        35        45        55
PRERBCS   MASMISSSAVTTVSRASRVQSAAVAPFGGLKSMTGFPVKKVNTDITSITSNGGRVKC  MQW

D5::8     MASMI   AVTTVSRASRVQSAAVAPFGGLKSMTGFPVKKVNTDITSITSNGGRVKC  MQW

D5::10    MASMM     TTVSRASRVQSAAVAPFGGLKSMTGFPVKKVNTDITSITSNGGRVKC  MQW

D5::13    MASMI        SRASRVQSAAVAPFGGLKSMTGFPVKKVNTDITSITSNGGRVKC  MQW

D5::18    MASMM             VQSAAVAPFGGLKSMTGFPVKKVNTDITSITSNGGRVKC  MQW

D5::27    MASMI                      GGLKSMTGFPVKKVNTDITSITSNGGRVKC  MQW

D5::51    MASMI                                              GGRVKC  MQW

D5::60    MASMM                                                        W

D37::53   MASMISSSAVTTVSRASRVQSAAVAPFGGLKSMTGFP               GRVKC  MQW

D33::52   MASMISSSAVTTVSRASRVQSAAVAPFGGLKSM                  GGRVKC  MQW
```

Fig. 2. RBCS transit peptide deletions. The amino acid sequence of the wild-type RBCS transit peptide is shown in single letter code in the top line. The nature of the amino acid deletion for each of the derivatives are shown below preRBCS and are designated by numbering the codons that are fused in the formation of the deletion

shown in Figure 3 A. In a parallel experiment, binding efficiences of these modified polypeptides were analyzed (Figure 3 B). The most striking result from these experiments comes from the comparison of the import and binding data (Figure 3 C). We were unable to find a deletion that would effect translocation but have no effect on binding. In every case, the level of import generally reflected the efficiency at which that polypeptide bound. A deletion polypeptide in which almost all of the transit peptide (D5 : : 60) was removed showed no detectable import and a very low level of binding (which probably represents nonspecific binding). In the case of D37 : : 53 and D33 : : 52, minor bands with the apparent molecular weights of 15,000 and 13,000 were observed. The 15,000 dalton polypeptide corresponded to the size expected if an intermediate is formed by a two-step processing event. In a separate experiment, chloroplasts were fractionated into soluble and membrane components, prior to electrophoretic analysis. Both of these minor bands were in the soluble fraction, further supporting the hypothesis that import and processing are independent events.

Two general conclusions can be drawn from these results. First, we find no evidence for sequences that are exclusively required for the process of translocation across the envelope membranes, in contrast to the many deletions that effect binding. Second, since deletions in the amino and carboxy terminal regions of the transit peptide resulted in similar reductions in the levels of binding, we conclude that either binding does not simply involve a short sequence of amino acids or if it does, then binding is strongly affected by the nature of distal sequences.

Recently, two RBCS transit peptide deletions have been tested *in vivo* using NPT fusions (Kuntz *et al.,* 1986). Deletion of the amino acids cys-

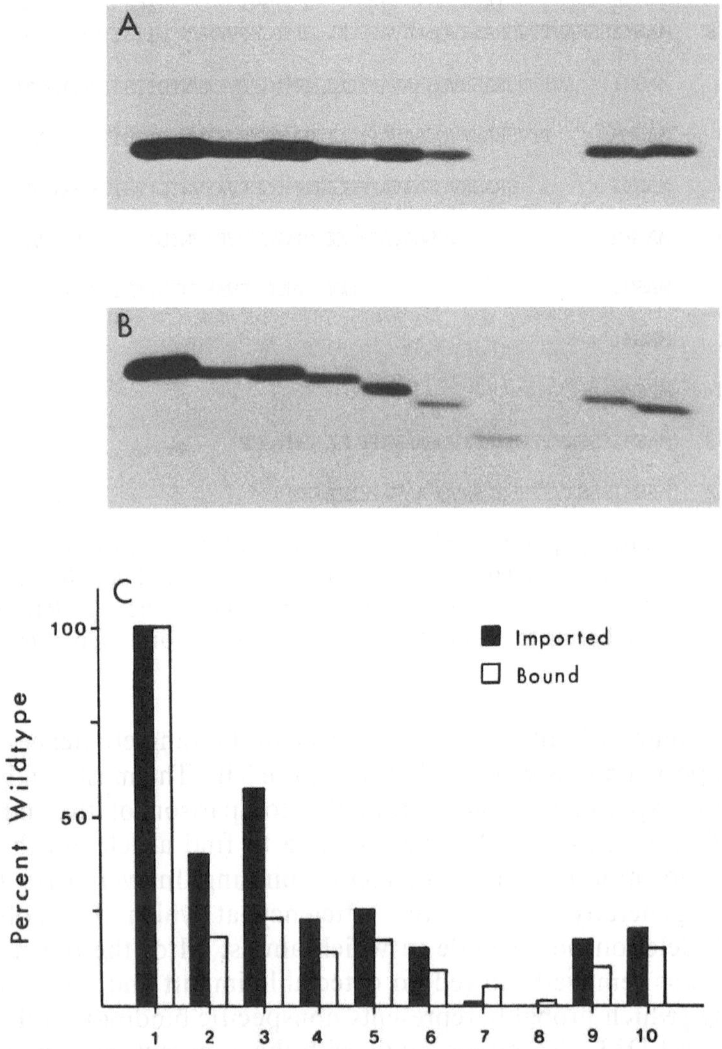

Fig. 3. In vitro import and binding of preRBCS and deletion derivatives. Pea chloroplasts were incubated in the light with translation products for 60 min. Intact chloroplasts were reisolated and analyzed by sodium dodecyl sulfate-polyacrylamide gel electrophoresis/fluorography. Lanes 1, preRBCS; 2, D5::8; 3, D5::10; 4, D5::13; 5, D5::18; 6, D5::27; 7, D5::51; 8, D5::60; 9, D37::53; 10, D33::52. A. Import characteristics of preRBCS and deletion derivatives. Chloroplasts were protease treated after incubation with translation products. B. Binding characteristics of preRBCS and deletion derivatives. Nigericin was added to the incubation mixture, at a final concentration of 400 nM, prior to the addition of translation products. C. The quanitation of the results in A and B. Values represent percentile of preRBCS

teine and methionine at the juncture between the the transit peptide and the mature protein, had no effect on the ability of this construct to be imported or processed. Therefore, the processing enzyme appears not to recognize the actual site at which cleavage occurs, but sequences on the amino-terminal side of the cleavage site. This is in contrast to the conclusion drawn from the experiments in which carboxymethylation of the precursor polypeptides completely blocked processing resulting in the suggestion that this cysteine residue was important for correct processing of the precursor polypeptide (Robinson and Ellis, 1984b). The second construct, containing a deletion of the 17 amino acids at the carboxy-terminal end of the transit peptide appears to interfere with proper import and processing of the NPT fusion polypeptide. Isolated chloroplasts from transformed tobacco plants contained three prominent forms exhibiting NPT activity, the major component comigrating with the bacterial NPT II activity. The other two forms were of intermediate size. All three forms cofractionated with the chloroplast membranes, in contrast to the wild-type fusion polypeptide which was found exclusively in the chloroplast stroma. The two intermediate forms were partially degraded by trypsin treatment indicating that at least part of the polypeptide is exposed to the outer surface of the chloroplast. This latter observation is in contrast with the *in vitro* data presented above. It is unclear whether this difference is a result of the slightly different nature of the two deletions (D37::53 versus D41::57) used in the two experiments or due to the fact that each construct contained a different "body" polypeptide.

It is of interest to compare the results from chloroplast import studies with related results obtained for mitochondria. Two mitochondrial presequences have been modified and analysed for import behavior. A series of deletions were constructed in the presequence of ornithine transcarbamyl transferase (Horwich *et al.*, 1986). Removal of the six carboxy-terminal amino acids result in wild-type levels of import. A series of amino-terminal deletions progressively reduced the level of import and an internal deletion of 15 amino acids completely abolished import. The most interesting results however, were the observations that substitution of one arginine (position 23) to glycine caused a complete loss of import while substitution at the other three arginines only moderately reduced the import levels. Moreover, if other amino acids such as asparagine, alanine or lysine were substituted at position 23, increasing levels of import were observed (approximately 20, 45 and 80%, respectively). These results strongly suggest that the overall conformation of the presequence is an important factor for the efficient import of mitochondrial proteins. The deletion results are similar to those obtained with preRBCS. It would be interesting to know the effect of these presequence deletions on binding to de-energized mitochondria. In contrast to these results, deletion analysis of the presequence of cytochrome oxidase subunit IV has shown that only the first of 12 of the 25 amino acids in the presequence are needed for efficient import of dihydrofolate reductase into mitochondria (Hurt *et al.*, 1985). Larger deletions in which only the first seven or nine amino acids of the presequence remain are not imported and do not bind to de-energized mitochondria.

VIII. Future Prospects

Currently, little is known about either the mechanism by which proteins are translocated across chloroplast membranes or the specific requirements and limitations operating on the targeting signals. As discussed above, there is indirect evidence for proteinaceous receptors on the outer surface of the chloroplast envelope. If there are specific protein receptors, then how many are there and are they divided into classes directing protein translocation to different internal locations? Are there additional proteins in the chloroplast membrane which mediate transport, for example proteins which form pores? What is the role of ATP in the translocation of proteins? Are there phosphorylated precursor intermediates, or is ATP used indirectly in the import process? Are the energy requirements of chloroplast and mitochondrial import as fundamentally different as the evidence from inhibitor studies suggest?

Despite our lack of knowledge on how proteins are translocated across chloroplast membranes, great success has been obtained in targeting foreign polypeptides into chloroplasts by simply fusing the transit peptide to the amino-terminus of a foreign peptide. The nature of the signals contained within these transit peptides is unknown, but the tools by which these signals can be decifered are available and are currently being employed. It appears that transit peptides contain at least two sets of information; organelle specific signals and signals which direct proteins to the correct internal organelle compartment. Recently, it has been proposed that mitochondrial transit sequences (presequences) consists of a simple linear array of short matrix-targeting domains, stop transfer domains, and cleavage sites (Hurt and van Loon, 1986). By varying the order and number of these domains and cleavage sites, specific signals can be constructed to target a protein to a given mitochondrial compartment. At present, no clear picture exists on how internal sorting occurs in chloroplast protein transport. This is due in part to the limited number of different transit peptides that have been sequenced. Does the chloroplast transit peptide also contain a linear array of signals that specify the internal localization? If so, will specific mitochondrial domains (for example stop transfer sequences) subsititute for the corresponding chloroplast domain? Answers to these and many other questions will provide greater understanding of the basic process of targeting proteins into chloroplasts and enhance our ability to genetically manipulate these organelles for the improvement of agronomic plants.

IX. References

Aoyagi, K., Bassham, J. A., 1985: Synthesis and uptake of cytoplasmically synthesized pyruvate, Pi dikinase polypeptide by chloroplasts. Plant Physiol. **78**, 807—811.

Berry-Lowe, S. L., McKnight, T. D., Shah, D. M., Meagher, R. B., 1982: The nucleotide sequence, expression and evolution of one member of a multi gene family encoding the small subunit of ribulose-1,5-bisphosphate carboxylase in soybean. J. Mol. Appl. Genet. **1,** 483—498.

Broglie, R., Coruzzi, G., Lamppa, G., Keith, B., Chua, N.-H., 1983: Structural analysis of nuclear genes coding for the precursor to the small subunit of wheat ribulose-1,5-bisphosphate carboxylase. Bio/Technology **1,** 55—61.

Cashmore, A. R., Broadhurst, M. K., Gray, R. E., 1978: Cell-free synthesis of leaf protein. Identification of an apparent precursor of the small subunit of ribulose-1,5-bisphosphate carboxylase. Proc. Natl. Acad. Sci., U.S.A. **75,** 655—659.

Cashmore, A. R., 1983: Nuclear genes encoding the small subunit of ribulose-1,5-bisphosphate carboxylase. In: Genetic Engineering of Plants, Kosuge, T., Meredith, C. P., Hollaender, A. (ed.), p. 29—38. New York: Plenum Press.

Cashmore, A. R., 1984: Structure and expression of a pea nuclear gene encoding a chlorophyll a/b binding polypeptide. Proc. Natl. Acad. Sci., U.S.A. **81,** 2960—2964.

Cashmore, A., Szabo, L., Timko, M., Kausch, A., Van den Broeck, G., Schreier, P., Bonhert, H., Herrera-Estrella, L., Van Montagu, M., Schell, J., 1985: Import of polypeptides into chloroplasts. Bio/Technology **3,** 803—808.

Chitnis, P. R., Harel, E., Kohorn, B. D., Tobin, E. M., Thornber, J. P., 1986: Assembly of the precursor and processed light-harvesting chlorophyll a/b protein of *Lemna* into the light-harvesting complex II of barley etiochloroplasts. J. Cell Biol. **102,** 982—988.

Chua, N.-H., Schmidt, G. W., 1978: Post-translational transport into intact chloroplasts of a precursor to the small subunit of ribulose-1,5-bisphosphate carboxylase. Proc. Natl. Acad. Sci., U.S.A. **75,** 6110—6114.

Cline, K., Werner-Washburne, M., Lubben, T. H., Keegstra, K., 1985: Precursors to two nuclear-encoded chloroplast proteins bind to the outer envelope membranes before being imported into chloroplasts. J. Biol. Chem. **260,** 3691—3696.

Coruzzi, G., Broglie, R., Lamppa, G., Chua, N.-H., 1983: Expression of nuclear genes encoding the small subunit of ribulose-1,5-bisphosphate carboxylase. In: Structure and Function of Plant Genomes. Ciferri, O., Dure III, L. (eds.), p. 42—59. New York: Plenum Press.

Coruzzi, G., Broglie, R., Edwards, C., Chua, N.-H., 1984: Tissue-specific and light-regulated expression of a pea nuclear gene encoding the small subunit of ribulose-1,5-bisphosphate carboxylase. EMBO J. **3,** 1671—1679.

Dobberstein, B., Blobel, G., Chua, N.-H., 1977: *In vitro* synthesis and processing of a putative precursor for the small subunit of ribulose-1,5-bisphosphate carboxylase of *Chlamydomonas reinhardtii.* Proc. Natl. Acad. Sci., U.S.A. **74,** 1082—1085.

Dunsmuir, P., 1985: The petunia chlorophyll a/b binding protein genes: a comparison of Cab genes from different gene families. Nucl. Acids Res. **13,** 2503—2518.

Ellis, R. J., 1981: Chloroplast proteins: synthesis, transport and assembly. Annu. Rev. Plant Physiol. **32,** 111—137.

Ellis, R. J., 1983: Chloroplast protein synthesis: principles and problems. Subcell. Biochem. **9,** 237—261.

Flugge, U. I., Wessel, D., 1984: Cell-free synthesis of putative precursors for envelope membrane polypeptides of spinach chloroplasts. FEBS Lett. **168,** 255—259.

Gantt, J. S., Key, J. L., 1986: Isolation of nuclear encoded plastid ribosomal protein cDNAs. Mol. Gen. Genet. **202,** 186—193.

Gasser, S. M., Daum, G., Schatz, G., 1982: Import of proteins into mitochondria. Energy-dependent uptake of precursors by isolated mitochondria. J. Biol. Cehm. **257,** 13034—13041.

Grossmann, A. R., Bartlett, S., Schmidt, G. W., Chua, N.-H., 1979: Post-translational uptake of cytoplasmically synthesized proteins by intact chloroplasts *in vitro.* Annals N.Y. Acad. Sci. **343,** 266—274.

Grossman, A., Bartlett, S., Chua, N.-H., 1980: Energy-dependent uptake of cytoplasmically synthesized polypeptides by chloroplasts. Nature **285,** 625—628.

Grossman, A. R., Bartlett, S. G., Schmidt, G. W., Mullet, J. E., Chua, N.-H., 1982: Optimal conditions for post-translational uptake of proteins by isolated chloroplasts. J. Biol. Chem. **256,** 1558—1563.

Hay, R., Bohni, P., Gasser, S., 1984: How mitochondria import proteins. Biochim. Biophys. Acta **779,** 65—87.

Highfield, P. E., Ellis, R. J., 1978: Synthesis and transport of the small subunit of ribulose bisphosphate carboxylase. Nature **271,** 420—424.

Horwich, A. L., Kalousek, F., Mellman, I., Rosenberg, L. E., 1985: A leader peptide is sufficient to direct mitochondrial import of a chimeric protein. EMBO J. **4,** 1129—1135.

Horwich, A. L., Kalousek, F., Fenton, W. A., Pollock, R. A., Rosenberg, L. E., 1986: Targeting of pr-ornithine transcarbamylase to mitochondria: definition of critical regions and residues in the leader peptide. Cell **44,** 451—459.

Hurt, E. C., Pesold-Hurt, B., Suda, K., Oppliger, W., Schatz, G., 1985: The first twelve amino acids (less than half of the pre-sequence) of an imported mitochondrial protein can direct mouse cytosolic dihydrofolate reductase into yeast mitochondrial matrix. EMBO J. **4,** 2061—2068.

Hurt, E. C., van Loom, A. P. G. M., 1986: How proteins find mitochondria and intromitochondrial compartments. Trends Biochem. Sci. **11,** 204—207.

Karlin-Neumann, G. A., Kohorn, B. D., Thornber, J. P., Tobin, E. M., 1985: A chlorophyll a/b — protein encoded by a gene containing intron with characteristics of a transposable element. J. Mol. Appl. Genet. **3,** 45—61.

Karlin-Neumann, G. A., Tobin, E. M., 1986: Transit peptides of nuclear encoded chloroplast proteins share a common amino acid framework. EMBO J. **59,** 9—13.

Kloppstech, K., Meyer, G., Schuster, G., Ohad, I., 1985: Synthesis, transport and localization of a nuclear coded 22-dk heat-shock protein in the chloroplast membranes of peas and *Chlamydomonas reinhardtii.* EMBO J. **4,** 1901—1909.

Kohorn, B. D., Harel, E., Chitnis, P. R., Thornber, J. P., Tobin, E. M., 1986: Functional and mutational analysis of the light-harvesting chlorophyll a/b protein of thylakoid membranes. J. Cell Biol. **102,** 972—981.

Kolansky, D. M., Conboy, J. G., Fenton, W. A., Rosenberg, L. E., 1982: Energy-dependent translocation of the precursor of ornithine transcarbamylase by isolated rat liver mitochondria. J. Biol. Chem. **257,** 8467—8471.

Kuntz, M., Simmons, A., Schell, J., Schreier, P. H., 1986: Targeting of protein to chloroplasts in transgenic tobacco by fusion to mutated transit peptide. Mol. Gen. Genet. **205,** 454—460.

Lamppa, G. K., Morelli, G., Chua, N.-H., 1985: Structure and developmental regulation of a wheat gene encoding the major chlorophyll a/b — binding polypeptide. Mol. Cell. Biol. **5,** 1370—1378.

Leutwiler, L. S., Meyerowitz, E. M., Tobin, E. M., 1986: Structure and expression of three light-harvesting chlorophyll a/b — binding protein genes in *Arabidopsis thaliana.* Nucl. Acids Res. **14,** 4051—4064.

Lin, Z.-F., Lucero, H. A., Racker, E., 1982: Protein kinase from spinach chloroplasts. I. Purification and identification of two distinct protein kinases. J. Biol. Chem. **257,** 12153—12156.

Lubben, T. H., Keegstra, K., 1986: Efficient *in vitro* import of a cytosolic heatshock protein into pea chloroplasts. Pro. Natl. Acad. Sci., U.S.A. **83,** 5502—5506.

Mazur, V. J., Chui, C.-F., 1985: Sequence of a genomic DNA clone for the small subunit of ribulose-bis-phosphate carboxylase-oxygenase from tobacco. Nucl. Acids Res. **13,** 2373—2386.

Mishkind, M. L., Wessler, S. R., Schmidt, G. W., 1985: Functional determinants in transit sequences: Import and partial maturation by vascular plant chloroplasts of the ribulose-1,5-bisphosphate carboxylase small subunit of Chlamydomonas. J. Cell Biol. **100,** 226—234.

Ohlrogge, J. B., Kuo, T. M., 1985: Plants have isoforms for acyl carrier protein that are expressed differently in different tissues. J. Biol. Chem. **260,** 8032—8037.

Pfisterer, J., Lachmann, P., Kloppstech, K., 1982: Transport of proteins into chloroplasts. Binding of nuclear-coded chloroplast proteins to the chloroplast envelope. Eur. J. Biochem. **126,** 143—148.

Pichersky, E. Bernatzky, R., Tanksley, S. D., Breidenbach, R. B., Kausch, A. P., Cashmore, A. R., 1985: Molecular characterization and genetic mapping of two clusters of genes encoding chlorophyll a/b — binding proteins in *Lycopersicon esculentum* (tomato). Gene **40,** 247—258.

Pichersky, E., Bernatzky, R., Tanksley, S. D., Cashmore, A. R., 1986: Evidence for selection as a mechanism in the concerted evolution of *Lycopersicon esculentum* (tomato) genes encoding the small subunit of ribulose-1,5-bisphosphate carboxylase-oxygenase. Proc. Natl. Acad. Sci., U.S.A. **83,** 3880—3884.

Robinson, C., Ellis, R. J., 1984a: Transport of proteins into chloroplasts: Partial purification of a chloroplast protease involved in the processing of imported precursor polypeptides. Eur. J. Biochem. **142,** 337—342.

Robinson, C., Ellis, R. J., 1984b: Transport of proteins into chloroplasts: The precursos of small subunit of ribulose bisphosphate carboxylase is processed to the mature size in two steps. Eur. J. Biochem. **142,** 343—346.

Robinson, C., Ellis, R. J., 1985: Transport of proteins into chloroplasts: The effect of incorporation of amino acid analogues on the import and processing of chloroplast polypeptides. Eur. J. Biochem. **152,** 67—73.

Schleyer, M., Schmidt, B., Neupert, W., 1982: Requirement of a membrane potential for the posttranslatinal transfer of proteins into mitochondria. Eur. J. Biochem. **125,** 109—116.

Schmidt, G. W., Devillers-Thiery, A., Desruisseaux, H., Blobel, G., Chua, N.-H., 1979: NH_2 — terminal amino acid sequence of precursor and mature forms of the ribulose-1,5-bisphosphate carboxylase small subunit from *Chlamydomonas reinhardtii.* J. Cell Biol. **83,** 615—622.

Schmidt, G. W., Bartlett, S. G., Grossman, A. R., Cashmore, A. R., Chua, N.-H., 1981: Biosynthetic pathways of two polypeptide subunits of the light-harvesting chlorophyll a/b protein complex. J. Cell Biol. **91,** 468—478.

Schmidt, R. J., Gillham, N. W., Boynton, J. E., 1985: Processing of the precursor to a chloroplast ribosmla protein made in the cytosol occurs in two steps, one of which depends on a protein made in the chloroplast. Mol. Cell. Biol. **5,** 1093—1099.

Schmidt, G. W., Mishkind, M. L., 1986: The transport of proteins into chloroplasts. Ann. Rev. Biochem. **55,** 879—912.

Schreier, P. H., Seftor, E. A., Schell, J., Bonhert, H. J., 1985: The use of nuclear encoded sequences to direct the light-regulated synthesis and transport of a foreign protein into plant chloroplasts. EMBO J. **4,** 25—32.

Small, L. S., Gray, J. C., 1984: Synthesis of wheat leaf nitrate reductase de novo following induction with nitrate and light. Eur. J. Biochem. **145,** 291—297.

Smeekens, S., van Binsbergen, J., Weisbeek, P., 1985a: The plant ferredoxin percursor: Nucleotide sequence of a full length cDNA clone. Nucl. Acids Res. **13,** 3179—3194.

Smeekens, S., De Groof, M., van Binsbergen, J., Weisbeek, P., 1985b: Sequence of the precursor of the chloroplast lumen protein plastocyanin. Nature **317,** 456—458.

Soll, J., Buchanan, B. B., 1983: Phosphorylation of chloroplast ribulose bisphosphate carboxylase-oxygenase small subunit by an envelope-bound protein kinase *in situ.* J. Biol. Chem. **258,** 6686—6689.

Stiekema, W. J., Wimpee, C. F., Tobin, E. M., 1983: Nucleotide sequence encoding the precursor of the small subunit of ribulose-1,5-bisphosphate carboxylase from *Lemna gibba* L. G-3. Nucl. Acids Res. **11,** 8051—8061.

Timko, M. P., Kausch, A. P., Hand, J. M., Cashmore, A. R., Herrera-Estrella, L., Van den Broeck, G., Van Montagu, M., 1985: Structure and expression of nuclear genes encoding polypeptides of the photosynthetic apparatus. In: Molecular Biology of the Photosynthetic Apparatus. Steinback, K. E., Bonitz, C., Arntzen, C., Bogorad, L. (eds.), p. 381—396. Cold Spring Harbor Laboratory.

Tobin, E. M., Wimpee, C. F., Karlin-Neuman, G. A., Silverthorne, J., Kohorn, B. D., 1985: Phytochrome regulation of nuclear gene expression. In: Molecular Biology of the Photosynthetic Apparatus. Steinback, K. E., Bonitz, S., Arntzen, C., Bogorad, L. (eds.), p. 373—380. Cold Spring Harbor Laboratory.

Tumer, N. E., Clark, W. G., Tabor, G. J., Hironaka, C. M., Fraley, R. T., Shah, D. M., 1986: The genes encoding the small subunit of ribulose-1,5-bisphosphate carboxylase are expressed differentially in petunia leaves. Nucl. Acids Res. **14,** 3325—3342.

Van den Broeck, G., Timko, M. P., Kausch, A. P., Cashmore, A. R., Van Montagu, M., Herrera-Estrella, L., 1985: Targeting of foreign protein to chloroplasts by fusion to the transit peptide from the small subunit of ribulose-1,5-bisphosphate carboxylase. Nature **313,** 358—363.

Van der Mark, F., Van den Briel, W., Huisman, H. G., 1983: Phytoferritin is synthesized *in vitro* as a high-molecular-weight precursor. Biochem. J. **214,** 943—950.

Wasman, C. C., Reiss, B., Bartlett, S. G., Bonert, H. J., 1986: The importance of the transit peptide and the transported protein for protein import into chloroplast. Mol. Gen. Genet. **205,** 446—453.

Westhoff, P., Jannson, C., Klein-Hitpass, L., Berzborn, R., Larsson, C., Bartlett, S. G., 1985: Intracellular coding sites of polypeptides associated with photosynthetic oxygen evolution of photosystem II. Plant Mol. Biol. **4,** 137—146.

References for Table I.

(1) Aoyagi and Bassham, 1985; (4) Cashmore *et al.,* 1978; (9) Chua and Schmidt, 1978; (13) Doberstein *et al.,* 1977; (17) Flugge and Wessel, 1984; (18) Gantt and Key, 1986; (22) Grossman *et al.,* 1982; (24) Highfield and Ellis, 1978; (41) Ohlrogge and Kuo, 1985; (45) Robinson and Ellis, 1984a; (50) Schmidt *et al.,* 1981; (51) Schmidt *et al.,* 1985; (64) Van der Mark et al., 1983; (66) Westhoff *et al.,* 1985.

References for Figure 1.

(2) Berry-Lowe *et al.*, 1982; (3) Broglie *et al.*, 1983; (5) Cashmore, 1983; (6) Cashmore, 1984; (12) Coruzzi *et al.*, 1984; (14) Dunsmuir, 1985; (29) Karlin-Neumann *et al.*, 1985; (32) Kohorn *et al.*, 1986; (35) Lamppa *et al.*, 1985; (36) Leutwiler *et al.*, 1986; (39) Mazur and Chui, 1985; (43) Pichersky *et al.*, 1985; (44) Pichersky *et al.*, 1986; (49) Schmidt *et al.*, 1979; (55) Smeekens *et al.*, 1985a; (56) Smeekens *et al.*, 1985b; (58) Stiekema *et al.*, 1983; (60) Timko *et al.*, 1985; (61) Tobin *et al.*, 1985; (62) Tumer *et al.*, 1986.

Subject Index

MIX
Papier aus verantwortungsvollen Quellen
Paper from responsible sources
FSC® C105338

If you have any concerns about our products,
you can contact us on
ProductSafety@springernature.com

In case Publisher is established outside the EU,
the EU authorized representative is:
Springer Nature Customer Service Center GmbH
Europaplatz 3, 69115 Heidelberg, Germany

Printed by Libri Plureos GmbH
in Hamburg, Germany